DELUSIONS IN
SCIENCE & SPIRITUALITY

"Susan Martinez has gifted us with another brilliant work. In a nonthreatening and nonjudgmental manner, Martinez methodically and masterfully presents her thesis that almost everything that we have been taught about our species' evolution, our planet, and our cosmic origins is totally wrong. With balanced displays of wit, wisdom, and intelligence, Martinez provides the reader with the 'rise of knowledge from unseen worlds' and deftly demolishes the 'delusions' in conventional and orthodox science and spirituality. This is a must-have book for anyone who wonders about our true history on this planet and the interconnectedness of all life. We highly recommend it."

BRAD AND SHERRY STEIGER, AUTHORS OF
REAL ENCOUNTERS, DIFFERENT DIMENSIONS,
AND OTHERWORLDLY BEINGS

"Martinez, trained in the theories of elite universities, has turned the tables on their ill-rooted assumptions. She demonstrates how their conclusions do not match reality. This volume crowns her triad of books that upheave today's illusionary views about humanity's origins, place on Earth, and role in the universe. Scholars and other readers thirsty for a new multidimensional cosmology will react to Martinez's work as a parched straggler seeing an oasis. One will be intellectually nourished and find stepping-stones to a better set of answers to our present mysteries of life."

PAUL VON WARD, AUTHOR OF
WE'VE NEVER BEEN ALONE
AND THE SOUL GENOME

"Martinez avoids going too deeply into scientific jargon, making *Delusions in Science & Spirituality* palatable for a wide audience. Great food for thought and a wonderful reference work, too. I highly recommend this book. It is liberating!"

JAMES WEBSTER, AUTHOR OF
LIFE IS FOREVER

"Martinez takes no prisoners."

MICHAEL TYMN, AUTHOR OF
RESURRECTING LEONORA PIPER

DELUSIONS IN
SCIENCE & SPIRITUALITY

The Fall of the Standard Model and the
Rise of Knowledge from Unseen Worlds

SUSAN B. MARTINEZ, Ph.D.

Bear & Company
Rochester, Vermont • Toronto, Canada

Bear & Company
One Park Street
Rochester, Vermont 05767
www.BearandCompanyBooks.com

Text stock is SFI certified

Bear & Company is a division of Inner Traditions International

Library of Congress Cataloging-in-Publication Data
Martinez, Susan B., author.
  Delusions in science and spirituality : the fall in the standard model and the rise of knowledge from unseen worlds / Susan B. Martinez, Ph.D.
      pages cm
  Summary: "Debunks cherished theories of mainstream consensus and reveals the deeper mysteries of the science of the unseen" — Provided by publisher.
  Includes bibliographical references and index.
  ISBN 978-1-59143-198-5 (paperback) — ISBN 978-1-59143-779-6 (e-book)
  1. Knowledge, Theory of. 2. Science—Philosophy. 3. Science—Social aspects. 4. Sprituality. I. Title.
  Q175.32.K45M32 2015
  501—dc23
                                                            2014034565

Printed and bound in the United States by Lake Book Manufacturing, Inc.
The text stock is SFI certified. The Sustainable Forestry Initiative® program promotes sustainable forest management.

10 9 8 7 6 5 4 3 2 1

Text design by Virginia Scott Bowman and layout by Priscilla Baker
This book was typeset in Garamond Premier Pro with Myriad and Papyrus used as display fonts

To send correspondence to the author of this book, mail a first-class letter to the author c/o Inner Traditions • Bear & Company, One Park Street, Rochester, VT 05767, and we will forward the communication, or visit the author's website at **earthvortex.com** or contact the author directly at **poosh8@gmail.com**.

For Dan Winner,
Creative genius, friend, humanitarian

Nine-tenths of existing books are nonsense; and the clever books are the refutation of that nonsense.

BENJAMIN DISRAELI

Conventional wisdom has a way of being wrong.

DAVID RAUP, *THE NEMESIS AFFAIR*

CONTENTS

ACKNOWLEDGMENTS

I would like to thank Jon Graham for having faith in my ideas and helping to make the publication of this book possible; and many thanks to Mindy Branstetter, a godsend from editorial heaven. My gratitude also to Torri Beck: Who says librarians are plain?

Thou hast quickened my members to be dissatisfied by the old revelations.

OAHSPE, BOOK OF KNOWLEDGE

When Death Is Not Death: A Parable

Master Abu-Ishak Chisti once told of a certain man who was believed to have died, and was being prepared for burial when he revived. The man sat up, but he was so shocked at the scene surrounding him that he fainted. He was put in a coffin, and the funeral party set off for the cemetery. Just as they arrived at the grave, he regained consciousness again, lifted the coffin lid, and cried out for help.

"It is not possible that he has revived," said the mourners.

"But I am alive!" shouted the man. He appealed to a well-known and impartial scientist and lawyer who was present.

"Just a moment," said the expert. He then turned to the mourners, counting them. "Now, we have heard what the alleged deceased has had to say. You fifty witnesses tell me what you regard as the truth."

"He is dead," said the witnesses.

And so he was buried.

Are the experts still burying the truth?

THE FIX IS IN

Myth has its charms, but the truth is far more beautiful.

J. Robert Oppenheimer

It is said that the first casualty of war is the truth.* There is a quiet intellectual war going on; the naked truth is a fatality in the battle for scientific and philosophical supremacy. Since this book is about the "ongoing
search for fundamental farces"† let me open with this. Around the time
I began writing *Delusions in Science and Spirituality,* a congressman
from my own state (Georgia), Paul Broun, who is also a medical doctor,
publicly announced, "All that stuff I was taught about evolution and
embryology and big bang theory—all that is lies straight from the pit of
hell," adding, "Global warming is a trick." Broun, as it happens, sits on
the House Committee on Science, Space, and Technology.

Although my political views are probably far from Rep. Broun's,
I, too, have had my fill of "fundamental farces." They are everywhere
and they are squarely in the way of truth. Until they are cleared off the
table, we cannot move forward. In these pages we will confront and
debunk the grandest myths of our time—albeit the current alpha dogs

---

*Winston Churchill's famous words reflected his belief that in time of war the truth is so
precious that it must be guarded by a posse of lies.
†This was the double-entendre subtitle of Steve Mirsky's article "Anti-Gravity" in *Scientific American* (2013, 86), reporting on the outspoken congressman's opinions.

in human knowledge—theories so self-congratulatory, so cosseted, so well-inked and oversold as to become Fact. Today's science magazines are one big commercial for the Standard Model (SM). As a result, everyone believes in evolution, ice ages, global warming, and so on. These solutions—unctuously labeled "attractive," "elegant," "sophisticated," "robust," "muscular"—are shot down tomorrow. Meanwhile, the alternatives to these moribund theories are ignored and despised, even though "the improbabilities of today are the elementary truths of tomorrow," as sagely declared by Charles Richet, 1913 Nobel Prize winner in medicine, in his book *Thirty Years of Psychic Research* (2003).

Scientific debate? Academic freedom? Not really. Turf wars and quarrelsome factions are only internecine, infighting, that is, within the Standard Model.

Scrambling to break records—as if knowledge were a race or a contest—the mainstream offers us "progress" in science in the form of

a particle that travels *faster* than light
the *largest* map ever made of dark matter
the *biggest* structure in the universe
the *most massive* star ever seen
the *most powerful* gamma ray burst ever observed
the *largest* color image of the universe
the *brightest* supernova ever recorded
the *most massive* black hole
the *earliest* version of *Australopithecus*
the *first* hominid to use fire
the world's *oldest* jewelry, *oldest* wine cellar, *oldest* pair of pants
the *hottest* summer on record
the *biggest* underwater volcano

If this record-breaking game is really progress, whom does it enlighten? Besides, some of these "breakthroughs" are less smashing than their blazing headlines suggest. In 2012, for example, archaeologists found evidence of 6,700-year-old corn in Peru, which was trumpeted as the oldest ever found in South America, 2,000 years earlier than previously thought. But as we will see, maize is quite a bit older in that region than 6,000 or 7,000 years. Headlines have even been made with the supposedly oldest turd in America: "startling evidence" from a cave in Oregon, these coprolites (feces) are hailed as comprising the oldest evidence of human presence in the Americas, at 14,300 years old. But not really; that date is still a conservative figure, considering that 50,000-year-old flake tools have been found in our hemisphere (much more on all this in chapter 8).

In the seventeenth century, the giants of science set the pace for us moderns by reducing the universe to fixed secular laws: Religion was hokum, it was science that should be trusted. And by the nineteenth century, God was driven out altogether, no longer needed to explain nature. Though some toyed with a certain double truth, trying to combine science and religion, intellectual confusion reigned, and even the physicist and mathematician Sir Isaac Newton "showed signs of the 'split mind,'" according to historian Lloyd Moote in *The Seventeeth Century*. Also critiquing medieval cosmologists, author and activist Arthur Koestler found them still wed to Ptolemy's geocentrism (see chapter 5). Koestler remarks that "They knew that the sun governed the motions of the planets, but at the same time closed their eyes to the fact." As Koestler saw it, the secret appeal of the Earth-centered (geocentric) system lay in "the fear of change, the craving for stability . . . in a disintegrating culture. A modicum of splitmindedness and doublethink was perhaps not too high a price to pay for allaying the fear of the unknown" (1959, 73, 76). Might this analysis apply as well to the twenty-first century?

SPLIT MIND OR DOUBLETHINK

Today, as we approach the denouement of our vaunted age of information (and disinformation), we're getting very close to truth time. Old paradigms, old regimes, are falling, and with them, old doctrines. On the cusp of this time of change, split mind or doublethink once again prevails. Carl Jung, for example, was split down the middle, as far as the spirit world was concerned. Even though he himself was a "sensitive" and had many paranormal experiences, he still maintained, "I cannot accept evidence for the independent reality of spirits" (Ebon 1978, 116); (much more on Jung in chapter 6). And how's this for doublespeak? According to one of today's leading cosmologists, Lawrence Krauss: "We've been so successful that the questions we're asking now are so deep that they may remain unanswerable for some time to come: and maybe forever. We don't understand this model that we have. It's completely inexplicable" (quoted in Pendick 2009, 48).

Doublethink also has science swinging from overly cautious on the one hand to outrageously speculative on the other. By turns, the expert comes off as the confident know-it-all, then without warning switches to the disingenuous confessor of ignorance. Split mind has science saying ever so humbly, "We have a lot to learn, we are still a work in progress. In fact, we don't really know what mechanism drives such-and-such . . ." Yet in the next breath, we hear the same scientists elevating the Standard Model into unassailable fact, impregnable to alternatives. Unassailable, for example, is the theory of evolution: "The real battle is over," declared the sci-fi author Isaac Asimov, concerning the debate over Darwinian evolution (1971, 165). "Evolution is quite simply the way biology works" (Hayden 2009, 42). In other words, if you disagree, you are simply ignorant of the facts.

Split mind also has people agreeing to two contradictory statements, as well as hedging their bets. Equivocation is part of the toolkit, affording an escape hatch, damage control, deniability. To give a single example: after studying Mrs. Leonora Piper, America's foremost turn-of-the-twentieth-century psychic-medium, psychologist/philosopher William James remained a skeptic, at least publicly. Yet, hedging, he wrote "I am persuaded by the genuineness of her trance and . . . believe her to be in possession of a power as yet unexplained" (James 1903, *Varieties of Religious Experience*) (that power being spirit communication; see chapters 6 and 7). Though it is generally assumed

Our friend Thomas Hayden (quoted above on the infallibility of evolution), proudly reported the blitz for Charles Darwin's two hundredth birthday in 2009: a plethora of lectures, exhibits, and festivities; in England, Darwin's face now graces the special two-pound coin; a five-day celebration at Cambridge; similar events in the United States; special exhibits at the Smithsonian, including one that shows "how orchids evolved and adapted according to Darwin's theory" (Hayden 2009, 42). To me, all this hoopla proves only that today's standing doctrines owe their fame and glory not as much to truth as to exposure—a constant, relentless barrage—publicity per se.

that Professor James did more than any other researcher to advance the survival hypothesis (the belief in the immortality of the soul), the

Fig. I.1. Mrs. Leonora Piper, the subject of Tymn's latest book,
Resurrecting Leonora Piper.

truth is he probably did more than any other person to impede it. In his popular book *The Varieties of Religious Experience,* James didn't even mention Mrs. Piper. What's more, "He continually beat around the bush on the survival issue," according to the spiritualist author Michael Tymn, who thinks James was more of a believer than he let on, but lacked the courage to admit it publicly for fear of damaging his reputation (Tymn 2013, 3–4). Off-duty, the experts might well say or think that the conservative consensus view of the SM is baloney. We are not above selling our souls for a place in the scientific sun.

Split mind is a natural outcome of a society that is trained to compartmentalize; most of us, with our own noncommittal, fragmented lives and mixed beliefs, are likewise casualties of split mind. And with so much specialization in disciplines and the workplace—which is another kind of fragmentation, called "hyperspecialization" by cosmologists—the rigid separation of departments of knowledge puts us in a somewhat precarious state. The Big Picture and warning signs of potential danger can be easily missed. The more hyperspecialized we are, the less we can expect a harmonious, "thinking" whole. The more territorial and competitive we are, the more pettiness we display over "jurisdiction," the sooner we cook up a recipe for disaster, as the run-up to 9/11 and its bureaucratic tangles proved to be. ⮑

Overspecialization, moreover, is the mother of isolationist orthodoxies and of insufferably technical language, as well as the progenitor of special interests, jealousies, and hostility, but, most of all, the worm's eye view. Albert Einstein opined, "It is not enough to teach man a specialty. Through it he may become a kind of useful machine . . . resembling a well-trained dog . . . but not a harmoniously developed personality. . . . Overemphasis on the competitive system and premature specialization . . . kill the spirit on which all cultural life depends" (1954, 66).

Specialization bars the Everyman from partaking in the fruits of those various labors. As a result, the sacred cows of science are received by a numbed public almost indifferently, reflexively. Accepted as established fact, the SMs have become public relations darlings, media darlings, handled by expert puppet masters with more style than substance, more showmanship than stripe, more face than fact. The brilliant American historian and social critic Christopher Lasch observed back in the 1970s:

> The master propagandist uses circumstantial evidence in a matter-of-fact tone* along with accurate details, to imply a misleading picture of the whole. . . . An educated public . . . cherishes nothing so much as the illusion of being well informed. . . . The more technical and recondite, the more convincing it sounds. Hence . . . the obfuscatory jargon of pseudo-science [with its] aura of scientific detachment—calculatedly obscure and unintelligible—qualities that commend it to a public that feels informed in proportion as it is befuddled! (Lasch 1978, 76–77)

In one of his enjoyable tirades, independent scholar, inventor, and author James Churchward lit into this obfuscatory jargon, without which pseudoscience could not survive: "The more technology, impossible to understand, the better it is, for here is a bluff for the public with no possible comeback" (Churchward 1968, 164). I read in a March 2013 issue of *Time* about the head of the International Monetary Fund, Christine Lagarde, who doesn't care to "speak in the opaque language of the IMF's economic policy wonks and sometimes has to interrupt a meeting to say 'Stop it. You've lost me. You have to use simple terms that people out on the street will understand, because otherwise you are just talking to yourselves.'" Indeed, the sooner we get rid of this pernicious habit, the better. It is only the proprietary voice speaking—the voice of hyperspecialization, along with its sugar-daddy special interest. Concerning the latter, Einstein once remarked incisively, "Private capitalists inevitably control the main sources of

---

*One irresistible example: "The debate is over," announced Arnold Schwarzenegger in the *New York Times* regarding global warming. "We know the science. We see the threat, and we know the time for action is now" (Marshall 2005). All bluster; rather than preventing a disaster, the global warming campaign is inviting one (see chapter 4).

information (press, radio, education). It is thus . . . quite impossible for the individual citizen to come to objective conclusions" (Einstein 1954, 157).

Have I dared to call today's SMs pseudoscience? I can tell you only that the further along I got in the research for this book, the more I found the experts, often enough at the cost of the taxpayer, explaining things that never happened! Big bang, evolution, Ice Age! Never happened. Read on.

Are not their doctrines trembling on their foundation?

OAHSPE, BOOK OF OURANOTHEN

I will be quoting freely from one of my "bibles," Oahspe (1882), especially from its Book of Cosmogony and Prophecy, which sets aside a good part of our present philosophy on the nature and structure of the universe, beginning with the "attraction" of gravitation (expanded on in chapters 1 and 5). In these pages, our itinerary goes from science (chapters 1–4) to psyche/self (chapters 5–7), to society (chapters 8–9). Even if Isaac Newton was (as I will argue) in error concerning the law of gravity, he was right about one thing—inertia. Objects (and ideas!) in motion tend to stay in motion; this is also known as the status quo: the idée fixe.

INERTIA OF THE MIND

Dead knowledge is the danger. It is the peculiar danger of scholarship, of universities.

ALFRED NORTH WHITEHEAD,
DIALOGUES OF ALFRED NORTH WHITEHEAD

According to Arthur Koestler,

The inertia of the human mind and its resistance to innovation are most clearly demonstrated not, as one might expect, by the ignorant mass—which is easily swayed once its imagination is caught—but by professionals with a *vested interest* in the monopoly of learning. Innovation is a twofold threat to academic mediocrities: it endangers their oracular authority, and it evokes the deeper fear that their whole, laboriously constructed intellectual edifice might collapse. (my italics) (Koestler 1959, 427)

It is only when we shed the materialist blindfold, which pervades today's leading doctrines (sworn at the thigh to a secular framework—godless and purposeless), that the forces of the nonphysical world come into play—answering, in time, the dead ends in psychology, philosophy, even medicine, climatology, and the hard sciences. I believe the powers in the unseen world will explain all the so-called mysteries of science and history, including the origin of mind. In fact, they are only mysteries when we refuse to study Es, the world beyond, unseen but potent, the ever-missing something.

In strict materialism and its secular model, matter (corpor) is asked to explain everything. But it can't. A Force, which is in the unseen, rules over all. Oahspe's Book of Cosmogony and Prophecy, a science manifesto that pulls no punches, declares that

> man has ever sought the cause [of phenomena] in corporeal things; he builds up certain tables and diagrams, and calls it science. . . . He searches for explanation by anything under the sun that is corporeal [tangible], rather than search in the subtle and potent, unseen worlds.
>
> OAHSPE, BOOK OF COSMOGONY AND PROPHECY 6:12

I find our modern ideas swayed by either fanatical secularism, which has earned the moniker "scientific fundamentalism," or its opposite, fanatical occultism (chapters 5 and 7). To the "sci mat" (scientific materialist), the heaven world, as Dr. Marvin E. Herring's cartoon suggests, is simply part of irrational thinking. But it is a patent falsehood (if not a basic piece of propaganda) that belief in a higher power or the invisible

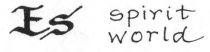

Fig. I.2. Sign of Es, all that is beyond. All that is is in the unseen. Es is the root of such words as is, essence, *and* Essene.

"You poor deluded man! Thank God, I'm an atheist!"

Fig. I.3. Fanatical secularism. Cartoon by Marvin E. Herring.

realm entails the abdication of reason. Did Kepler, Galileo, Newton, Pasteur, Faraday, or Descartes—all believers—lose their reason? No, religion is not a superstition, as atheists say. It is a very different outlook.

Charles Darwin and Sigmund Freud (chapters 2 and 6) came up with perfect solutions to satisfy the secular, God-free paradigm. As a result, theorists have ever since been trying to account for incorporeal things in corporeal terms, like asking the brain to explain the mind, or asking dark matter to explain gravity. The scientific materialist would reduce all, including human consciousness, to physical laws and chemical reactions, acknowledging neither soul, nor unseen realm, nor purpose. Spiritually illiterate, sci mat does not see the faces, only the vase (see fig. I.4.). And in this cynical age, the keepers of the Standard Model say

Fig. I.4. In white is seen a vase. In black are the profiles of two faces. The famous Rubin vase was created by a psychologist to illustrate the bias of perspective.

that even if we do have a soul, it is not relevant to science; let's just call it the Unconscious (chapter 6). Yet to critics like Richard Webster, Freud's unconscious is an elaborate and "complex pseudo-science" (1995, 438).

Would it be too overwhelming for science to grant the reality of a spiritual realm (Es) over which corpor—or human devices—has no known control?

THE UNSEEN

What is essential is invisible to the eye.
ANTOINE DE SAINT-EXUPÉRY, *THE LITTLE PRINCE*

Much of what exists is, frankly, in the unseen. Who can see gravity? Nothing in black holes can be observed (see chapter 1) . Nor can we see subatomic particles or "dark energy" or the void into which the galaxies recede. The quantum world itself is a "vast porridge" where nothing is fixed or measurable. And the more we chase down the "particle" (a thing), the more it appears to be a "wave" (an action). Matter and energy, we have come to accept, are ultimately interchangeable; therefore, the seen world (matter) is an aspect of the unseen (energy, force). $E = mc^2$. Today (paradoxically, thanks to theoretical physics), the unseen is coming into its own; much of the universe lies beyond what we can observe. Quantum physics, without intending to, has brought us to the doorstep of Es. But to the strict materialist, elements (visible, tangible) still govern forces (rather than the other way around), and forces (being unseen) are immaterial. Yet elements of themselves have no force whatsoever, as discussed in chapter 1.

George Morley, who was hierophant of the British Church of Kosmon at Surrey, England, once said in a lecture:

> Today, with all our civilization and learning, we have no higher conception of the Great Architect of the Universe and the brotherhood of man than had they of the ancient world. The grand philosophies we have built have toppled to the ground, and we are left like a rudderless boat upon the troubled sea of speculation. . . . In many cases the result has been to extinguish faith altogether and drive men into a kind of fatalistic and selfish materialism. (Morley 1962)

Today, as we move imperceptibly from the hectic age of information to the age of maturity and understanding (Kosmon), forces unseen (like gravity and dark energy)—as opposed to "things" in the seen world—are becoming better known. Sooner or later, enlightened science will recognize the long-sought mechanism of things, the all-embracing twin processes of condensation and dissolution—forces, not elements—which make and unmake worlds.

Science today upholds the secular paradigm by teaching that the physical world created itself and everything else. We are taught that quanta and the stars as well as all species and culture are self-creating, self-organizing entities. Forget the Great Architect. No Creator need apply. No Higher Intelligence, we are taught, is involved, only the spontaneous emergence of structures, systems, and order—not unlike the theory of spontaneous generation that was scotched more than 150 years ago. This self-creating universe not only banishes the Great Spirit to the unimportant realm of "belief," it also ignores the actual mechanism behind creation: vortexya (the dynamo, or force field, underlying

Fig. I.5. Two images of a spiral: (top) spiral nebula and (bottom) artist's rendering of the twist that engulfed Tokyo in 1923 after the Great Kanto Earthquake (see also figs. 1.9 and 1.10, pp. 59 and 60)

condensation, as discussed in chapters 1 and 4). They say, for example, that the massive energy of a hurricane self-organizes into a huge weather system (see appendix A). But as we will come to see, it is the power of the vortex that drives the hurricane and moves the planet. We recognize that power by its shape—the corkscrew as seen in the spiral nebulae of galaxies or the spiral pattern of a shell. The line of sight to the Sun is also a spiral; and the Earth's course through the galaxy is helicoidal, a screwlike path. Michio Kushi called it the "Spiral of Life."ᴌ

Causation, the very heart of science, has fallen on hard times. Symptoms are celebrated, root causes lost sight of. As seen in these chapters, no actual cause of a big bang (chapter 1), of evolution (chapter 2), of ice ages (chapter 3), of global warming (chapter 4), of planetary influence (chapter 5), and of civilization itself (chapter 8) is known. Sunspots, for example, are a symptom, not a cause. Do sunspots affect terrestrial life? No! This eleven-year cycle is of the Earth system, not the Sun. Sunspots are a sign of that cycle, not its cause (the subject comes up again in chapters 3, 4, and 5).

Another example: Do methane and carbon dioxide *cause* global warming (chapter 4)? Many scientists, like Roy W. Spencer, Ph.D., speak of "a natural increase in the CO_2 . . . as a *result* of warming. Note that this is opposite of causation in the theory of human-made global warming. . . . Just as in the case of clouds and temperature, we are once again confronted with the question of cause versus effect" (my italics)

Spencer goes on to explain:

> Since there is less cloud cover over the Earth in unusually warm years . . . the argument went, the warming *caused* less cloud cover, which allowed more sunlight in, which enhanced the warming. . . . But how did the researchers know that the warmer temperatures caused a decrease in cloud cover, rather than the decrease in cloud cover causing the warmer temperatures? Well, it turns out they didn't know. (my italics) (Spencer 2010)

This is Spencer's specialty; he is an atmospheric scientist. "We now have published evidence of decreases in cloud cover *causing* warmer temperatures, yet it has gone virtually unnoticed" (my italics). Spencer challenges those who say that human-made global warming has prompted an increase in El Niño effects in recent years.

> I think it is much more likely that causation is actually operating in the opposite direction: more frequent El Niños . . . [explain] the

warming in the late twentieth century. The general issue of cause-versus-effect is at the core of many mistakes that have been made in the interpretation of how the climate system works. (Spencer 2010, 21, 72, 101, 128, 154)

What is the cause of pole shift (a.k.a. magnetic reversal)? No one knows, but plenty of guesses are offered: Some say the earth's liquid core generates the magnetic field and may initiate its flips (Mara Grunbaum). It is suggested that since the core can slip somewhat in place, the poles can wander, even completely reverse; though how that happens no one knows. Alternately, it is the buildup of ice at the poles that caused them to flip. Then again, it could be plate tectonics, that is, isostacy. Or "nutation" (a word referring to axial wobble, causing radical displacement of the planet's axis of rotation). Or is it a slippage of Earth's solid crust over the molten interior, changing the polar location? Or perhaps some large-body impact jolted the Earth enough to reverse its polarity. Or pole shift happened when Earth entered the photon belt of the Pleiades (Von Ward 2011, 8). Or it may be the "result of the way in which the Earth's magnetism is generated" (Gribbin and Plagemann 1974, 53). Indeed, "little is known for sure about how or why the field flips from north to south . . . there is no good explanation" (Raup 1986, 183). I think the rolling of the Earth, herein called "oscillation" (chapters 3 and 4), will help solve this problem.

We are inevitably faced with the problem of closed doors and outright censorship. There is a fine line between the sin of omission and the sin of obstruction. The students and public are only allowed access to scientific information that is harmless to a tiny handful of powerful scientific tyrants. Although the twenty-first-century brain trust, our intelligentsia, is global, the myths challenged in this book are maintained largely by the usual cabal of white, male, English-speaking gentlemen. Conventional wisdom comes to us by way of "the fragile assumptions and cliquish associations upon which presumed truth is often built" (White 1980, 112).

It seems the more fragile the assumption, the greater the *chutzpah*. Today's experts make a point of saying that evolution is not a theory; it is a fact. Despite attempts by these intellectual imperialists to treat the battle as won, "actually the questions remain wide open" (Taylor, *The Great Evolution Mystery,* 1983). Science, avers another observer, can be "as dictatorial as the most fanatical ecclesiastical organizations" (Bros 2008, 34). Equal opportunity does not apply to, say, big bang, evolution, or global warming, whose findings come to us *ex cathedra* as policies (not truths). Thinking outside the box is not part of the job description for these professionals.

A curator at one of the world's top museums,* who happened to keep an open mind during a cosmology controversy, was fired on the spot and forced to clear out his office immediately. In other words, the cost of admission to the club, the winning team, is loyalty to the paradigm. These oaths of loyalty to the SM—and to all sci mat, for that matter—leave the honest and noble servant of truth out in the cold. To question a single brick in today's cosmological edifice would endanger it all, and "This is the threat that keeps most astronomers from looking for a flaw in the chain" (Acheson 2008, 153). They really cannot afford to take the plunge.

Instead of examining the flaw, standard operating procedure is to call it a mystery or an "anomaly," a fluke, a freak, even when it could well be a key factor. Young scientists coming up are not even exposed to such "anomalous data," while alternative views that might

---

*At the time of the Velikovsky controversy, *The Daily Princetonian* (February 1964) editorialized on "the personal vituperation, deliberate misrepresentation of facts, off-hand misquotations, efforts at suppression of the books containing the theories, and the denial of the right to rebut opponents in professional journals that the independent scholar Immanuel Velikovsky encountered. . . . Far more was going on than mere challenge to established ideas . . . theories held because of the vested interests they represent" (Velikovsky 1965, xii). Faced with threats of boycott, Macmillan, Velikovsky's original publisher, relented, turning over *Worlds in Collision* to another publisher. Macmillan was blackmailed by the cabal that threatened to cut off their textbook market.

explain them are contemptuously blackballed. They are, essentially, taboo. ⋏

CONCEALING THE EVIDENCE

The halls of academia are very much like the New Inquisition. They have not yet burned people at the stake, but they have thrown people in jail and destroyed careers.
PHILIP COPPENS, *THE LOST CIVILIZATION ENIGMA*

Investigative journalist Philip Coppens, for his part, avers that "lost civilizations" (chapter 8) are not really lost; they are "excluded *on purpose*. By a consensus view embedded within the walls of academia; this has grown like a cancer." An entire series of paradigm-shaking artifacts, evidence of a lost technology (extremely ancient lenses, for example) belonging to such excluded cultures, ends up buried inside "the walls of various museums." Too much of this "anomalous" evidence (pointing to a high civilization in the Stone Age) is "scattered in various museums" (Coppens 2013, 275, 248).

Paleolithic men, said French researcher Robert Charroux, were familiar with masonry and lived in large towns with streets, artisans, and probably even hairdressers, though much of the evidence for this is "kept hidden in the back rooms of museums." Mothballed. "Who has ever seen the six fine specimens [from La Vaulx] that disappeared into the Museum of Saint-Germain-en-Laye?" asks Charroux. Many other artifacts bearing particularly "complex and skillfully executed carvings have disappeared." The prehistoric library of Lussac-les-Chateaux goes back to the Magdalenian, probably more than fifteen thousand years ago. These stone books, which contradict the SM's conservative chronology, were long sequestered in Paris at the Museum of Man; even today, the most interesting parts "have never been shown to the general public" (Charroux 1971, 28, 85, 119, 48). The same can be said of the wonderful finds at Glozel or India's Tirvalour Tables with much "too early" astronomical knowledge, long sequestered in Paris and possibly destroyed (Charroux 1971, 117).

A few battle-weary archaeologists, defeated by this "cheating by concealment" (a.k.a. "discreet fraud") have come forward, reporting that evidence that contradicts the SM has been thrown away, stolen, or wrapped in burlap and plaster in the back rooms of museums. Loads of "anomalous" artifacts have vanished into storage bins, if not into thin air. "Major caches of archeological material are handed over to the Smithsonian, only later to disappear down the memory hole," laments author Richard Dewhurst, concerning "the suppression of hundreds of 'out-of-context' finds, all submitted to the museum in naive ignorance of the museum's official policy of suppression of alternative perspectives" (2014, 12, 229). My colleague, author Patrick Chouinard, in his book *Forgotten Worlds,* runs down a particular artifact that challenges "accepted scientific wisdom" and was taken by the Smithsonian and subsequently lost: "a royal example of how the establishment continues to suppress theories and discoveries" that don't match the paradigm. Meanwhile Chouinard reminds us of those blond mummies found in China that were "consigned to the dusty reaches of the Xinjiang museum at Urumchi. . . . Their anomalous nature . . . led this discovery to be intentionally buried by the Communist regime for almost twenty years" (Chouinard, *Forgotten Worlds,* 2012).

In Australia, maverick archaeologist Vesna Tenodi tells the sorry tale of skeletal remains proving preaboriginal races that were deliberately destroyed. In their place, the experts served up "intellectual kitsch . . . the fabrication of Australian prehistory for political purposes." Tenodi has "a thick folder of their responses to my work, consisting mainly of threats of legal action . . . 'We'll take you to court, our lawyers will destroy you!'" (Tenodi 2013, 15–16).

A simpler tactic is mere dismissal: Evidence that can reasonably account for "anomalies" may be discounted as "long since discredited," an obsolete theory. A certain "defensive unanimity" among scientists (noted by Corliss 1980, 37) is the bulwark that keeps alternatives at bay, sidelining all opponents. This consensus, this tendency to agree on an interpretation, even when equivocal, has a name: *groupthink*.

THINKING OUTSIDE THE BOX

Consensus is a political concept, not a scientific one.

JOHN KAY

The models may be agreeing now simply because they're all tending to do the same thing wrong.

ROY W. SPENCER,

THE GREAT GLOBAL WARMING BLUNDER

I must admit that I can afford to write and think outside the box. I am not an academic; I left that world many, many years ago. Institutionally filtered knowledge was not my cup of tea. Books written by "dissidents" outside the box have been boycotted or savagely reviewed, the targets labeled pseudoscientific, frauds, failures, mentally incompetent, "loony," their ideas "absurd," claptrap. No one in the know takes them seriously. They are losers, liars, charlatans. Their credibility must be destroyed by any means possible—defame, disgrace, burn them at the (career) stake.

And hit where it hurts: the pocketbook. Bottom line: Funding, after all, is in the hands of "politicized decision makers" (Mitton 2005, 3). Any defiance of the SM is career suicide, throwing out any chance of getting your research approved and underwritten. Money talks. The truth has no price. Get on the "climate change" bandwagon, for example, and you're funded, you're published. "To even suggest that [global warming] may not be the entire story, is to face harsh consequences: loss of grant funding [or] . . . [the] inability to publish one's data and views" (Schoch 2012, 279).

Employment itself is at stake. I recently had the opportunity to chat with a manufacturing executive who at one point had very publicly spoken out against global warming. He regretted it. It did not go well for him. I asked him to please expand on his claim that it was a government hoax, but he would not, saying, "I want to keep my job." That was the end of the conversation. Another critic of global warming once asked an expert: Why don't we hear from those scientists who doubt the dangers of carbon dioxide (greenhouse gas)? "It's the money!" replied the scientist. "Twenty-five billion dollars in government funding has been spent since 1990 [this was in 2006] to research global warming." If climate change boiled down to simple and natural fluctuations, "there wouldn't be much money to study it" (Stossel 2006, 204).

The same goes for cosmology: "Unless you work for the Big Bang theory [chapter 1] you will not get academic funding" (Lerner 2004, 20.) When astronomer Sir Fred Hoyle dared to buck it, his "opponents deployed enormous resources to wrong-foot him" (Mitton 2005, xvii). One of Hoyle's colleagues in "continuous creation" (the best alternative to the big bang theory) was the observation-astronomer Halton Arp. After publishing his findings, Arp's status plunged, his work rejected and ridiculed by the astronomy establishment. Finally, denied telescope time, Arp took early retirement and moved to West Germany. It was a classic case of "theory rul[ing] over observation, like the Ptolemaic astronomers who refused to look through Galileo's telescope," laments science writer Eric Lerner. As cosmologist Hannes Alfven commented lightheartedly about that great Italian scientist's censorship and house arrest, "Galileo was just a victim of peer review" (Lerner 1991, 228, 53).

Peer review, in a word, is the process that decides what to fund and what to print. But on what basis? Everyone knows the review is empaneled to prevent alternative views from getting into open court. "The objectivity of science," as geologist Robert Schoch forthrightly stated, "is a myth. . . . Submissions to [high-status scientific journals] are subjected to the peer review system. The reviewers act as censors . . . guarding the status quo." While honors are heaped on those advocating the SM, "competing theories are marginalized. . . . Dissenting views . . . must be suppressed . . . detractors locked out of jobs, publication outlets, and grant funding" (Schoch 2012, 122–25).

In an open letter to e-mail subscribers, scholar and theorist John Feliks, editor of the online *Pleistocene Coalition News,* wrote in a similar vein, "Ours and my experience with peer review is pretty bad. . . . The peer review comments [on one of his papers] did not match in any way the high level of those I received openly from leading researchers . . . I do not have respect for peer review in anthropology, nor should anyone." Feliks talks about the SM "selling the public another ape-man [see chapter 2]. That's what you do in physical anthropology . . . [warrant] an ape turning into a man. Ulterior motives behind peer review, and personal or special interest, run those publications."

Oh, not that everyone bucking the system is right. Not at all. There is enough pigheadedness to go around. Few are enlightened, but there is no escaping that the professions have essentially become monopolies, each its own little fiefdom. Big bang monopolizes cosmology: everything we don't understand about cosmogenesis—blame it on the one big bang. Evolution monopolizes anthropology: everything we don't understand about human beginnings—blame it on Darwinian natural selection and mutations. "Ice ages" monopolized climate study of the past: everything we don't understand about ancient geology—blame it on glaciers. Global warming monopolizes today's climatology: everything we don't understand about weather—blame it on human-caused warming. The unconscious monopolizes the science of psychology: everything we don't understand about the mind—blame it on the unconscious.

PART ONE: SCIENCE

I have divided this book into three parts: Science, Self, and Society.

Today's favored scientific theories are essentially arguments from authority, all sacred cows—big bang, evolution, Ice Age, global warming—each one operating in an atmosphere of pressure-driven consensus, each pumping out boilerplate explanations. All these chapters are about change, our theories of change, influence, triggers. Concerning the question of change: I don't believe the universe has changed much (chapter 1); nor do I believe that humanity has "mutated" from one species to another (chapter 2); neither does the Earth change a great deal from warm to cold (chapter 3); nor does the Earth switch from cold to warm (chapter 4). I think that the instability of society itself has found its way into our theories, which have come to disdain the immutable, that which changes but little. Bewildered by the pace of change in modern society, theorists have mistakenly posited change where there is no change, as well as acceleration where there is no acceleration. In chapter 1, I challenge astrophysics's notion of accelerating expansion of the universe. Chapter 2 questions the supposition that *Homo sapiens*'s evolution accelerated at the Great Leap Forward (ca. forty thousand years ago). Chapter 4 confronts the claim that global warming is accelerating; and chapter 8 disputes the dictum that culture accelerated six thousand years ago, resulting in the birth of civilization.

But what I *do* take into consideration in chapters 1 through 4 is the age and aging of a planet, especially Earth, which are indeed critical factors. Touching on this question of age, chapters 1, 2, and 3 address supreme scientific mysteries of the past, clearing up some of the reasons for each "mystery." What was the cause of the dinosaur extinction, for example, or of ice ages, or of Mars's dessication? A mystery? Not really. The simple factors of age and aging are rescued from oblivion to decipher these "mysteries."

So-called mysteries also arise because of the unresolved and disputed question of design, purpose, and plan. Of course, the sci mats—cosmologists and evolutionists (chapters 1 and 2)—undervalue it, while occultists—astrologers and reincarnationists (chapters 5 and 7)—overvalue it! Evolution, ruling out intelligent design, teaches instead that the great diversity of life is due to arbitrary, random deviations from a norm. In this view, the minuscule chance of *accidental* design features (itself an oxymoron, for design implies plan) somehow adds up to superb system and order. This blind random process, "a giant lottery" in biologist Michael Denton's words, "is one of the most daring claims in the history of science" (Denton 1986, 149).

Are we truly pawns of chance? If, under the SM, the universe itself has no purpose (chapter 1), and if the human race also has no purpose (chapter 2), why should we as individuals have any purpose? Purposelessness or randomness, though it has no explanatory power whatsoever, has been seized by scientism as the answer. For example, agriculture (chapter 8) supposedly began with "accidental" sprouting of seeds in garbage heaps: the same "accident" occurring in Asia, America, the Middle East, and Egypt!

Another example: when Paleolithic megaliths are discovered (under the sea or in unexpected locations), the verdict is that—being "too early" for the SM—they are merely random formations shaped by nature (geofacts, not artifacts), even when great causeways and geometric structures indicate a site of archaeological importance.* Randomness is further invoked in the quantum view that events in nature are analogous to a game of chance (the so-called uncertainty principle). For instance, it must be "by chance that the earth has a moon, that we have day and night, that we have a sequence of seasons, that we have oceans and water, atmosphere and oxygen . . ." (Velikovsky 1965, 7).

Isn't it remarkable that so many of today's so-called sciences rest not only on crazy chance but also on a kind of bizarre, magical thinking? Such is the abracadabra of the big bang,[†] in which the entire universe came about in one inconceivable explosion—creation of everything out of nothing. Wow! That notion bothers people like astrophysicist Eric Lerner: "The contradictions of quantum mechanics are swept under the rug . . . [while] articles written about them tend to conclude, 'Isn't the universe bizarre?' . . . Quantum mechanics introduces magic into the heart of science. . . . The basic principle of causality is abandoned" (1991, 354). Human evolution (chapter 2),

---

*Natural water erosion, for instance, was invoked to account for (to *dis*count, really) the ancient underwater ruins stretching south from the coast of Okinawa to the island of Yonaguni, a total of eight prehistoric sites sporting grand boulevards, majestic staircases, and magnificent archways. A book on the excavation by Masaki Kimura, a Ryukyu University professor of geology (with chapter titles like "Discovery of a Civilization Lost in the Sea" and "A Utopia Sunk in the Pacific Ocean"), has not been translated into English. The saga of these suppressed Japanese finds reeks of "a conspiracy of managed information" (Frank Joseph 1997, 200).

†Speaking of bizarre and magical thinking, did you know it was Edgar Allen Poe, the great fantasist, who, in a story titled "Eureka," first proposed the big bang?

we are also taught, came about thanks to random mutations (genetic accidents), in such a way that the brain magically and "suddenly became efficient . . . [a] leap from animalistic being to homo sapiens [that] took place overnight. A miracle? Miracles just don't happen" (von Däniken 1974, 80).

In the absence of true mechanism, and with randomness taking the stage, so-called scientists have turned to phony crises to solve the enigmas of science and history. The incredible, haphazard, and implausible—cataclysmic thinking—now become the accepted explanations (as explored in chapters 1, 3, 4, and 8). With no other explanation in sight, alarmist thinking steps in and, thanks to the violence of our age, the notions of big bang, survival of the fittest, glacial devastation, and menacing global warming take the spotlight. "Such visions and dreams of world destruction," says psychiatrist Anthony Storr "are frequent precursors of schizophrenic episodes" (1996, 90).

Is science having a nervous breakdown?

Or is it all a gimmick? Have we been fashioning terrifying theories to grab the attention of the public with its famous two-second attention span? The catastrophe card is the equivalent of wild conspiracy theories. I have already touched on gradual condensation as the real mechanism behind world making. But now catastrophists are saying, "Planets, once thought to form gradually out of coalescing dust and gas, are now viewed as the survivors of a violent process of collision and accretion, the winner in a Darwinian competition to build up enough gravity to control one's own orbit" (Adler 2006, 46). Whew. The paragon of the ultraviolent twentieth century's embrace of violent science, physicist and cosmologist George Gamow, wrote, "In days long past, our planet was born from the Sun, its mother, as the result of a brief but violent encounter with a passing star" (1948, 1). Wild asteroids crashing to Earth, assert today's science gurus, caused everything from ice ages to mass extinctions. The dinosaurs, say peer-reviewed articles, were wiped out by a rock six miles in diameter that crash-landed here 65 million years ago. Serious articles in the most prestigious journals tell us that the Moon was formed out of a thrilling, sci-fi–worthy collision between Earth and another large object: "About 4.5 billion years ago, a rogue protoplanet . . . slammed into Earth and blasted a huge amount of debris into space . . . some of the debris clumped together to form the moon"* ("Gravity's Pull" 2003, 4). Our forebear Cro-Magnon man disappeared twelve thousand years ago, not due to any spectacular natural disaster (as numerous authors claim), but thanks to prosaic backbreeding (see chapter 2).

> *[For the media], bad news is good news. . . . The alarmists' publications get all the press.*
>
> ROY W. SPENCER,
> *THE GREAT GLOBAL WARMING BLUNDER*

Rogue objects and interplanetary exterminators sell so much better than mundane processes like gradual heat loss (chapter 4) or gradual accretion that formed the Moon (chapter 1). Or exquisitely slow magnetic reversal versus headline fodder like instant pole flip, or "somersault" (chapters 3 and 4).

Smacking of hype, the sensational big bang makes people smile. It's entertaining. To many, it is just a big joke. "Earth's Explosive Origins Revealed," scream the headlines. Scientists are also saying that our planet "swallowed" smaller planets; and that the Sun "flung out" particles; and that supernovae "sprayed" stuff from which the solar system formed, followed by a "shock wave," and so on. "Without such violent mixing," one article concludes, "Earth might not have come to exist" (Keats 2013, 41). Thus does so-called empirical science sound a bit more like science fiction every day, offering "slam," "swallow," and "shock" in place of the mechanism of vortexya, wherein the force field conceives planets by an orderly process and without benefit of extraneous calamities.

Indeed, very few axioms are needed to understand the universe.

---

*More likely, the Moon formed by gradual accretion as a vestige of the condensation that formed the Earth itself.

Think of Occam's razor: Look for the simplest, broadest, and most systematic explanation, then substantiate it with observation and impartial documentation. Science hasn't come to it quite yet, but it's there: vortexya. In this new/old science, we will learn to distinguish "cosmic" stuff from stuff much closer to home, that is, in our own geomagnetic field. In this connection, let's consider geologist Charles Lyell's warning not to recur to extraordinary agents, such as cosmic projectiles or other dreamed-up catastrophes, which, growled another critic, are "an easy way to explain great events . . . [they] are the mainstays of people who have very little knowledge of the natural world" (Cohen 1973, 146).

In my view, paroxysms made to serve as a quick scientific fix sprout directly from the cynicism, disconnected thinking, and decadence of our times. The bangsters, for example, say the universe is decaying. Is that so? Lerner has wisely alerted us to such tainted formulas, given the intimate relationship between the ideas dominant in cosmology and those of society at large. "The Big Bang's golden age in the seventies," he cautioned, "corresponds to the end of the postwar boom and a new decade of growing pessimism" (1991, 163). From a strictly scientific point of view, Fred Hoyle found the conceptual difficulty of the universe's sudden origin insuperable.

Unconvinced by the alarmist interpretation, chapter 3 spotlights how theorists have made the very slow and steady oscillation of the globe into a sudden and catastrophic pole shift. Magnetic reversal was not at all a sudden or threatening event; quite the contrary, it was an extremely slow and gradual balancing mechanism. Chapter 4 then takes up the global warming "catastrophe" that supposedly threatens our civilization.

In these pages, I ask the reader to consider—in addition to the first (axial) and second (orbital) motions of the Earth, the third (oscillation) and the fourth (c'vorkum) motions of the Earth—in order to complete our knowledge of earth science, as recognized by the ancients and mentioned in the Book of Wars:

The Lords took on corporeal forms and talked and reasoned with mortals, especially regarding the stars, the Moon, and the Earth: teaching the four motions of the Earth: axial, oscillaic, orbitic, and vorkum; the plan of the hissagow [solar phalanx]; and the cycles of the Earth; the cycles of the sun; and the vortices that move them all.

OAHSPE, BOOK OF WARS 37:7

Ice age, big bang, global warming, evolution—they are all clichés that most people take with a healthy grain of salt. There is a "deep-seated distrust of expertise" among ordinary people (Pollack 2003, 37). They tend to believe the experts as little as they do the politicians and the propagandists. ᴧ

PART TWO: SELF

Here we will move into the reigning schools of thought claiming to discover the secrets of human nature, personality, how the mind works, even what constitutes the soul. The veritable chaos of theories—especially of the new age bent—only underscores why we must understand the spirit world correctly. It is not enough to simply believe in it; we need to know how it works. It's time for a little housecleaning in the new age. The hocus-pocus of astrology is untangled in chapter 5, while the illusory Freudian and Jungian unconscious is unmasked in chapter 6, and the black art of reincarnation is exposed in chapter 7. In each of these popular doctrines we find the underlying premise to be little more than a glorified idea, a kind of intellectual sleight of hand, which incorrectly has us all predestined and predetermined from the start.

Chapters 6 and 7 also challenge the view that ideas or instincts reside in our genes, having been put there as survival mechanisms; these chapters also lay bare our disastrous divorce from the soul. Both the theory of the unconscious and that of reincarnation have grossly misrepresented the spirit of humanity. Finally, these two chapters probe the common thread of the human psyche—a tangled thread—which we unravel simply by differentiating the mortal life from the immortal life. Part two inevitably deals with misguided and sloppy occultism, which has disfigured our thinking and besmirched our dignity.

PART THREE: SOCIETY

Was the Fertile Crescent in the Near East the true cradle of civilization? Should we believe today's esoteric protohistorians who say that a highly evolved civilization (Atlantis) was inundated and destroyed by massive floodwaters at the end of the last ice age, the sages of that high culture finding refuge in high mountain areas? I think that is imagined history. What we do need to pursue is the lost race theory in order to make sense of the advanced civilizations of the Paleolithic as well as the irresistible similarities between them.

The book closes with a look at democracy (chapter 9) and the paradigm shift moving us into the future.

Out of the Box

In Carl Jung's famous OBE (out-of-body experience), he was in a coma in 1944, hospitalized after a heart attack. Drifting far above the Earth, his free mind saw an etheric temple carved out of a huge block of stone. It was an NDE (near-death experience), for the nurse saw him surrounded by a halo of light (typical of those near death). Jung's spirit drew closer to the temple in Etherea, and it dawned on him that inside he would find the meaning of life. But a spirit-form appeared before him and insisted he return to Earth. Describing his dominant emotion later, Jung said he felt depressed and furious at being back in his body and in this world of "boxes."

Today, on the cusp of great change, the mandarins of science still cling to those boxes and with them the reins on this age of information (and disinformation) which is nonetheless sweeping full bore into the coming age of understanding. It is a time in which we shall make a giant U-turn, as we improve our perception of rotation, gravity, condensation (not explosion), and the unseen world that powers it all!

Don't listen to all the garbage your teachers fill your poor student head with. But get your degree anyway . . . to have the piece of paper, but you can forget everything you've learned.

LEIF DAVIDSEN

PART ONE
SCIENCE

Abbreviations

bya = billion years ago   byr = billion years old

mya = million years ago   myr = million years old

kya = thousand years ago   kyr = thousand years old

CHAPTER 1

BIG BANG

Or the Universe Has Always Existed

Who knows for sure, who can proclaim here,
Where it originated, from whence this creation came?
"THE SONGS OF CREATION" IN THE *RIGVEDA*

Though the cosmos may indeed have eternal laws, there seems to be no permanence to *man's* laws, prematurely labeled "scientific fact." The universe's geometry, for example, was once thought to be elliptical, then

Physicists were awed, astounded, and devastated when the ultimate origin of the Big Bang was revealed.

Fig. 1.1. "Everything we see could have emerged as a purposeless quantum burp in space." Such are the delirious sound bites we hear from proponents of the "godless particle" as discussed in this chapter. "Creating 'stuff' from 'no stuff' seems to be no problem at all" (Krauss 2012, 5). But it is a problem . . . Cartoon by Marvin E. Herring.

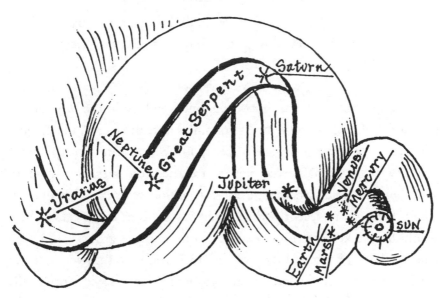

Fig. 1.2. The cyclic coil. Adapted from Plate 47 in Oahspe. The solar system is not flat in this diagram. Neither are the requirements of a flat universe satisfied by "dark matter" (as discussed below).

hyperbolic, then spherical, then concave, then convex. In the latest draft of cosmology, it is flat. Which one is right? An infinite universe, without boundary, as discussed below, could hardly be flat. Is today's standard model (SM) any exception to the entire history of science, which admits the overthrow of orthodoxy from one stage to the next?

OLD GALAXIES

Nothing is more curious than the self-satisfied dogmatism with which mankind at each period of its history cherishes the delusions of the finality of its existing modes of knowledge.

ALFRED NORTH WHITEHEAD

Galileo bested Aristotle, Copernicus toppled Ptolemy, Kepler superseded Copernicus, and Newton's microphysics was trumped by Einstein's relativity and Bohr's quantum mechanics, which make up the current SM of field theory. Now, with all due respect to hardworking scientists, we are compelled to put aside today's planetary views, if they are wrong.

Like the big bang, which says that about 13.7 bya* (maybe as much as 20 bya), the universe was hot-wired. But even here, at the very start of the model, there is a problem. Astronomers have discovered much older galaxies, some that must have taken 100 byr to form. Astrophysicist Fritz Zwicky, who pioneered dark matter, neutron stars, and supernovae, thought some stars were a million billion years old. Astronomer Brent Tully, more recently, discovered supercluster complexes, which would have taken a trillion years to form, definite grounds for challenging the big bang and its birthday 13.7 bya. It was, incidentally, in the late 1940s that the late, great astronomer Fred Hoyle sarcastically and offhandedly came up with the term *big bang* while speaking on the radio. We will meet the colorful Professor Hoyle later.

So how does the big bang work? In one split second the whole she-bang flashed out from a superhot "something" smaller than an atom—a speck far smaller than a proton. Is this science fiction I am hearing or sober cosmology? Well, if you are going to replace the Creator, it better be sensational.

In its first moments, the secular pundits go on to report, the big bang spewed matter helter-skelter, creating a featureless primordial soup of subatomic particles. Three minutes later, simple atomic nuclei began to synthesize quarks. These first bits of primordial matter began to "pull together" to form the first atoms. A short 300,000 years later, fast-burning blue stars transmuted the big bang's pristine hydrogen, helium, and lithium into heavier atoms.

Decreasing in density after the bang, all this matter, in time, somehow formed huge "clumps," which became galaxies. That's a big "somehow": The details are unknown. To explain this seeming miracle of self-creation, big bangsters say that gravity then began pulling together the clumps. This is what supposedly continues to hold the planets in place. Twelve bya the

---

*The big bang sets the maximum distance from which we can receive information to 13.7 light-years. Are we still the center of the universe?

swelling universe began to subside, making it possible for gravity to pull matter together out of those primordial seeds. This then is offered as the genesis of the first stars and planets: the denser regions presumably pulled in nearby material by virtue of gravity. This is how all those primordial fragments came together in just the right way to form galaxies.*

Thus have the building blocks of life, according to our science gurus, formed and organized themselves into self-sustaining units. But can things really form all by themselves? Although scientific materialism says yes without hesitation, other cosmogonies do not allow it. The nineteenth-century physicist James Clerk Maxwell, for one, mathematically disproved Pierre-Simon Laplace's nebular hypothesis, which had the solar system condensed from a cloud all by itself. Not only would the Sun require a faster spin, but that nebular material would have actually pulled apart, rather than come together.

Nowhere in etherea [deep space] is there a solution of corpor [matter] sufficient to put itself in motion or to condense itself, or to provide the road of its travel.

OAHSPE, BOOK OF COSMOGONY AND PROPHECY 2:12

As we go on, we will offer the principle of condensation, not dissolution or explosion, powered by vortexya (the cosmic whirlwind) as the key to astrogenesis. Do we really understand the physical mechanism in outer space that compresses gas into a nice planetary ball? We ask this because a cloud of hydrogen must be sufficiently compressed, in the first place, for gravity to dominate it. Gas molecules out in space do not just clump together. What would compress them? Science has no answer, least of all the big bang model, which assumes the universe started out as smooth (homogeneous) expanding gases with no organizing principle at all. Gases are not wont to pull together; on the contrary, they gradually move outward.

*See chapter 2, on evolution, where theorists propose the same ineffable power of nature to organize itself, evolving plants, animals, and humanity, à la Charles Darwin.

How could the stars, planets, and galaxies evolve from floating gases? "Gravity," explains my friend and colleague James M. McGill, "is not a sufficient mechanism to do this. In outer space, the gas is millions of times more expansive than the critical compressed size needed for gravity to hold it as a stable star . . . Vortexya must be the cause of compressing interstellar gas into a ball that creates stars and planets" (studyofoahspe.com). Vortexya, which is roughly the winding action of the electromagnetic field (EMF), is the long-sought world-maker, as discussed further below and in the chapters to come. Parsimoniously, it also gives us the "theory of everything."

Who formulated and approved that do-it-yourself universe, anyway?

A great flurry of cosmological theorizing culminated in the roaring twenties with Edwin Hubble's telescopic observations of a universe that seemed to be expanding. It was around this time that the Russian mathematician Aleksandr Friedmann and the Belgian priest G. H. Lemaître arrived at the mutual opinion that the universe was indeed expanding from a denser, earlier state. Yet it bothered some scientists that the church's doctrine of creation ex nihilo was what had inspired the Roman Catholic Lemaître to invent the big bang. The leap out of nothingness looked like a scientific confirmation of the first few sentences of Genesis; this led to the approval of the big bang theory by Pope Pius XII in 1951.

Is space really an expanding entity, forever dragging the galaxies away from each other? Although Einstein's general theory of relativity was used as a framework for Hubble's discovery of alleged expansion, Einstein's own gut reaction was negative: his theory of general relativity does not posit expansion. Neither could the great Jewish sage accept quantum theory's overall abandonment of determinism and causality, an event occurring with no conceivable cause and no clear mechanism. "The absence of a mechanism," observed cosmologist David Raup, "is often used as a weapon against research conclusions that we don't like" (1986, 109). But since we do like the big bang, we overlook its indeterminacy—and its other flaws. In fact, before Hubble's Big Mistake, most physicists, including Einstein, were inclined to believe that the cosmos had always existed. Einstein doubted both big bang and black holes throughout his life, always looking for a better, more consistent, explanation.

While Hubble's "law" contends that galaxies have been spreading out ever since the big bang, "Hubble's constant" denotes the rate of that expansion, for the galaxies appear to be rushing away, apparently still fleeing the horrid big bang. Then, in 1998, analysts concluded (counterintuitively) that the expansion, rather than slowing down, was accelerating! Faint distant galaxies were thought to be flying away from the Earth faster than brighter nearby ones—presumably establishing the picture of a universe that keeps growing, carrying galaxies farther apart—faster.*

But if the galaxies are flying away from everything else, wouldn't space, sooner or later, become empty? We are, in this scheme, asked to wrap our minds around statements like the following: "We cannot say with any real certainty that the rest of the universe will disappear in the far future. Although accelerating expansion will cause distant galaxies to recede from view . . . the Milky Way will merge with its neighbors" (Krauss 2008, 12). And how does the popular theory of colliding galaxies fit in with an expanding universe? According to the SM, galaxies, pulled about by the gravity of other galaxies, may have their own trajectories across the "balloon"; this "local" movement explains how galaxies can collide, and how the Andromeda Galaxy is moving toward us. But that movement toward us may only be apparent. What if the rushing away of galaxies is actually an optical illusion? "Cosmic expansion may be just a mirage" (Pendick 2009, 51). This is where c'vorkum, the fourth motion of Earth, comes in—the movement of the entire solar family through space.

---

*However, distant galaxies, it is explained, seem to move apart only because the space between them is increasing—just as raisins (the metaphor of choice) move apart in a rising loaf of bread, or as painted dots expand on an inflated balloon. Galaxies, in other words, are not really "moving away," but only give that impression, since they are part of the allegedly expanding fabric of space-time. The universe's expansion, then, does not actually push bodies into new territory; rather the space-time grid itself is inflating. At least according to the SM's "inflation" theory.

C'VORKUM

The earth is not in the place of the firmament as of old.

OAHSPE, BOOK OF JEHOVIH 4:20

Perhaps it is Earth's movement through the firmament, this c'vorkum, that makes it look as if certain stars are moving away from Earth— something like the landscape seen rushing away from the window of a passing train, or the oncoming landscape (Andromeda) that appears to be rushing headlong at us. *C'vorkum* is an ancient term referring to the pathway taken by the solar system in its movement through the galaxy; that is, the orbit made by the Sun with all its planets. This is the "etherean roadway" illustrated throughout Oahspe. (C'vorkum will be addressed again in chapter 4 as an unforeseen factor in climate change.)

Even if the universe *is* expanding, Swedish astrophysicist Hannes Alfven and others have pointed out that it does not necessarily require a mythological big bang to explain it. "There are actually a half-dozen competing explanations" (Lerner 1991, 278–79). One of them is Alfven's plasma theory; another might be Einstein's equations involving solutions that do not need to draw on the "singularity" thesis (a single point of origin).* Too, there are analysts who wonder, "What if the bang resulted from a collision between universes?" In which case it would not require the phenomenon of inflation (expansion). Yet others say the universe is actually contracting, with no origin in time. Still others say only a part of the universe is expanding. The only thing expanding here is a fog of doubt.

Has any mechanism for expansion ever been found? Even Alan

---

*Since most of our universe is hidden from view, astronomers admit that the big bang is "still largely unexplained" (Davis 2002, 298). The event is referred to as a singularity, providing "in all truth, no insight into what might have powered the bang itself" (Greene 2012, 23). Singularities (stuff that cannot be explained by the SM) did not sit well with Einstein. Among them: the notion that everything started from an infinitely dense dot. Where did the dot come from? No one knows, and the guessing game begins. Nobel Prizes are handed out for the most loudly applauded and distinguished guesses.

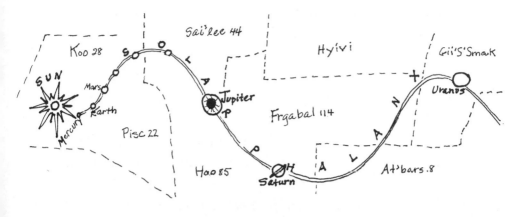

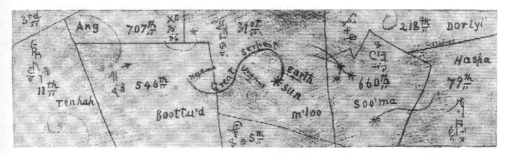

Fig. 1.3 C'vorkum. Top: Travel of the Serpent in the Roadway of the Firmament (adapted from Oahspe Plate 36). The orbit of the Great Serpent (the solar system) takes 4.7 million years. Bottom: Location of the Great Serpent during the second nine thousand years after humanity's creation (adapted from Oahspe Plate i089).

Guth, the man who identified expansion in 1979, admits that its mechanism remains unexplained. To fill that gap, the SM postulates a hidden dimension, with something called antigravity (a.k.a repulsion) at work, a drive that pushes all things apart. To make that idea work, something called "dark energy" is called into existence—a mysterious force that pushes the universe apart at an ever-faster rate. Dark energy is presumed to have emerged 5 bya against its opposite, the force of gravity, when the cosmos kicked into high gear and began accelerating. This, true or not, completes the SM of the universe.* ꙡ

---

*Why am I not surprised that the big bang theory—a theory entailing immense violence and the antagonism of two great forces—came along right after the planet's first world war?

When it was found that distant stellar explosions (type Ia super-novae) were slightly dimmer than expected, the "explanation" was that the expansion of the universe was accelerating. And the cause of this acceleration was dark energy. Since exploding stars have a known, intrinsic brightness, they were used as "standard candles" to determine the star's distance and velocity, thus calculating the rate of acceleration. Critics, however, say that supernovae have been altogether misinterpreted and only create the illusion that the expansion of the universe is speeding up. When the Hubble telescope photographed one of these exploding stars in 1997, astronomers jumped to the conclusion that it confirmed the idea of space uniformly filled with an invisible form of energy that creates a mutual repulsion between objects normally attracted to each other by gravity. This new energy, as a property of space, would not only serve to counteract gravity but also to explain the acceleration noted or assumed by cosmologists. If dark energy is a constant vacuum energy (inherent in space), it is then reasoned that it would cause acceleration, because what it does is push galaxies apart.

The entire argument smacks of circularity. Cosmologists, hedging their bets, modestly add, "A universe of questions still remains" (Stone 2007, 12). Such qualifiers thread throughout these arguments: "Exploding stars . . . show that the cosmic expansion *may* be accelerating: a sign that the universe *may* be driven apart by an exotic new form of energy" (my italics) (Hogan 1999, 46). More to the point, the apparent dimness of these distant supernovae could just as well be due to demagnification caused by gravitational lensing; or possibly they are farther away than their redshifts suggest (see the discussion of redshift, page 65); or perhaps cosmic dust screens out some of the light. Indeed, lately they have been reporting that our universe is "dustier" than we thought, spreading huge amounts of space-dust and gas in every direction, which could redden a supernova's light. "This finding is important because astronomers use the brightness of certain stellar explosions . . . to study the universe's expansion" (Krauss 2008, 22). ᴧ

DARK ENERGY

*If the size of everything . . . evenly expands, distant objects
only appear to be redshifted.*
ERIC LERNER, THE BIG BANG NEVER HAPPENED

The cosmic light neither attracts nor repels.
WALTER RUSSELL, THE SECRET OF LIGHT

And what exactly is this "dark energy" that was so conveniently discovered to validate expansion and repulsion? Most of the universe, say
cosmologists, is composed of invisible energy (hence, labeled "dark"),
about which little is known. "When it comes to dark energy," confesses
Lawrence Krauss, a theoretical physicist and cosmologist, "we know
that it exists, but we don't know anything about it" (quoted in Pendick
2009, 48). Yet it is "just what we need to explain the repulsive gravity
. . . [that] pushes every galaxy away from every other, driving the expansion to speed up" (Greene 2012, 23).

But here is the irony: even though the SM has discovered this
"weird" dark energy—really *re*discovered it, for it is almost the same as
Aristotle's quintessence and the nineteenth century's "ether," though
neither of these posited expansion or acceleration—it nonetheless contradicts other parts of the SM. How so? Its density is too low and its
energy too great, which are just the properties of the despised classical ether. The problem for the SM is that it requires a certain average
density of mass, but space turns out to be much more rarified than we
supposed. (Ether is rarified.) Scientists bemoan "the spectacular failure
of attempts to explain [this]. . . . When the astronomers deduced how
much dark energy would have to permeate space to account for the
observed cosmic speedup [acceleration], they found a number that no
one has been able to explain. . . . The dark energy density is extraordinarily small" (Greene 2012, 23)—a figure of .138, with 120 zeros
after the decimal point! Doesn't this tell us the "speedup" idea (and
its chain of inferences) is crumbling? Maybe "expansion" is actually

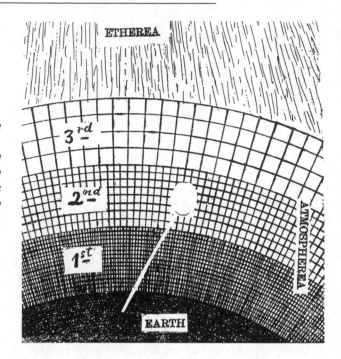

Fig. 1.4. Oahspe diagram of rarification. Cross-hatching indicates comparative densities, seen as more rarified in the upper regions.

slowing down. Maybe it is constant. Maybe it is not expanding at all.

Meanwhile, when physicists calculate the value of this dark energy, it works out to be more than a hundred orders of magnitude too large, indicating "a repulsive force strong enough to rip all matter apart in an instant" (Stephen Battersby 2005, 6). All these developments, as some frankly see it, promote "a fundamentally irrational physical universe" (Lerner 1991, 362). One Nobel laureate in dark energy says (if you can believe this): "I have no clue what dark energy is" (Marsolek 2012, 62).

Let the clueless experts read up on the classical ether or even recent astrophysics, which finds an "ethereal energy" threading through empty space. For science now concedes space is not a vacuum, after all; it is filled with fields of force whose medium must be the imponderable ether that was so contemptuously disregarded after the famous Michelson-Morley experiment* more than a century and a quarter ago. When skywatchers in 1998 declared that the expansion of the universe was accelerating,

*Published in 1887, the Michelson-Morley experiment attempted to detect relative motion of matter through the stationary luminiferous aether. —*Ed.*

they explained this by invoking dark energy, a "kooky energy" (Hogan 1999, 45) that supposedly repels rather than attracts. Well, they found the right thing for the wrong reason. It is not "new," "weird," or "exotic." This kooky energy was known to Aristotle, Paracelsus, James Clerk Maxwell, Sir Oliver Lodge, Sir William Crookes, Thomas Edison, and many others.* Only they called it "ether." ⋀

To explorer and naturalist Alfred Russel Wallace, it consisted of "minute vibrations of an almost infinitely attenuated form of matter . . . a recondite force . . . [with] a power of motion as rapid as that of light or the electric current." Edison, for his part, called electricity "an etheric force" and, presaging today's EVP (electronic voice phenomenon), he spent many years trying to tap this force in order to develop a "soul-telephone" that could be used as a telepathic channel between the living and the dead, something to improve on the "trumpets" of the spirits. Even Einstein's cosmological constant was "like the ether," for it "endows the void with an almost metaphysical aura . . . [perhaps] as far-fetched as angels sitting on the head of a pin" ("Special Report" 1999, 53). But ether is not so far-fetched; it is merely rarified, corpor in solution, sublimated beyond the reach of any physical instrument. We can't nail it down.

Physicist John A. Wheeler once remarked that the ultimate constituent of the universe is the "*ethereal* act of observer-participancy" (my italics), referring to the virtual particles that inspired the Heisenberg uncertainty principle: stuff shows up because we are looking for it. Hence, "objective" reality fades away. It was this subjectivity, this evanescence of the ethereal wave that had made us think of space as insubstantial, a vacuum. If anything was going on in interstellar space, the best we could do was call it "quantum fluctuations." Einstein said, "The ether, if it exists, cannot be detected." This is exactly what bothered science—these "virtual particles" that pop in and out of existence too quickly to be detected, thanks to ether's low amplitude and high vibrational frequency. This is precisely the

*For Crookes, Lodge, and Wallace, the spirit world of ether was the ground of energy. Ether, to Heraclitus, was the soul of the cosmos.

behavior of corpor in solution. It all seemed like a ghost world,* these spectral particles now sought after in sensitive, costly, and dangerous underground experiments. ᴧ

Perhaps we have been trying too hard to capture it; it will not obey the laws of ordinary atoms. Think of the billions of dollars we would save by eating crow and humbly accepting the evanescent, inaccessible ether once and for all. The goal of today's costly fusion experiments is to find the hypothetical Higgs boson particle, millions of times smaller than the nucleus of an atom. So far, $10 billion have been spent on the particle accelerator at CERN (the European Organization for Nuclear Research), called the Large Hadron Collider (LHC). This smashes protons together in a seventeen-mile-long tunnel beneath the Swiss-French border, the collisions presumably "mimicking" the big bang. Primordial black holes, it is also thought, were created during the big bang. But are we playing with fire? "The concern is that micro black holes . . . if produced at the LHC, could be a recipe for a worldwide disaster" (Williams 2008, 23, 59). And although it is supposed that artificial nuclear fusion will "mimic" the big bang, this is an expensive assumption.

> *Scientific discoveries of major importance still to be made . . . are not hidden deep in the atomic nucleus . . . not so cunningly concealed that they require billion-dollar particle accelerators . . . to unravel them.*
>
> RICHARD MILTON, *ALTERNATIVE SCIENCE*

Just in time for the temporary closing of the LHC in 2013, scientists announced they had found the "magnificent" Higgs boson God particle. Read on, and find that *"they believe"* they have found the "elusive particle which *many theorize*" (my italics) started the universe. The data, we are told, "strongly indicates" it. Trumpeting this research as a

---

*The term *ghost world* refers to the solvent of corpor, also known as *ether* or *ethe*. Ether can readily pass through solids. "Dark matter particles are hard to spot since they can pass through ordinary matter as if they weren't there" (Lemonick 2013, 15).

"strong contender for the Nobel Prize,"* scientists nonetheless "stopped short of saying conclusively that it was the same particle. . . . We still have a long way to go" (Heilprin 2013).

"Most of us," admits science writer Bob Berman, "are already bored with today's mind-numbing list of particles . . . bits of flotsam" that the Large Hadron Collider has discovered. But "to what end?" asks Berman. "Bosons, mesons, pions, kaons, anti-quarks, J-particles: how much of this can we handle?" Will these particles really give us the theory of everything? "Any day now," Berman answers sarcastically, though with this insight, "navigating the particle swamp will keep Europeans occupied for decades, a vast improvement over their traditional warfare pastime of bygone centuries" (Berman 2009, 16).

It is driving scientists crazy, allowing invisible, immaterial, undetectable stuff to fill all space! Yet that is what it does—it pervades all things and is the solvent of all things. Ether (ethe) is the very pith of the unseen world.

There are two known things in the universe: ethe [unseen] and corpor [seen], the former is the solvent of the latter. Earth substance is soluble in ethe, the great etherean firmament being a dense solution of corpor.

OAHSPE, BOOK OF COSMOGONY AND PROPHECY 2:1–4

Ethe holds corpor in solution; this is the condition of atmospherea and of the etherean regions beyond. And when a portion of this solution is given a rotary motion it is called a vortex. Indeed, ethe is more subtle, more sublimated, when axial motion of the vortex is

---

*In October 2013, that prize was indeed awarded to Peter Higgs and François Englert for their role in postulating the Higgs boson particle.

swifter, i.e., in the higher regions. Nor is a vortex a substance or thing of itself, any more than a whirlwind is, or a whirlpool in the water. As a whirlpool cannot exist without water, or a whirlwind exist without air, so a vortex cannot exist without the etheic solution.

⋏   OAHSPE, BOOK OF COSMOGONY AND PROPHECY 3:13

The same concept was taught by Carl Friedrich von Weizsäcker: that all the planets were born from a cloud of dust, while the galaxies built themselves around the heavy elements in a series of whirlwinds.

The old, discarded ether theory, now enjoying a comeback, is needed to resolve the conundrums of physics. The space vacuum, which is not really "smooth," is a seething mass of virtual particles, a kind of "foam" or "particulate ether," as suggested by physicist John Bell in an interview with cosmologist, astrobiologist, and physicist Paul Davies: Bell argued that the ether was wrongly and arbitrarily rejected by the Michelson-Morley experiment, purely on philosophical grounds, that philosophy being strict materialism: what is unobservable does not exist. Nevertheless, the Michelson-Morley experiment "does not exclude the existence of an ether . . . [which] is necessary for the propagation of electromagnetic and gravitational forces without acting at a distance" (www.lightenergy4.homestead.com). Space, in other words, is not passive; the Faraday/Maxwell electromagnetic field confirmed that there *was* a force out there affecting electrical masses. To account for it, these great nineteenth-century theorists posited a medium pervading the whole of space—ether. But with no mechanism to explain it, or eyes to behold it, or instruments to measure it, it was judged unsound and rejected by traditional science.

The SM, in addition, suffers from two mutually incompatible theories: relativity and quantum mechanics. Science has been unable to reconcile Einsteinian physics—which seeks certainty, order, and majesty in the universe—with quantum mechanics, which settles for random behavior of particles under a principle of uncertainty. As far as Einstein was concerned, "God didn't play dice with this universe." It wasn't a crapshoot. It wasn't chance. It wasn't random. The uncertainty, you might even argue, is our own, not the universe's!

Frustrated by the increasingly paradoxical theories spun off quantum physics, Einstein once quipped, "I would rather be a cobbler than a physicist." Since his time, things have only gotten worse. Today's cosmologists, as Eric Lerner sees it, are building "bizarre towers of ad hoc hypotheses . . . something Einstein, the lover of simplicity and beauty . . . would never have tolerated. . . . The Big Bang . . . has flunked every test, yet it remains the dominant cosmology; and the tower of theoretical entities and hypotheses climbs steadily higher" (Lerner 1991, 162).

Inquiring minds have asked: if the big bang begins All and Everything with a super-violent explosion out of nothing, where did that tremendous energy come from? And how (contradicting the second law of thermodynamics) did it make order out of chaos? Do we really think the majesty of worlds came from one colossal, mindless, purposeless detonation?

> *You either have to say there have been monstrous coincidences . . . or there is a purposive scenario to which the universe conforms. . . . The question as to whether the universe is a product of thought . . . that is my personal opinion.*
>
> FRED HOYLE

The possibility that the world happened by chance, someone once posited, is less than the possibility of *Webster's Unabridged Dictionary* springing fully formed from an explosion in a print shop. "A causal basis has been lost sight of," admitted Sir Arthur Eddington, remarking on quantum physics's dead end. The big bang, after all, is without a known cause, as is the origin of matter itself. Not a few scientists are "uneasy with a universe that . . . came from nowhere" (Mitton 2005, 118).

The abyss, the void of uncertainty, enshrined in the twentieth century as a "principle," only masks the precarious state of current thinking. Neo-Darwinian evolution (chapter 2) also invokes it: The supposed incompleteness of the fossil record reflects "uncertainty." Global warming invokes it, too. Yet in his own way Darwin warned against it: "It is those who know little and not those who know much, who so positively assert that this or that problem will never be solved by science" (Hayden 2009, 48). Haven't many "unknowables" of the past become known? Today the unseen world awaits discovery or, should I say, disclosure—bumping up the "unknowable," the uncertain, to the knowable.

The imponderable ether or the "virtual particles" of Werner Heisenberg's uncertainty principle are no less real or "uncertain" just because they fail to conform to the standards of a material (seen) and mechanistic world. Instead, let us better understand that world. When Heisenberg found that it was experimentally impossible to chart an electron's velocity and position at the same time, he concluded that we cannot know the electron's future. It is uncertain. Besides, the spin of a photon is not definable until you actually measure it. But should we label this "uncertainty"? Or move on to study the ether in its own right?

Quanta (filling in for ether) are not "things"; hence they are labeled uncertain. We could only say that a particle had the *probability* of being in a particular location. "Hard science" dead-ended right there. Heisenberg and Niels Bohr—rather than embracing the unseeable ether—rejected it and pronounced that reality is at best run by chance. Einstein abhorred this, holding out for a model of reality describing "things themselves and not merely the probability of their occurrence" (Einstein 1954, 276). Nevertheless science chose to elevate, even glorify, this philosophy of chance and doubt by extolling the "principle" of uncertainty: "Solutions, as quantum reality teaches, are a temporary event. . . . In this new world, you and I make it up as we go along. . . . We must live with the strange and bizarre . . . [and] stay comfortable with uncertainty and confident of confusion's role. . . . We will have to muddle our way through. But in the midst of the muddle . . . [we go] deeper and deeper into a universe of inherent order" (Wheatley 1994, 151). Writer and consultant Margaret Wheatley's last two sentences nicely illustrate split mind, as discussed in the introduction. This is also the mantra of Stephen Hawking. "I'm glad that our search for understanding will never come to an end, and that we will always have the challenge of new discovery. Without it we would stagnate," he told *Discover* (2005, 65). Of course he is glad about that—indeterminacy keeps the ball rolling at physics headquarters—just as theorists in other fields, including climatology and psychology, welcome the slick escape clause with open arms. Now I think I understand what is meant by "people who thrive on chaos."

A further example might be the questionable "feedback" mechanisms in world climate. Roy Spencer thinks "it is this persistent uncertainty that keeps the fear of catastrophic global warming alive. As long as scientists can claim there is a great uncertainty about feedbacks, then there is always the possibility . . . that we have already pushed the Earth past the point of no return" (Spencer 2010, 71, 102).

Uncertainty comes to the rescue in psychology as well:

> The modern discovery of the unconscious shuts the door forever. It definitely excludes the illusory idea . . . that a man can know spiritual reality. In modern physics, a door has been closed by Heisenberg's principle of indeterminacy, shutting out the delusion that we can comprehend an absolute physical reality. . . . The unconscious can only be approximately described—like the particles of microphysics—by paradoxical concepts. What it really is, we shall never know, just as we shall never know this about matter. (von Franz 1964, 253)

The same noncommittal vagueness masquerading as reasoned expertise is implanted in Carl G. Jung's archetypes, which are pretentiously called "primary possibilities." To Jung, the journey into the mind was an uncertain path.

What a sorry state we are in, with our various sciences wimping out to closed doors and giant question marks. No wonder we are so insecure. We need certainty! We deserve it. We can achieve it. It is a patent falsehood that we shall never have it. Staying "comfortable with uncertainty" is not only disingenuous and cowardly, it is a green light for slippery moral relativism and all its integrity-destroying sophistry!

Particle physicist Niels Bohr searched for cosmological revelations under a microscope. Would the very small reveal the very large? Based on particle physics, the big bang model, as I see it, has flagrantly overworked the microcosm/macrocosm analogy (as further discussed in chapter 5). "The microworld is not a Lilliputian version of the macroworld; it is something qualitatively different" (Paul Davies, quoted in Groothuis 1986, 96). The atom is a metaphor, not a mirror, of the universe. Should we not study a thing for what it is, not what it is like, and leave metaphors to the bards? Who is to say the universe behaves just like a tiny particle wiggling under a microscope? Scientists nonetheless are using laboratory data on nuclear fusion to corroborate the big bang: experiments being conducted at setups like LHC (see page 44) are trying to prove something that allegedly happened billions of years ago. It's worse than the tree falling in the forest. Better to go from knowables to the unknown rather than the other way around.

What then are the knowables? Let's consider the general uniformity of arrangements and elements in the cosmos. The Moon is constituted of materials similar to those on Earth: iron, stone, clay, water. Comets are made of earthly substances like water, oil, gold, stones. The universe is homogenous, the same wherever we look; chemical elements are roughly equivalent throughout the galaxy. But wait. There's a problem. If matter spread out evenly throughout the universe, it makes the blind fate of a huge explosive scattering most improbable. Neither does the big bang give the early universe enough time to reach thermal equilibrium, meaning there should be differences in energy at different places. A violent bang, in short, would have to be fine-tuned to an absurd degree to secure the homogeneity we find out there. The radiation known as cosmic microwave background (CMB), supposedly released after the bang, is incredibly uniform. Well, that is why the concept of inflation was invented, referring to an extremely rapid expansion of "everything" in the early moments of the bang, driven by dark energy that permits an expanding gas of evenly distributed atoms.

Cosmic microwave background radiation (CMB) is assumed to be the remnant, the "lingering echo," of the big bang, preserved from the time when the cooling universe was less than 1 myr. Not all agree with this neat explanation. Perhaps CMB emanates from more recent structures "instead of being related to an event that happened billions of years ago" (Kreisberg 2012, 49). CMB could be coming out of massive stars or from nearby galaxies whose energy is absorbed by interstellar dust. Indeed, Fred Hoyle's team thought CMB came from the center of active galaxies as part of their continuous creation model (see page 63). Although CMB was celebrated back in the sixties as the slam dunk for the big bang, its "static" actually comes from all directions in the heavens. And it is too smooth to account for the helter-skelter explosion or for matter to have "clumped" together into celestial bodies. How could stars and planets have been forged out of this smoothness, this evenness? They couldn't, says Eric Lerner. There simply was not enough time for such large-scale structures to form, nor was there enough matter. The plasma cosmology model has an alternate explanation of these structures, as well as of the abundance of light elements, the absorption of radio waves by the intergalactic medium, and the formation of spiral galaxies and quasars.

Even spiral nebulae in the far reaches of space are distributed uniformly. Why was the early universe so orderly? An implausible circumstance, say physicists. That's when these geniuses invented the "multiverse," the mind-bender that would have us believe that other universes were disorderly and chaotic, but ours, being orderly, survived and thrived, thanks to its tidiness. Just a lucky break. *Happenstance* is one word these anti–intelligent design scientists use to explain it. For example: "I think there is something that needs explaining," said Martin Rees, England's astronomer royal and chief explainer.

While people of faith accept a well-designed universe simply as a token of the Great Architect, the agnostic scientist always needs an explanation. And so the latest explanation is "M-theory," postulating a "multiverse" composed of billions of universes, which, by chance, run on different physical laws. To the atheistic explainer, the physics of "our" universe is so exquisitely orderly, the precision so "uncanny . . . [that it is] unlikely to an absurd degree." Being so unlikely or odd (an overwhelming improbability), it must be a fluke of the multiverse, occurring by sheer force of numbers, in the same sense that snapping

Fig. 1.5. There is no intrinsic conflict between science and religion; it is only the webs we weave. Cartoon by Marvin E. Herring.

up all the lottery tickets guarantees buying the winner. Order, then, is a kind of spectacular, fortuitous accident. Is this scientific thinking? Or antireligious twaddle?

Labeling a theory "metaphysical," admits Martin Rees, "is a damning put-down," because metaphysics is not science and its notions cannot be proven or disproven. However, Rees argues that the multiverse "genuinely lies within the province of science," even though he concedes in the next breath that (1) the concept remains speculative, (2) the premises upon which many multiverse calculations rest are "highly arbitrary," and (3) "the theory hangs on assumptions . . . about the physics of very dense states of matter." Indeed! Although Rees (in 2000) thought that "within the next twenty years we may be able to put the multiverse on a firm scientific footing . . . ," M-theory, according to Brad Lemley, seems "little more compelling than the conjured-by-God hypothesis," a toss-up. M-theory also seems to make us (sapient beings) alone in the vast universe. After all, what are the chances of this "overwhelming improbability" (Lemley 2009, 66–69) happening twice, or multiple times?

GRAVITY

But let us return to the one and only universe we know, and its gravity. The textbooks say the Moon is held in orbit by gravity; the entire solar system is supposedly held together by gravity. Allow me to quote the ever-candid James Churchward in *The Second Book of the Cosmic Forces of Mu* on the matter: "I cannot find that any of our scientists have attempted to explain what the force of gravity is that carries matter down, or to show either where or how it originates." Gravity, contends the SM explainer, is a force of attraction between massive bodies; nevertheless "its source is a mystery" (*Scientific American*, May 2007, 104).

Apparently just as mysterious as where the energy of the big bang came from. Theory simply cannot explain the powerful force that inflated the universe. Even the Large Hadron Collider "may be unable to shed light on these cosmological quandaries" (Stephen Battersby 2005, 6).

But gravity should not be a mystery; it is only misdefined and conveniently invoked as the force that made the stars condense out of the expanding debris of the big bang. Isaac Newton said every particle in the universe attracts every other particle with a force that depends on their mass and the distance between them. But that can't be right. My understanding of gravity involves no such thing, no such attraction.

The Book of Cosmogony and Prophecy (chapter 1) boldly asserts: "That which thou callest gravitation is no attraction between bodies." What then? I believe that planetary science cannot be mastered without an understanding of the nature of force fields; in plain language, gravity is the *centripetal* force of that field. And we are calling that force vortexya. "The only force capable of turning the scattered gases [of early Earth] into a spherical form is a Centripetal Force . . . bringing her into form, out of chaos. . . . In the great nebula of Andromeda . . . the lines which the gases are taking are easily followed, showing them to be working *toward a center*. . . . All centripetal forces produce vortices" (my italics) (Churchward 1968, 11–13). As long ago as 1665, the Italian theorist Alfonso Borelli suggested that a centripetal force kept objects from leaving the Earth. That's gravity.

Fig. 1.6. A cross section of an advanced nebula. The
great centripetal force pushes the gases toward the
center. After Churchward (1968, 10).

PUSH NOT PULL

Newton didn't quite get it right. Gravity is really a field, like a magnetic field.

ALBERT EINSTEIN, *IDEAS AND OPINIONS*

Therefore gravity is a push, not a pull. No pulling force has ever been found to account for the 14.7 psi pressure at sea level. That force must originate in the Earth's electromagnetic field (its vortex), not inside the Earth. According to adherents of the SM, "We still do not know what causes this magnetic field, and according to our best knowledge of the properties of the Earth's interior, it should not be there at all! [The idea that] the source of terrestrial magnetism is situated in the central iron core can hardly stand up . . ." (Gamow 1948, 94). Gravity is nothing more than the downward push or cascading of the field's vortexian power. Nor do we need to postulate attraction or magnetism to explain why planets stay in their places. Their own swift rotary motion keeps them there.

The vortexian whirligig not only makes but also sustains worlds: In the beginning of a planet, corporeal solutions are propelled toward

"𝕿𝖍𝖊 𝖋𝖔𝖗𝖒𝖆𝖙𝖎𝖔𝖓 𝖔𝖋 𝖙𝖍𝖊 𝖜𝖔𝖗𝖑𝖉 𝖇𝖊𝖌𝖆𝖓 𝖜𝖎𝖙𝖍 𝖆 𝖛𝖔𝖗𝖙𝖊𝖝."

—Anaxagoras of Ionia
ca. 430 BCE

ANAXAGORAS CLAZOMENIVS
PHILOSOPHVS.

Fig. 1.7. Anaxagoras of Ionia, ca. 430 BCE.

the center, and when the center is sufficiently dense, it manifests as light and shadow. Then, after vapor and vortex have carried the proto-world forth, it condenses into a solid and takes its place as a newborn planet. Vortex approximates the EMF; here is your theory of everything. Indeed, Faraday-Maxwell's "field" anticipated the all-embracing science of vortex, the force field that transmits every relevant energy—including gravity—and that powers the rotation of the planet.* It is the great spiral, the whirlwind of creation, that puts and keeps planets in place.

Things do not fall to the Earth because of some imagined magnetic "pull"; they are driven down by the centripetal force of the vortex. It is a fallacy that the apple rushes to the Earth just as iron filings rush to the magnet. Magnetism is not the principle behind gravity. Vortexya is. Moreover, instead of gravitation acting across empty space between planet and Sun, Descartes had space filled with matter, jostling particles, which formed whirlpools, vortices, sweeping the planets around the Sun. The planets, in short, were whirled around by vortices in an all-pervading ether. Descartes repudiated magnetism and any form of action at a distance. Though I think he hit the nail on the head, his celestial mechanics has been rejected by the keepers of the SM: "He took . . . a step backward by explaining magnetism and gravity as whirlpools in the ether. . . . Filling all space with monstrous eddies and vortices" (Koestler 1959, 501).

James Churchward, constitutionally outside the box and therefore ahead of his time, said it all in his fundamental observation (almost a hundred years ago) that "the earthly *elements* could not pull the apple down. It requires a *force*. Elements are not forces, but they are permeated with forces" (my italics). Exactly the same understanding is found in the Book of Cosmogony and Prophecy.

What is called corporeal substance, which hath length, breadth and thickness, remaineth so by no power of its own. Corpor has no

---

*We have not yet discovered how to use pure vortexian energy as a source of power, but I believe we will do so in the years to come (see Martinez, "Power from the Nightside," in Kenyon 2008).

power in any direction whatsoever, neither attraction of cohesion, nor attraction of gravitation; nor has it propulsion, but of itself it is inert in all particulars ... Attraction existeth not in any corporeal substance as a separate thing. Nor is there any substance of gravitation. These powers are the manifestation of vortexya.

OAHSPE, BOOK OF COSMOGONY AND PROPHECY 1:45, 1:27, 6:11

Newton, despite himself, thought it "inconceivable that inanimate brute matter should ... affect other matter ... as it must do, if gravitation ... is inherent in it.... That gravity should be innate ... [and acting] at a distance ... is to me so great an absurdity, that I believe no man ... can ever fall into it" (quoted in Koestler 1959, 339).

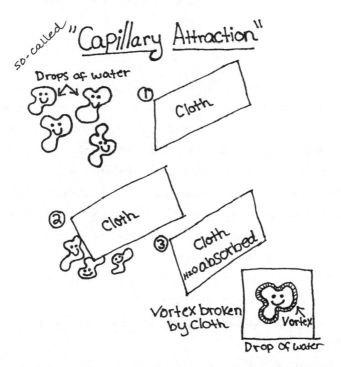

Fig. 1.8. "Capillary attraction" is a misnomer. A drop of water has no attraction for a piece of cloth, nor a piece of cloth for a drop of water.

When the cloth approaches the drop of water, it breaks the vortex thereof, and the water goes into divisible parts into the cloth.

OAHSPE, BOOK OF COSMOGONY AND PROPHECY 1:28

Planets, we must realize, have no attraction or repulsion whatsoever in relation to one another. Rather, the force involved lies in the vortex, the electromagnetic envelope encasing worlds and galaxies and atoms. Science does not see it (yet): "Quite honestly, our understanding of . . . magnetic fields is as uncertain as our knowledge of how life began," Donald Goldsmith freely confessed in *The Hunt for Life on Mars* (1997). Masses (elements, corpor) do not, cannot, attract matter; but vortices (force fields) do, just as the whirlwind attracts particles to its midst. ↻

WHIRLWIND

As the whirlwind gathereth up dust, and driveth it toward a center, so is the plan of [the] universe.

OAHSPE, BOOK OF INSPIRATION 1:23

James McGill explains it like this: "Within the vortex, which is the etheic solution in rotary motion, lies the secret of creation. Atoms, molecules, moons, planets, stars, and galaxies are round in shape because matter is driven together at the center and formed into a ball by the [spiral] force of the vortex" (www.studyofoahspe.com).

Hannes Alfven, whose plasma theory was developed out of the physics of electromagnetism, also sees the universe formed in this manner, that is, by the whirlwind of magnetic fields, concentrating matter into planets and stars by virtue of vortices of energy. It is nothing less than the universal corkscrew, the swirling bathtub drain, the funnel cloud of the tornado, the spiraling hurricane, the whorled volcanic slurry, the helical DNA, the structure of quartz, as well as all the plants like hops that curl around a stake.

It is also the screwlike path of our planet's journey across the galaxy, the spiral movement of the stars, the coiled nebula,* even the ocean wave that curls into itself. Alfven's magnetic vortex filaments, like the

---

*Those with "etheric sight" (chapter 6) have even seen the spirit leave the dense body in a spiral movement.

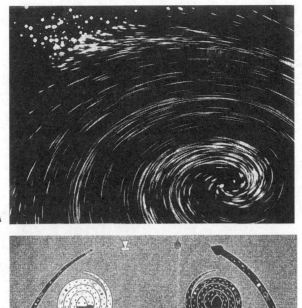

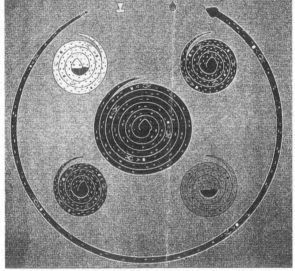

Fig. 1.9. (A) Black whirlwind. Thus does Apache cosmogenesis describe the Beginning as holding only darkness, water, and cyclone (i.e., the twisting power). Indeed, their Creator, Hactcin, formed the first animal by whirling it around rapidly in a clockwise motion. The whirling icon, sometimes in swastika form, remains a prominent symbol in Southwest Native American life. Emblem of the creation, it everywhere symbolizes the life of the universe and nature's power.

(B) The spiral of life is seen in a Navajo sand painting.

Behold, I made a whirlwind in Etherea, hundreds and hundreds of millions of miles across, and it driveth to the center a corporeal world from that which was unseen.

OAHSPE, GOD'S BOOK OF BEN 2:15

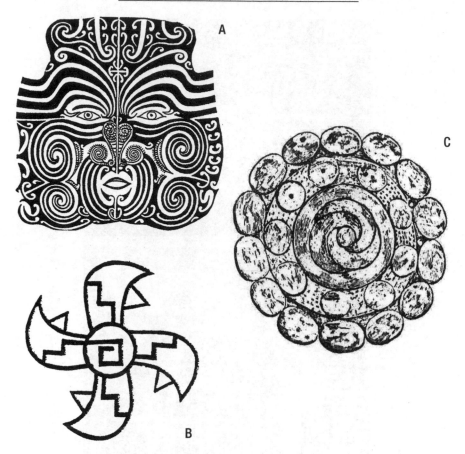

Fig. 1.10. Cosmological knowledge is imprinted in ancient forms of art. The cosmic spiral can be seen in the art of New Zealand, India, the Americas, the Pacific, Crete, Mesopotamia, Scandinavia, Asia Minor, Samarkand, China, Russia, Africa, Malta, and Northwest Europe—often on megaliths—aptly interpreted as a diagram of electric energy or life-enhancing power. (A) A Hawaiian image. (B) Swastika. The gammated cross in both Old and New Worlds symbolized nature's powers. (C) Gorget from Tennessee mounds, showing a spiral.

centripetal force field, stream toward galactic cores. Starting in the 1960s, space probes confirmed Alfven's model, as did the discovery that the rotation of Jupiter's magnetic field, described as twisting plasma vortices, produces electric current. Alfven's opinion of the big bang? "A myth" (Lerner 1991, 48).

The same vortexian principle of twist underlies James Clerk

Einstein postulated the existence of torsion (twisting) fields in 1931. Author David Wilcock, referring to biophoton research, writes that the source field creates these virtual protons "when there is rotating, vortex movement within the field itself" (Wilcock, 2011, 243). A spiraling electromagnetic wave (called a torsion wave) travels through the ether, following the path of a perfect phi-spiral. As such, "the fundamental wave . . . is a torsion wave. . . . Every movement . . . leaves its mark on the ether in the form of torsion waves. . . . Stars emanate torsion wave-like energy, the result of the spinning of the stars . . . and all atoms are in fact torsion wave generators. . . . In implosion physics . . . spiraling is towards the nucleus of the atom . . . toward the zero point of the electromagnetic vortex" (www.slideshare.net/UltimateWaterInc/the-torsion-wave-is-the-structure-of-light-energy-dynamics).

Maxwell's revolutionary unification of electricity, magnetism, and light, all of which result from the whirlpool motion of fields. Inspired by Maxwell's concept of electromagnetism, Einstein then replaced force at a distance with fields acting locally: "The field concept indeed displaced the concept of action at a distance" (Einstein 1954, 74). All known forces diminish with distance. As far as Hoyle was concerned, the force worked by direct action, "at no distance in the four-dimensional sense. Strictly, the phrase 'action-at-a-distance' should be changed to 'action-at-no-distance'" (quoted in Lindsay 1971, 422).

In the enlightened work of Faraday and Maxwell, the facts of nature—electricity, magnetism, light, optics, and even gravity—were finally linked under a common umbrella, a single reason for being: the field, unifying the disparate facts of science under one grand, comprehensive law, the theory-of-everything (TOE), a single axiom, "an utterly simple idea" (Einstein 1954, 74). That axiom, vortexya, allows light, heat, magnetism, and electricity to all be one and the same thing under different conditions.

Vortexya manifest in magnetic flame is called ELECTRICITY

Vortexya manifest in hydrogen-oxygen combo is called THUNDER

Vortexya manifest in drive is called GRAVITY

Vortexya manifest in phosphorous is called LIGHT

Vortexya manifest in mineral (iron) is called MAGNETISM

Vortexya manifest in actinic force is called HEAT

Vortexya manifest in east-west currents is called WIND

Vortexya manifest in overlapping currents is called CYCLONE and
WATERSPOUT

Vortexya manifest in discord is called CLOUD and HAIL

Vortexya manifest in transverse needles is called COLOR

Vortexya manifest in absorption is called EYE

Vortexya manifest in animals is called INSTINCT

Though rejected, Maxwell's model would be used not only by Edison to electrify the world, but also by Einstein, whose equation $E = mc^2$ came almost directly from Maxwellian physics. Although the Michelson-Morley experiment in the nineteenth century supposedly proved that the ether does not exist, quantum physics still needed something to fill space, a propagating medium, its extra gravity absolutely necessary to keep galaxies from flying apart! Science refused to accept Maxwellian physics, if only because it required an all-pervading ether of space, which was not detectable and does not emit any radiation. It is the wild card of cosmology. Quantum physics, which promptly took its place, would brook no invisible ether, yet reinvented it under a new name, "dark matter"—invisible stuff (subatomic) that just has to be there.

If truth in science is established by objective experimentation,* the big bang cannot pass muster. The bangsters say that primordial forces organized themselves. This miracle cannot be tested or falsified. Neither

---

*Quantum theory itself has apparently made scientific objectivity a thing of the past, for adherents of this theory say we cannot study particles as separate from ourselves or as stable "things." Our own act of observation, as noted above, is inherent in the process. That is, certain particles come into focus only when we observe them. In the quantum world, things do not exist without us.

Scientists say they've found signs of dark matter

Scientists working on the Cryogenic Dark Matter Search experiment say they've uncovered three promising signs that could point to the discovery of dark matter. The researchers, who presented their findings at the American Physical Society meeting, are part of a larger group of scientists, including those at the Large Hadron Collider, who are searching for the elusive particle. **BBC**

Fig. 1.11. Dark matter.
Cartoon by Marvin E. Herring.

can a fictitious big bang explain asymmetrical particles and fields. "The foundations of fundamental physics and its whole tower of symmetry are tottering" (Lerner 1991, 341).

Continuous creation is the most compelling alternative to the big bang. Creation, in this view, is not the work of a moment. Immanuel Kant said that "millions and whole myriads of millions of centuries will flow on during which always new worlds and systems of worlds will be formed." Now, the SM assumes that the deeper we look into the universe, the older it should be. "But that is not the case. No matter where we look . . . old stars and young stars appear everywhere. . . . Creation is happening at every single level, not just at one single point fifteen billion years ago; there was no big bang" (Kreisberg 2012, 49).

If creation proves to be continuous, we will need to elevate condensation, the very opposite of the bangers' explosion, to the status of world-maker. It is the aggregation of stuff (background material) that is the organizing principle. In Fred Hoyle's continuous creation, matter is produced in the nuclei of galaxies by condensation. Indeed, astrogenesis is more in keeping with the congregation of matter than with any wild dispersal or tearing apart. The Chinese have an origin myth of how the universe was formed from condensed vapor, "an idea very similar to the scientific reality of star formation and cosmic evolution" (Chouinard 2012, 215). The two concepts of explosion and condensation, as Hoyle made clear, are contradictory. Stellar systems, in the combined work of astronomers Hoyle, Geoffrey and Margaret Burbidge, and William Fowler, could produce all the known elements. The most common ones (helium, carbon, oxygen, nitrogen) were built up in the stars by process of fusion.

Continuous creation also implies infinity, worlds without end! Only in the last few decades have we begun to find these other worlds, some similar to Earth, with a typical galaxy containing hundreds of billions of stars. The "expanding" universe, or the "multiverse," is, really, the infinite universe. Infinity is a fundamental premise of modern mathematics. A finite universe is actually the Dark Ages view, just as "the basic assumption of the medieval cosmos—a universe created from nothing—[is] now the assumption of modern cosmology" (Lerner 1991, 162).

Fig. 1.12. Design of the finite universe, as pictured by medieval philosophy.

FROM BEGINNING TO END

He is forever bringing together and forever dissolving and dissipating worlds without number.

OAHSPE, OURANOTHEN'S THIRD DECLARATION, 21

I make and dissipate everlastingly . . . beyond number made I them [worlds].

OAHSPE, PLATE 26

For some reason today's science, led by figures such as Stephen Hawking, feels compelled to impose a beginning on the cosmos, and an ending as well. But not all scientists think along these lines. Hannes Alfven's plasma cosmology sees an infinite universe in both time and space, like a circle with no beginning and no end.

> *The universe as a whole has no end, and it may have had*
> *no beginning.*
>
> ANDREI LINDE

Although the SM says the universe is about 30 billion light-years in radius, it is also known that it stretches out farther than we can see, "and it is older than we know, possibly infinite and eternal" (Acheson 2008, 156). Enter Halton Arp, one-time observational astronomer at the Mount Wilson and Palomar observatories. Arp boldly and heroically challenged the reigning view, asserting that galaxies are born hither and yon, and that expansion is questionable.

A MEASURE OF AGE

There are many worlds, new coming into being every day.

OAHSPE, THE LORDS' THIRD BOOK 2:10

Halton Arp's work on "redshift" exposed the big bang as a chain of inferences, a theory invented to explain the supposedly expanding universe and why the galaxies seem to be moving away from each other. It was the faulty redshift assumption of expansion that led to the big bang hypothesis in the first place. Expansion would stretch the wavelength of light passing between galaxies; *redshift* refers to longer wavelengths. Light is shifted toward the red end of the spectrum because of its motion away.

Mathematician and physicist Christian Doppler, in the nineteenth century, noted a spectrum shift toward the red when a source of light moves away from the observer. Thus the light from a star that is receding is seen at a lower frequency; that is, redder. High redshift, then, means the star is far away; low redshift means it is nearby. It is thought, in other words, that redshift is due to motion away from the viewer: the faster it recedes, the greater the redshift. Comparison is made to a train whistle sounding a lower pitch as it moves away.

It was then assumed that fainter redshifted galaxies are more distant, and the correlation made between increasing redshift and decreasing luminosity was assumed to be due to velocity. Fainter should be not only farther away but also moving faster. Finally, it was deduced that the relationship between velocity and distance of those receding galaxies was based on expansion and acceleration.

Back in the 1930s, however, not everyone thought redshift was caused by expansion; even Hubble cautioned against accepting this view. And in the following generation, Halton Arp dared to point out that redshifts are not necessarily due to the putative expansion of the universe. Some scientists say the universe is not expanding at all, and that redshift is created as light travels through space. As photons travel, their energy declines. "This effect could explain the high redshifts of some quasars, since light traveling through the outer atmosphere of the quasar could be redshifted before leaving it" (Lerner 1991, 428).

Dim quasars can have high or low redshift; these objects might be much closer than the distances calculated by assuming their redshifts are due to the expansion of the universe. Quasars might actually be close matter traveling at a high velocity, having been expelled with great speed by nearby galaxies. This contradicts the bangsters who say quasars are extremely distant.* Besides, other scientists have found galaxies with high redshift that are older than the big bang!

In the late sixties, improved observation of quasars (brilliant, compact objects thought to be extremely distant) led to the idea that these objects might be much closer than originally thought. Local quasars, according to Fred Hoyle and Geoffrey Burbidge, could have been expelled at high speeds from the cores of disturbed galaxies. Since this could blow the whole big bang theory out of the water, "Geoff and Fred were often disturbed in the dead of night by the telephone . . . a mysterious voice . . . threaten[ing] them with the ruin of their scientific reputations" (Mitton 2005, 292–95).

Their steady-state theory argued that high redshift could easily be interpreted as nearby matter traveling at a great velocity. It was also pointed out that when the position of two extragalactic objects is very close, but their redshifts very different, both redshifts could not possibly be due to distance or the expansion of the universe. Indeed, when Halton Arp observed the closeness of a high-redshift quasar and a low-redshift galaxy, it indicated that some other explanation was in order. Meanwhile, Margaret Burbidge was collecting further examples of galaxies and quasars in close proximity. Even connected objects (so-called double stars, which may have been quasars) were found with different redshifts! Her husband Geoff then declared that quasar redshifts do not necessarily indicate immense distances; rather, these quasars could be local objects ejected at very high velocities from the nuclei of galaxies. Perhaps, thought Geoff, "galaxies beget galaxies."

---

*On a similar note, scientists have only recently discovered nearby gamma rays, previously thought to originate only in deep space. These smaller bursts, called terrestrial gamma-ray flashes (TGFs) "occur much closer to home . . . [indeed] in Earth's atmosphere" (Palus 2011, 18).

Embedded in Oahspe are passages showing nebulae as the birth-place of stars. For example:

> Ahead lay the swamps of Ull, where seven corporeal stars [had been] dismembered a billion years ago, now set with a'jian fields, and forming nebulae . . .
>
> OAHSPE, BOOK OF FRAGAPATTI 3:11

In other words, it looks like the universe is an enormous recycling center. In which case, an exploded or "dismembered" star scatters its elements into space and, later, under conditions of density (a'jian), it condenses into new worlds (nebulae). In a similar paradigm, Hoyle's group viewed the solar system as formed not from big bang material, but from a dismembered supernova.

But here is the important part: Halton Arp figured out that redshift is probably due to neither distance nor velocity, but to age. Astronomer Ray Lyttleton's view—that redshift might be a function not of expansion, but of gravitational field—could also be tied into age, the age of the field, the vortex—the EMF. "The electric and magnetic field," reasoned Einstein, "could not be conceived independent of the state of *motion* of matter, which was considered the carrier of the field" (my italics) (Einstein 1954, 74). That's right; and its motion is a matter of how fast it is rotating, which, in turn, is a measure of age.

> *The ultimate source of a quasar's immense power is rotational energy.*
>
> ERIC LERNER, *THE BIG BANG NEVER HAPPENED*

So what we have is the brightness of a planet or galaxy proportionate to its motion—that is, its rotational velocity—which depends on its age.* With aging comes a weaker EMF, slower rotation, and diminishing radiance. Research at Swinburne University in 2007 challenged

---

*For more on the direct relationship between rotation and age, see page 179 in chapter 4.

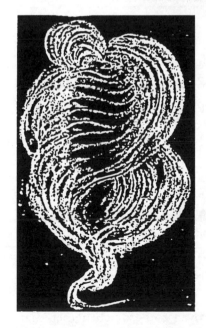

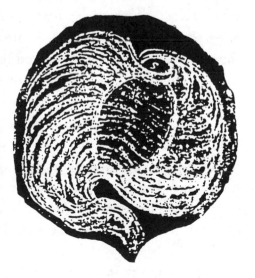

Fig. 1.13. Four images indicating the shape of a vortex as it ages.

"the fundamental assumption that the strength of electromagnetism has been constant through time. . . . They need to rethink their ideas . . ." ("Physics laws flawed," www.sciencealert.com/au/news/20071012 -16699-2.html). And, according to Princeton astronomer H. N. Russell, the physical properties of a star depend on its chemistry, mass, and age.

The age factor alone—egregiously neglected by the SM—could seriously undermine the current paradigm. Mars's great age is seldom mentioned, even though it is believed that its magnetism died out almost 4 billion years ago. Saturn and Jupiter might be the youngest members of our solar family. Young planets tend to be large and low-density. Gaseous Saturn, which is composed of light elements, is only one-eighth as dense as Earth. If velocity and turbulence are marks of a planet's wild beginnings, which I believe they are, these are young planets. Jupiter's great rotational speed (just nine hours) is a sign of youth. Saturn, too, rotates on its axis in less than ten hours. Saturn's satellite Dione orbits ten times faster than the Moon orbits Earth. Saturn's cloud belts are extremely turbulent, as are Jupiter's: swirls and eddies move through the tumultuous jet streams that roar around the huge planet.

In 1974 Jupiter's magnetic field almost destroyed Pioneer 10; the twist factor is enormous. Saturn's rings, in addition, are brilliant. Rings, belts, and satellites are features of young worlds; Saturn has the most satellites (twenty-one), and Jupiter is next with sixteen. And with its numerous belts, Jupiter is one of the brightest planets. Also indicating incipience is the "poisonous atmosphere" of Jupiter, as well as the presence of hydrogen and helium in the atmosphere of both Jupiter and Saturn: Earth's is almost gone. These are all characteristics of a vigorous young world. Compare the rotational velocities of Jupiter and Saturn to those of other planets: Mars takes 24½ hours to complete a rotation; Jupiter and Saturn each complete one rotation in less than 10 hours. "What causes this variability?" asks Velikovsky (1965, 6–7). Isn't it age, after all?

When redshift was measured, it proved to be much higher than that of the farthest known galaxies. It was so bright at such a distance

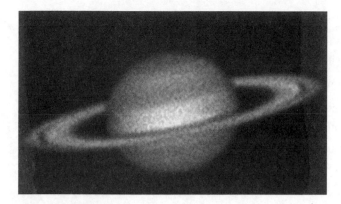

Fig. 1.14. Saturn

that no known mechanism could explain that much energy except a brand-new vortex forming a young, fiery world. Arp took this, and quasars, to represent "newly created matter"; the redshift yardstick for expansion was just a mistake. And if the yardstick is wrong, the whole chain of inferences based on it is wrong, including the big bang and an expanding universe. Higher redshift means younger age (of galaxy or quasar). Galaxies, for Arp, are born and then die (fade away).

Chain of Inferences

Most of twentieth-century cosmology, according to Amy Acheson, is based on a chain of theories. "If any link in the chain is wrong, the entire chain collapses. . . . [Everything] could become obsolete overnight" (2008, 153). Much of the chain, I notice, is composed of if/then syllogisms: *if* space is expanding, *then* the universe must have once been smaller, a speck that swelled outward. But then why is the expansion rate of the universe so perfectly fine-tuned for life? The answer, as we have seen, must be the notion of the "multiverse": a great if not infinite number of universes—some 10^{500}—each supposedly with a different cosmological constant as well as different laws of physics. Under this logic, there is bound to be one (ours) with a cosmological constant (expansion rate) so well suited to life: a luck-of-the-draw sort of thing.

But different physical laws? Albert Einstein wondered if nature did not always play the same game. I doubt that we are looking at different physical laws, only different densities, motions, and consistencies, and, of course, different ages. Invoking new physical "laws" to accommodate the inconsistencies of the SM and each new observation—and all for the sake of keeping the big bang theory afloat—is not exactly the scientific method. Not even close.

It has been supposed by imaginative minds that "The Big Bang could have been the result of a collision between our universe" and another one (Chown 2005, 16). A (very) hypothetical cosmic fuel, in this scheme, could have been concentrated in a tiny region, driving a stupendous outward rush. It is further hypothesized that this fuel would replenish itself, "power[ing] countless other bangs, each yielding its own separate, expanding universe" (Greene 2012, 23).

This recurring universe reminds me of nothing so much as the highly extravagant Hindu doctrines of (1) vast cosmic cycles of destruction and rebirth, (2) and the equally far-fetched doctrine of reincarnation. Madcap science says that if this inflation kept going, our universe may be just one of many that pop up as "new episodes of inflation begin." And in this reckless view, the big bang is not unique and our universe is "just one bubble in a vast froth of other bubble universes." Thus, according to collision theory, as our own bubble expands, it could "run into one of the others." We might even expect "an infinite number of collisions with other bubble universes" (Nadis 2012, 72).

I hate to burst your bubble, but I seriously doubt my universe started like that.

Unfortunately, Arp also paid a heavy price for his views. Denied telescope time and demoted by his fellows at AAP (the Association of Astronomy Professionals), he was forced to resign. He moved to Germany. Similar tactics were brought to bear against the Burbidges. Antibangers and friends of Hoyle, they were denied both telescope time and substantial research grants. "Academic politics," chortled Geoff. "It's all politics."

The late MIT mathematician Irving E. Segal did not go for the big bang, either. The universe is not expanding, he said, and redshift is not what they think. Instead, it can be explained using the principles of mathematical symmetry, a hidden symmetry of the universe that cannot be seen, but whose consequences can be observed.

Sci-fi wizard Isaac Asimov recalled the time in 1948 that Fred Hoyle and other astronomers tried to refute the big bang or any "beginning" at all. Their theory of continuous creation—being an apparent departure from the principle of conservation of mass and energy—was supposedly disproven. Said Asimov, "Now the theory of continuous creation is just about dead" (quoted in Frazier 1986, 301). His colleagues smugly agreed that this model, a.k.a. "steady state," is "past its sell-by date" (ca. 1965).

The Hoyle contingent, arguing that the universe has always existed much as it is today, saw new matter continuously creating galaxies. Even positing dark matter did not help the bangsters completely explain the formation of galaxies and stars. Geoff and Margaret Burbidge, working with Hoyle, showed how stars create elements continually by virtue of nuclear reactions. They thought single-proton atoms could eventually form everything in the universe. Stellar nucleosynthesis, not the vaunted big bang, could have produced the denser elements from hydrogen and helium. Under this theory, new matter comes not from a single source (the big bang), but from many sources all the time.

> *Big Bang failed when cosmic rays were shown to be produced in the present-day universe rather than the distant past.*
>
> ERIC LERNER, *THE BIG BANG NEVER HAPPENED*

Cosmic microwave background radiation (CMB) enthusiasts were among those claiming to have trumped steady state. The Hoyle theory was further invalidated, others claimed, when more distant regions of the universe were found to be different from our own. Some scientists, though, argued that those regions are not so distant. Yet others (Australian astronomers) said they are distant but not different.

Fred Hoyle, England's most renowned astrophysicist/cosmologist of the 1950s, once president of the Royal Astronomical Society and a prominent figure in thermonuclear synthesis, gallantly battled all those Cambridge bangsters. Even though they howled at Hoyle's public dismissal of the big bang, none of them "ever contradicted him in a scientific paper." Considered "almost as dangerous as Velikovsky" (Mitton 2005, 146–47), Hoyle was certain of "the perpetual bringing into being of new background material. . . . There is no end to it all. . . . Continuous creation can be represented by precise mathematical equations. . . . I cannot see any good reason for preferring the big bang idea. . . . There is not the smallest sign that such an explosion ever occurred." Indeed, he frankly declared: "Almost every development during the last fifteen years has taken the world along the wrong road" (Mitton 2005, 146–47).

From a certain perspective we can understand how the big bang came to be conceived in the same period (1930s to 1940s) that the power of the atom was mastered and released (wartime). Scientists in the 1930s were pulled away from academia and put to work on radar, weaponry, and splitting the atom. This was the period during which conscientious scientists began to worry that they were betraying their trust by producing weapons of mass destruction to drop on human beings. *That* was the big bang! August 1945—Hiroshima and Nagasaki.

Cosmology is getting more fantastic (the adjectival form of "fantasy") all the time. Outside the cosmologist's ever-shifting models, there is nothing to say the universe was ever in a different state. The whole field of space science has imperceptibly slipped away from truth-seeking. It's now engaged in the business of model proving, and it's getting more bizarre and violent with each passing year. More effort is made to accord with the theory than to accord with the facts.

The bangsters also seem to think they owe us a prediction of how the world will end, presumably in a big crunch, gravity eventually gaining the upper hand, the world caving in on itself. The universe, you see, will keep on expanding to the max, at which time it will start collapsing back into a black hole (the big squeeze).

Alternatively, the expanding universe will eventually pull everything apart, or the universe will just keep expanding forever and end in an "infinite scattering." On the other hand, if there is more matter in the universe than we think, gravity will stop expansion, and contraction back to a zero point (the cosmic egg) will occur, which will eventually produce another big bang! And so the cycle would repeat itself again and again. I hope my tax money isn't paying for this grandiose guessing game.

As we move now from chapter 1 (the origin of the universe) to chapter 2 (the origin of species), we see these two reigning theories, big bang and evolution, linked by a common theme, or rather a common assumption: the randomness of origins. Fred Hoyle, I might mention, did not believe in Darwin's theory of evolution; he doubted that random chemical reactions could have accidentally produced the complex molecules of even primitive life-forms. By chance? Not possible. The philosophy of mere chance and blind forces began, not with the big bang (1930s), but earlier still (1850s), with the Darwinian idea that things—whether stars and planets or plants and humans—form and improve and become more highly developed all by themselves, without the help of a Creator. By chance, not design. Atheism requires it.

OUT OF NOTHING

The regularity . . . and beauty of the sun, moon, and all the stars . . . are sufficient to convince us they are not the effects of chance. . . . When you see a dial or water-clock, you believe the hours are shown by art, and not by chance; and yet you should imagine that the universe, which contains all arts and the artificers, can be void of reason and understanding?

CICERO

This is one of the things I like least about the big bang: it teaches our children to believe in an irrational, hit-or-miss, arbitrary universe; a cosmos whose existence involves no intrinsic beauty, purpose, or plan; a universe with a convulsive, violent, unimaginably shocking genesis. If I had a choice, I would teach the young a universe of system and order, balance and harmony. Not by random came the worlds into being. Not by chance did humans come to walk the Earth.

> As I have shown system in the corporeal worlds [planets], know thou, O man, that system prevaileth in the firmament.
>
> OAHSPE, BOOK OF JEHOVIH 6:9.

Yes, we are teaching innocent, open minds that everything—all energy and matter—exploded into existence from nothing. Ex nihilo. A senseless, desultory, medieval universe, producing itself quite magically out of nothing. But "nothing" can't explode. And, since space didn't exist yet, either, all this happened nowhere. Every place came out of noplace. I see this secular framework, ironically, as even more magical and surreal than the oft-impugned biblical idea of creation, which at least gives us an origin in intelligence and love. Big bang and evolutionism are the nonbeliever's bible, established to eliminate Creator from the "scientific" picture. Yet without wisdom in the universe, our theories, one by one, stumble and fall, for they are desperate to explain the workings of it all without a Higher Power and forced to make nature run contrary to its own laws.

Richard Dawkins, one of today's chief sharpshooters protecting the House of Darwin and its requisite atheism, during a public debate with an Anglican argued that the fine design of the universe was simply a fact of nature. "Why bother with God?" he asked. The reply: "For the same reason you don't want to." Everyone laughed except Dawkins, who protested, "That's not an answer!"

His compatriot, Sir Arthur Eddington, once contended that "random disturbances" could have awakened the universe from its slumber. Random? Even Fred Hoyle, who was himself an atheist, had his disbelief badly shaken by the nature of the carbon nucleus, which to him seemed to have been designed by "some supercalculating intellect. . . . who monkeyed with physics, as well as with chemistry and biology. . . .

There are no blind forces worth speaking about in nature. . . . I think that all our present guesses are likely to prove but a very pale shadow of the real thing" (Mitton 2005, 117, xi).

Charles Darwin, though, would beg to differ. Let us move on then, to the most popular and most misguided of today's scientific SMs: human evolution!

CHAPTER 2

EVOLUTION

Darwin Trumped by the Nookie Factor

Crocodiles lounge in bliss while plovers pick leeches from their teeth, offering dental hygiene in return for food. The whale shark waits calmly as pilot fish swim in and out of its mouth, cleaning its teeth. Time and again, nature finds creatures, which normally would be their meal, becoming their partners. How could this symbiosis have evolved? What gradual steps of evolution or blind chance could have brought them into being? On what day did the crocodile decide it was to his benefit to entertain plovers? It is doubtful their feathered friends would have stuck around had they seen their cousins eaten alive by the fearsome crocodile. Darwinism, in short, cannot explain symbiosis. Such behavior never could have developed in a piecemeal fashion or through mere chance.

Not only is symbiosis, in all its charming varieties, found in the wild, but so are cases of selflessness, even sacrifice for others. This is what evolutionists call the "problem of altruism." But altruism is not a problem, unless your theory of life happens to require innate selfishness and an unremitting "struggle for existence"—every man for himself—driven by tooth-and-claw competition, which ultimately decides who has the advantage, the edge, who will monopolize the "niche." These are the clichés of the Darwinian dog-eat-dog world, an aggressive, opportunistic—even predatory—philosophy of organic life: the "war of nature."

Darwinism teaches us that the reward for being a slightly faster or more alert antelope is to survive over your slower, less savvy counterpart; as a result, over many generations, the whole antelope population will run faster "and with many such changes over time eventually become a new species. . . . Evolution, Darwin's 'descent with modification through natural selection' would have occurred" (Hayden 2009, 45). Is this really how different species came about? Who are we trying to fool with this glorified myth called natural selection? As biologist Lynn Margulis plainly sees it, "Natural selection eliminates and maybe maintains, but it *does not create*. . . . The laws of genetics show stasis, not change. . . . [N]atural selection was a process of elimination and could not produce all the diversity we see" (my italics) (Teresi 2011, 68–69).

However, John Rennie, former editor of *Scientific American,* protests such views: "On the contrary," he cites, "point mutations" (like bacterial resistance to antibiotics) and fruit fly research where, experimentally, specimens have freakishly grown legs from the head, to prove "that genetic mistakes can produce complex structures. . . . Natural selection . . . can drive profound changes in populations over time" (Rennie 2002, 80–82). Although this sort of argument (rhetoric, really) represents the majority view, I seriously doubt that we are all genetic mistakes, or that torturing fruit flies sheds any light on the origin of human beings.

I place more stock in the minority view as expressed by James Churchward: "Our scientists have been trying to build a castle in the air," stated Churchward after finding not "a single case where one animal is changing into another, such as the missing link between a reptile and a mammal. . . . What has been termed steps in evolution has been *mere physical modifications* . . . without, in any way, making the animal more complex . . . nor reflecting anything more than simple adaptations. . . . It is utterly impossible to make a living animal chemically more complex . . . because a chemical change means poison; poison means death" (my italics) (Churchward 1931, 111).

Darwin's theory of natural selection as the means of genetic change and the struggle for existence inevitably revolves around the idea of competition. Sure, you can find examples in nature (and human nature) of vicious rivalry and the cunning snare; yet some theorists think *relaxation* of competition is the factor that enabled new species to survive. When the German philosopher Friedrich Nietzsche learned of Charles Darwin's popular theory of an unremitting struggle for resources, he blandly remarked it was probably inspired by "overcrowded England." More to the point, Harvard anthropologist David Pilbeam, in his 1978 review of Richard Leakey's *Origin,* confessed: "Perhaps generations of students of human evolution, including myself, have been flailing about in the dark. . . . The theories are more statements about us and ideology than about the past."

Perhaps I can take this to mean that in a war-torn age like our own, "survival of the fittest" amounts to control of people, resources, territory, weaponry. Might is right. And with the overtly expansionist history of Western civilization as the backdrop of all our theories, the Darwinian narrative, which has creatures colonizing and exploiting the niche, is a fine charter myth for imperialism per se, not only explaining but somehow justifying the powers that be: almost a master race theory.

Not coincidentally, Darwinism is most vigorously supported in Britain and America. Indeed, author Richard Milton has commented that "no science programme has ever been shown . . . on British television questioning any aspect of the Darwinian mechanism, because it is the most strictly observed taboo subject. Censorship of anti-Darwinian ideas is rather more overt in the United States." Milton goes on to note the "sharp contrast" in attitude in Russia, for example. Yet he finds it "deeply disturbing . . . that such public debate of a taboo subject can take place in so recently a totalitarian society, but is effectively forbidden in tolerant, democratic Britain and America" (Milton 1996, 90).

If this paradigm of conquest and dominance is really the key to success—writ large and small in the animal kingdom—how come most apes are vegetarian? Most animals, for that matter, are cooperative, harmonious creatures who do not fight for food, sex, or lebensraum.

Nevertheless, evolutionists have long touted the competitive exclusion principle, which holds that two related species cannot coexist in the same habitat. The stronger or better, as Darwin maintained in *The Descent of Man,* "drives out its brutish ancestor. . . . Extinction at the hands of a successor is inevitable" (Darwin 2004). One species, then, the fittest one, preempts the niche. The rest disappear. But this is still guesswork. In fact, it is humbug, at least in Africa, where intensive excavation of the bones belonging to our "brutish ancestor," rather than revealing a neat succession of parent and daughter species, has found them living together!

This unfortunate discovery of coexistence, which still barely earns a footnote in the textbooks, not only twits the competitive exclusion principle, but worse, much worse, topples the whole evolutionary scheme of ancestry and descent. Darwinism should have stopped dead in its tracks in 1972 when Richard Leakey discovered Skull 1470 in Kenya, the cranium so different from all other known forms of early humans that it did not fit into any model of human evolution. Skull 1470's brain was, surprisingly, much bigger than *H. habilis*'s, even though 1470 was quite a bit *older.* Reverse evolution? Leakey had to conclude that there were several different kinds of early humans, of which we will soon hear more.

The problem presented by 1470 was resolved by retesting the volcanic tuff in the region. This exercise conveniently determined that 1470 was more recent, therefore an *H. habilis* type, thus avoiding an embarrassing reversal to the scheme of ever-improving evolution. But 1470 was too large to be an *H. habilis,* and he had much better chopping tools. The expected heavy bones or visor brow of that early type just were not there; besides, 1470's skull was too modern to be an *H. erectus* (let alone an *H. habilis*) and his cranial capacity was well beyond that of *H. habilis.* How could his moderate brow and flat face be ancestral to *H. erectus,* who is famous for his massive, barlike brow ridge and prognathous face? Not terribly likely that 1470 evolved into a more archaic form! Devolution?

Broca's area, a region of the brain linked to speech, which is found in anatomically modern humans (AMHs), was also present in 1470. Its leg bones, moreover, were almost indistinguishable from *H. sapiens.* Yet this man was a contemporary of *Australopithecus,* who is even older than *H. habilis.* But, dear reader, the solution to the problem is extremely simple: crossbreeding—that is, two quite different types coexisting and "exchanging genes." Only interbreeding gets us past this tangled web. Leakey himself declared: "Either we toss out this skull [1470] or we toss out our theories of early man . . . [for] it leaves in ruins the notion that all early fossils can be arranged in an orderly sequence of evolutionary change" (Leakey 1973, 820–28).

STRANGE BEDFELLOWS

Coexistence makes evolutionists very nervous.
MARTIN LUBENOW, *BONES OF CONTENTION*

The Big Picture reveals this: instead of australopiths (the earliest humans) being succeeded and replaced by *Homo habilis,* as the names and standard fossil sequence go, instead of *H. habilis* afterward being replaced by *H. erectus* (the next higher type) instead of *H. erectus* next being replaced by Neanderthal and Cro-Magnon, these different races of humans *coexisted* in Africa, sometimes side by side, sometimes in nearby cave systems. Rather than demonstrating the orderly parent-daughter relationship that Darwinism posits, all these different early humans in Tanzania, Kenya, and Ethiopia were contemporaries, neighbors, and also bedfellows. Humans, as we are coming to see, are the most promiscuous of primates; hence, a marvelous hybrid.

Instead of lowly hominid creatures gradually becoming human (through stages of development), instead of humans *changing* from one type of hominid to the next higher type, Neanderthal did no such thing. Neanderthal's brain was too large to link back to low-order (that is, small-brained) hominids. Harvard's leading paleoanthropologist in the early twentieth century, Earnest Hooton, called this a less credible miracle than the changing of water into wine. Neither was

Neanderthal anyone's predecessor, least of all *ours*. Harvard's William Howells found no evidence of Neanderthals going in the modern direction. To Marcellin Boule, Wilfred E. Le Gros Clark, and Louis Leakey (Richard's famous father), Neanderthal was merely some irrelevant distant cousin. There were traces of moderns *earlier* than Neanderthal, instantly disqualifying Neanderthal as our ancestor. In Europe, modern elements *preceded* Neanderthal at Swanscombe, Fontechevade, and Steinheim sites. Swiss archaeologists, after making skull comparisons, also concluded that Neanderthals did not give rise to living Europeans.

The same sort of situation applies to Western Europe's Grimaldi man, whose existence "in close association with the much more primitive Neanderthal form shows that Homo neanderthalensis could not be the ancestor of Homo sapiens, since both species were contemporary" (Boule 1923, 243). All told, there is no transition in Europe from Neanderthal to *Homo sapiens* (Cro-Magnon), the latter appearing on the scene not by evolution, but quite suddenly.

Neanderthal, analysts also concluded, was too "specialized" to become anything else, no less modern man. And how could he be the ancestor of us all if his bones were basically confined to Southern Europe (and Western Asia)? Or if "fully modern men . . . lived in sub-Saharan Africa, Australia, and other regions of the world, while the Neanderthals still inhabited Europe" (Felicitas Goodman 1981, 146)? Finally, Neanderthal could not possibly be our ancestor, because *H. sapiens* largely absorbed him, rather than evolved from him.

Linear or even branching evolution began a nosedive with the excavation of old Neanderthal caves (Skuhl and Tabun) in Palestine: here *Homo sapiens* and *Homo neanderthalensis* coexisted for 45,000 years—and crossbred. Hooton pointed out in *Up from the Ape* that the series was much too brief for evolution to have occurred. How could Neanderthal "have changed this rapidly into modern man . . . in so brief a space of time"? Besides, "a radical race mixture [would be] . . . exactly the sort of phenomena that are shown in the skeletal series from the caves of Skuhl and Tabun in Palestine." In fact, later workers used these Israeli finds to *eliminate* Neanderthal from sapiens phylogeny altogether, for here a modern type was actually the *predecessor* of Neanderthal man. On those slopes of Mt. Carmel, a mixed burial (studied since the 1930s) shook the family tree: Qafzeh Cave revealed an early *H. sapiens* mixed in with others who looked considerably less modern. Excavations from these Palestinian caves confirmed the coexistence—even cohabitation—of *H. sapiens* and Neanderthals; and with every variety of intergradation, crossbreeding and "signs of hybridization" (Day 1988, 125) came into view.

This fusion of types found in the caves of Tabun, Skhul, and Qafzeh shows modern and Neanderthal traits freely admixed. Most of these specimens have the trademark heavy torus (brow ridge), but many others are markedly modern in cranial vault, occiput, face size, and limbs. Despite obvious mixing, those who pointed out the hybridization of these two different races were silenced by a semantic ploy—labeling them Neanderthal*oid,* and arguing for a "transitional" stage of evolution (see page 91). The fact that the more anatomically modern humans (AMHs) at Qafzeh were using Neanderthal tools makes us suspect that the main result of these crossings was retrogressive; that is, the back-breeding of Cro-Magnons into Neanderthal stock (by which the latter inherited their big brains).

All those impossible reversals (such as *H. habilis* possessing more gracile skulls than Neanderthal or early Neanderthals being more AMH than later ones, or 1470 "becoming" more archaic) put the lie to evolution. This is a warning against facile theories of "progress," which, as we will see, are also falsely assumed by reincarnationists, arguing the soul's progress where we sometimes see civilization going backward and where we observe democracy's shadow of progress.

"The prognosis for Darwinism," wrote the distinguished and versatile cosmologist Sir Fred Hoyle thirty years ago, "is now very poor." We have, thought Sir Fred, "been bamboozled" (1984, 114). The handwriting on the wall came into sharper focus with recent twenty-first-century finds. Let's start with "hobbit," discovered in 2004; then move on to Denisova (X-Woman) announced in 2010; and finally to *Australopithecus sediba,* reported in 2011.

Indonesia's three-foot-seven-inch-tall, sixty-pound hobbit, also

Fig. 2.1. Artist's rendition of "hobbit." An astonishing composite of almost every hominid type, this creature, Homo floresiensis, *looked more like a "Heinz 57" than any imagined "transitional stage" of evolution.*

dubbed "Lady Flo" after her disinterment from a cave on the island of Flores, near Bali, broke all the rules. To begin with, her clan—thirteen additional specimens were recovered—had survived until a mere 12,000 years ago. This was way too recent for *Homo erectus,* who is supposed to have been replaced (edged out of the niche) much earlier.* Yet Lady Flo was indeed erectuslike: long arms, thick bones, hunched shoulders, short legs, large flat feet, bulging eyes, prominent brow, sloping forehead, and—the proverbial clincher—no chin.

---

*According to the SM, *Homo erectus* lived in Africa about 1.6 mya. But, wait. The race of *H. erectus* was previously thought to be no more than 500 kyr (Brace 1979, 5). And according to anatomist Charles Oxnard, he was only "a few hundred thousand years old" (Oxnard 1984, 332). Back in 1929, anatomist and anthropologist Sir Arthur Keith shelved *Pithecanthropus* no earlier than 220 kya! Now he is eight times older! How did that happen? In any case, the experts tell us that *H. erectus* went extinct 120 kya, or was it 400 kya? Gone from Africa, others say, 500 kya. Or was that 300 kya? In more recent books we are likely to hear that they vanished only 60 kya, which agrees with Le Gros Clark's date of *H. erectus* in Asia 60 kya. But old habits die hard and even current works, such as Brian Switek's 2010 book *Written in Stone,* put the *H. erectus* die-off at 200 kya. But Lady Flo is only 12 kyr, and she is erectoid.

The mystery only deepened with measurements of hobbit's brain, one-third the size of modern humans' brains, "chimp-sized" at 400 cc (cubic centimeters). Uh-oh, the brain of erectus, supposedly older than hobbit, was twice as big as Flo's, which certainly undermines the "principle" of evolutionary improvement and brain enlargement over time. If that wasn't enough to send evolutionism into a fatal spasm, Lady Flo's wrist and flaring pelvis aligned her with *H. erectus*'s predecessors: the long-extinct australopiths who supposedly died out millions of years ago! On top of all this confusion were Flo's respectably modern teeth and frontal and temporal lobes. Her neural connections were "highly evolved." Flo's people, no slouches, were skilled hunters with sophisticated tools. But all the confusion evaporates with the understanding that Flo was a hybrid of several races.

How can anyone believe the worn-out fairy tale that one species morphs into a different species? Have breeders ever turned a cat into a dog? The changes and "transitions" cited by anthropologists are a scientific illusion, a falsehood so big that we, dumbfounded, believe

Research: Ancient "hobbit" humans were unique species

Analysis of *Homo floresiensis* fossils reveals that the early humans were a unique species with brains larger than previously thought, according to research published in the journal *Proceedings of the Royal Society B*. Nicknamed "the hobbit" for its short build, the Homo floresiensis could be a descendant of *Homo erectus*, scientists say. The research adds to evidence that the *Homo floresiensis* was not a deformed modern human but rather its own species. **LiveScience.com**

Fig. 2.2. Hobbit. Cartoon by Marvin E. Herring.

it. There have been no changes: anthropologists admit the brain of *Australopithecus* did not increase in size or become more human for a million years. The *H. erectus* race also changed remarkably little, without any brain increase during their tenure on Earth. Incongruously, the early australopith known as "Lucy" (like Flo) had some modern aspects. Her hip structure was so refined as to make it hard for her to climb trees, and her family, the gracile australopiths, had bigger brains than the robust australopiths—who, indeed, are not as old. For gracile is older than robust in both South and East Africa.

The puzzle of *Australopithecus* appearing as two distinct kinds (gracile versus robust) now dissolves, if we factor in crossbreeding. No, evolutionary "selection pressure" did not change gracile into robust; that is a fruitless argument. East African robusts like Zinj (*Au. boisei*) were less humanlike, though younger, than the gracile australopiths. All these impossible evolutionary reversals are proof enough that Darwinian phylogenesis doesn't pan out. Such anachronisms in the robusts are only the result of *Australopithecus* retro-breeding with more archaic mates.

Sterkfontein fossil feet in South Africa proved to be quite tiny (under four inches long). The anklebone of these specimens, named "Little Foot," was extremely humanlike, even though they are the earliest known hominid in South Africa: Little Foot was judged to be 1 million years older than most of the other Sterkfontein finds (*Au. africanus*). But, again, the anachronistic difficulty of this specimen's too-early modernity evaporates if we allow it to represent one of the earliest cases of *Au. africanus* mixing with their AMH neighbors.

Paleoanthropologist Donald Johanson wondered how his famous Lucy (*Au. afarensis*) also got such modern feet. This tiny, baffling creature had caused one anthropologist to murmur: "This little midget will mess up everything," meaning the supposed run of step-by-step evolution from primitive to modern. Meanwhile, in the acclaimed Laetoli (*Australopith*) footprints of Tanzania, a small and "improved" foot was again seen on these little people (only four foot seven inches tall). That foot featured a rounded heel, an uplifted arch, and a forward-pointing big toe, typical of the modern foot. All it takes are a few of the right *H. sapiens pygmaeus* genes—the AMH race once known as the "Ihins" (see chapter 8).

But with the discovery of hobbit, experts were reduced to calling it everything from "pathological" to "bizarre" to "a cretin" to "a pygmy freak." The scientific dogfight over those bones led nowhere. All the blood, sweat, and tears of these "hobbit bone wars" might have been spared, had Lady Flo's obviously mixed morphology (seemingly contradictory skeletal features) been taken at face value. A wonderful composite of archaic and advanced features, Flo was just a happy hybrid. In other words, she didn't "evolve" from anybody. Her race—like all the races before and since—was, quite simply, crossbred, a mixed bag possessing the genes of *Australopithecus* and *Homo erectus* as well as moderns.

Mixed morphology, after all, is seen in early hominids across the board. Table 2.1 offers several examples.

TABLE 2.1. MIXED MORPHOLOGY

NAME(S)	LOCATION/ DATE GIVEN	MIXED TRAITS
H. ergaster, ER 3883, ER 3733	Koobi Fora, Africa	tall, early erectus; archaic features mixed with modern ones
H. habilis, handy man	Africa/ca. 2 myr	short and gracile, long arms; apparently an improved, upgraded australopith
H. heidelbergensis, Mauer man	Europe/130 or 750 kyr	pre-Neanderthal but modern dentition
Kabwe man	Broken Hill, Rhodesia/ 40 kyr or 400 kyr	tall and strong erectus-Neanderthal mix
Kanapoi, *Au. anamensis*	Kenya/4.2 myr	no chin but modern tibia
Krapina man, progressive Neanderthal	near Zagreb/28 or 130 kyr	"modernized" features
Laetoli, *Au. afar*	Tanzania/3+ myr	pygmy-sized, modern footprint (despite great antiquity)
Skhul	Mt. Carmel, Israel/40 or 80 kyr	modern crania and vocal tract combined with Neanderthal traits

NAME(S)	LOCATION/ DATE GIVEN	MIXED TRAITS
Sterkfontein, Mrs. Ples	South Africa/1 to 3 myr	australopith, but modern pelvis
Swanscombe, archaic sapiens	UK/120 or 225 kyr	thick-skulled, with "too-early" modern traits
Taung, *Au. africanus*, Dart's child	South Africa/2 myr	brain and teeth more modern than face
Turkana boy, KNM-WT 15000	Kenya/1.6 myr	"mosaic": primitive head on modern body

"X Woman," found in 2010, was the next bomb to drop on the house of evolution. Denisova woman (only part of her pinky finger was discovered, actually) was found buried in a cave in the Altai Mountains of southern Siberia. *Discover* magazine, noting the controversy occasioned by this forty-thousand-year-old unknown hominid (*hominid* meaning any humanlike creature), billed the story: "Pinkie Pokes Holes in Human Evolution." That sliver of bone yielded the DNA of a perplexing hodgepodge of modern *H. sapiens,* Neanderthal, and some other unidentified lineage. Although X Woman was instantly dubbed a "new species," thus garnering a standing ovation in the media, it looked more like just another case of crossbreeding. Indeed, analysts said Denisova must have descended from a "hybrid spawn." Three races were blended in her DNA. Why call that a new species? A new race, yes; but "species"?

The Fraudulent Coining of "New" Species

In my view, every single hominid in the fossil record is of the same species: *Homo sapiens*. The only differences are racial (subspecies). When they found "distinct" mtDNA in the Denisova woman, that sliver of bone instantly became a new species, even though the DNA differential could just as well reflect crossbreeding. Assigning new species to every new fossil (sometimes named after the person who discovered it or the sponsor of the excavation) is like having a star named after you. Frankly, there is little hope of hitting the news with mere varieties or races; only new species will do the trick. With Denisova, the glory went to the geneticists, the finger fragment supposedly marking an entirely new group of ancestral humans. Well, the more they dig, the more mixes they find, gratuitously calling them "species" and gloating over the ever-expanding catalog of human and prehuman taxa. Yet critics do ask: Is it possible that the scientists, who have given new species names to every early Homo find with significant differences, have made our family tree more complicated than it really is? (Gore 1997).

The fundamental design of the human cranium, after all, is the same from the earliest pithecanthropines (erectoids) right up to the modern races. Some features vary among recent races almost as much as between different fossil types. Therefore, we might just as well split up present-day humanity into several species, which would be a falsehood. Multiple human species—no; multiple races—yes. The fossil people have been trifling with us: a slightly different-shaped jaw and—presto!—a new species is born.

Even though the variation in early humans was considerable, it is still only variation. If the early heterogeneity within the human family had been properly understood, "many of the mistakes in anthropology could have been avoided" (Lubenow 2005, 112–13)—meaning these differences, even if minor, have led theorists (mistakenly) to take mere varieties as separate species, falsely multiplying the number of human taxa. For there was only one, albeit highly diversified, human species. And with relentless crossbreeding over time, the important differences have now been neutralized.

J.B.S. Haldane, the esteemed English biochemist and founder of the new synthesis (neo-Darwinism) said "new species may arise by hybridization" (quoted in Lovtrup 1987, 308). Well, in that case we don't need evolution at all.

Right on the heels of the X Woman splash, an African specimen hit the news (2011), its discoverers trumpeting yet another mongrel as a brand-new species and "game changer." It was easy enough to classify the Johannesburg find, named *Australopithecus sediba,* with Africa's well-known and extremely archaic australopiths. But one thing stood out: the shape of the front brain was modern in its organization. This was an outrageous combination, to find the otherwise oldest hominid type with a well-developed brain. Hyped as a "jaw-dropping find" and our long-lost ancestor,* sediba nonetheless raised the skeptic's flag. Its date, for one thing, was probably much more recent than claimed; indeed, the original date of 5.5 myr had to be changed to 1.97 myr. The too-rapid evolution of its advanced brain was also suspect, prompting nothing but fancy rhetoric to explain it: "vigorous experimentation," "fits and starts of evolution" (Hotz 2013, A2). But these pathetic sound bites failed to mask sediba's ineluctable "jumble of parts," literally a head-to-foot combination of australopith and much "later" types. Obviously a hybrid. No big deal. Certainly not a new species.

Given *Au. sediba*'s highly developed brain as well as longer legs and more modern hand grip, the specimens were labeled a rapid improvement from the lowly australopith. But, screaming hybrid, this South African specimen was no "bridge species" or evolutionary "transitional"—only the lucky recipient of some advanced genes—just like Denisova and hobbit! Breathlessly, reports on sediba contended that it is amazing what evolution has made out of spare parts, bits and pieces. But is it evolution that jumbles the genome? Or is it simply interbreeding, which certainly gives quick results, easily explaining the "too-rapid" evolution of sediba's brain (its teeth, nose, and pelvis were also too modern).

---

*After garnering all that media attention two years earlier, a 2013 redaction admitted that sediba may, after all, be a "dead end" (Hotz 2013, A2) in human evolution.

Research finds human-like traits existed in ancient hominids

An analysis of a collection of hominid fossils reveals that the ancient creatures had a mix of physical characteristics similar to humans and apes. The findings, published in six research papers in the journal *Science*, describe ape-like arms but human-like hands, among other traits. The fossils belonged to the *Australopithecus sediba* and were excavated in 2008 near Johannesburg, South Africa. While the study's researchers acknowledge they were unable to provide a direct link between the hominid and humans, other experts say the findings help support the process of early evolution.
Wall Street Journal

Fig. 2.3. Sediba.
Cartoon by
Marvin E. Herring.

We are always finding surprisingly modern bones where they should not be. . . . Human evolution may exist more in the minds of academics than in any location on Earth.

VINE DELORIA JR., *RED EARTH, WHITE LIES*

So-called bridge species, propped up as intermediates or transitional, are an illusion. We still have not found the specimens that presumably link *Australopithecus* to erectus to Neanderthal to Cro-Magnon.

> *For all these creatures the mystery is the same: why are there no transitional fossils leading up to them? . . . The fossils go missing in all the important places. . . . Most of the [so-called intermediates] are simply varieties, artificially arranged in a certain order to demonstrate Darwinism at work, and then rearranged every time a new discovery casts doubt upon the arrangement.*

FRANCIS HITCHING, *THE NECK OF THE GIRAFFE*

Was Africa's Laetoli Man a "transitional" hominid, linking archaic and modern anatomy? Laetoli apparently had no divergence of big toe in those famous footprints. Did this prove he was transitional? Not at all. How could they be transitional if they were contemporaries with their supposed ancestors, the australopiths? Nor are Laetoli's modern features modifications, only the result of blends—hybrids. The shortness of Laetoli Man (he stood only four foot three inches in height) came from the little people, the AMH Ihins; same with the modern foot. The Ihins (*Homo sapiens pygmaeus*) were contemporaries of early humans, and, though very different in type, they mixed with them. Gene exchange (the nice word for "nookie")—not evolution—gives us the "intermediate" types of the fossil record, the so-called transitionals of step-by-step evolution.

Bogus transitionals, trotted out to prove the metamorphosis of lower into higher hominids, are nothing but the hapless offspring of exogamous unions, which is the factor that produced the great variation so typical of interbreeding populations. The variability that appears at every turn simply resolves into the incessant crossbreeding of different stocks—nothing more. What I call the "mixing of races," say, of Neanderthal and modern, evolutionists call "transmutation" or "speciation," proven, they say, by specimens like those "intermediates" at Skhul. But, as we have seen, neither transitionals nor variability fit the bill at Israel's Qafzeh and Skhul caves. The bone people were obliged to confess that there were two different types of humans, two different races, in the Near East.

John Feliks, in a diatribe against the standard model, quotes directly from biology textbooks, which claim the fossil record provides clear evidence of "evolutionary transitions." But according to Feliks, these statements only "prove that the authors of a leading biology textbook either have no idea what they're talking about . . . or are participants in fraud. . . . Evolution is plagued by one fiasco after another" (Feliks 2013, 10–11). And how will advocates of continuous transition explain away the gaps, the missing hundreds of types that should be shading imperceptibly one into the next? Paleontologists openly admit that the transitional forms required by Darwinism have never been found. Here lie cognitive dissonance, split mind, sustaining the paradigm—even though it has been checkmated.

When we take a closer look, we can see that the *H. habilis* to *H. erectus* transition involves too extreme a change, especially in size; that is, too great an evolutionary leap. The very small body size of OH 62 (*H. habilis*) compared to the near giantism of some *H. erectus* specimens truly staggers the imagination.*

Please explain to me why, from around 300,000 BP (Before Present) (the supposed end of *H. erectus* time) to around 100,000 BP (beginning of modern man), no *H. erectus* fossils appear in the record, just where we should find them "transitioning." Where are the intermediates? Even Darwin worried that the record would not support his theory: Why is not every geological formation full of such intermediate links? How do such gaps and sudden big changes compare to evolution's insensibly fine gradations over immense spans of time? These unorthodox leaps and bounds are the very opposite of gradual evolution.

Yet that is just what we find in the record: fish swim into the fossil record out of nowhere, fully formed. So do the great apes. The mammals emerged abruptly. So did the invertebrates. They have no yesterday. The story is the same for chordates (the precursors of vertebrates) and the flowering plants, not to mention sudden new species at the Triassic-Jurassic boundary. Why were no ancestors ever found for the creatures of the famous Cambrian Explosion† 500 million years ago? The strata under the Cambrian are almost empty of animal fossils. We just can't find the forerunners of starfish, octopi, sea urchins, clams, or snails—only bacteria, worms, and primitive organisms like algae. An abominable mystery? "Darwin's big mystery," said Lynn

---

*It might be possible to explain the size and strength of certain erectoids by "heterosis" (hybrid vigor), as in Ngandong, OH9, Turkana, Weidenreich's "goliaths," and von Koenigswald's Meganthropus.

†The Cambrian explosion was the relatively short event beginning around 542 million years ago in the Cambrian Period, during which most major animal phyla appeared as indicated by the fossil record. — *Ed.*

Margulis, "was why there was no record at all before . . . 542 million years ago . . . and then all of a sudden you get nearly all the major types of animals. . . . There is no gradualism in the fossil record" (Teresi 2011, 67–68).

Even dinosaurs and mastodons materialized suddenly. And the milestones in human evolution are no different: *Homo erectus* and Neanderthal, it is grudgingly admitted, appeared abruptly on the scene. Ditto Cro-Magnon, the first paragon of fully modern humans, star of the acclaimed "sapiens explosion"—the big bang of evolution—out of nowhere. All these quantum jumps speak not of evolution. Something else is going on.* How is it that the very apex of evolution, the masterful work called humanity, with our incomparable brain, took the least amount of time to emerge? And why should there be a quickening of pace as the stages of evolution proceed? Does it make sense that simple species took millions of years to evolve but *H. sapiens,* by far the most advanced, managed to emerge in only a few thousand years? Even Sir Arthur Keith wondered how the extraordinarily complex human brain could have expanded in the relatively brief course of the Pleistocene. And how could the brain of *Pithecanthropus* (*Homo erectus*) "have evolved into the modern human form? I cannot credit such a rapid rate of evolution" (Keith 1929, 436). Well, the way to account for this is simply by looking at crossbreeding with a more advanced type—the little AMH Ihins. After all, there is no increase in *H. erectus* cranial capacity during the next half million years (after the initial 50 percent increase from *Australopithecus*'s). It stayed around 1,000 cc until the pithecanthrops were upgraded by crossbreeding with the moderns in their midst.

The solution given by the experts runs something like this: evolution sometimes proceeds at a faster pace for reasons yet to be discovered. Perhaps it was a sudden neurological change, an intellectually

---

*That something else is creationism, whereby the fossil record would and does indeed show creatures appearing fully formed, discussed in detail in chapter 8 of my book *The Mysterious Origin of Hybrid Man.*

advantageous mutation. Yet there is no trace in any fossils of such neurological developments. Well, then, maybe rapid progress was triggered by strong competition between different populations under severe environmental pressures.

Last time I checked, the theory of "competition" was circling the drain.

In any event, this jump was given a proper name: saltation or rapid branching or "punctuated equilibria." *Au. sediba*'s brain seems to have increased in size—or rather shape—relatively quickly. A similar jump happened again in the hominid line, approximately 600,000 years ago, another spurt in brain growth, a punctuational event. And again, we are told that some fortuitous mutation in the hominid genome about 50,000 years ago spurred cultural advances. It must have been some radical reorganization, a neurological change that allowed humans to develop culture and sophisticated behavior, some sort of intellectually advantageous mutation. But such a "burst of evolution" is actually a profound contradiction in terms, an oxymoron. The very definition of evolution entails minute changes by small degrees over a huge span of time. Every one of these "jumps" signals a new hybrid type, not evolution.

I am always amazed at the one-sided picture of evolution portrayed in the papers and journals, most recently in Matt Ridley's "A Relief to Darwin: The Eyes Have It" (2012, C4). Although Ridley refers to Darwin's correspondence with America's botany professor Asa Gray concerning the fearsome challenge presented to evolution by the complexity of the eye, the fact is, the Harvard botanist thought Darwin's theory could *not* explain the eye. (Ridley does not mention this.) Indeed, the eye's morphogenesis kept Darwin awake at night. It was, and is, frankly a black eye for evolution, its founder openly confessing in *The Descent of Man* that the eye's "inimitable contrivances . . . formed by natural selection, seems absurd in the highest degree" (2004).

Ridley also invokes the acclaimed name of Francis Crick (who discovered DNA together with James Watson) to clinch his argument, again unmindful that the Nobel Prize winner actually believed that DNA was far too complex to have evolved by random chance. How could the eye, "though it came to see, not be designed for seeing?" (Lovtrup 1987, 232). Not only was Darwin's blind chance (random mutation) a most improbable explanation for the eye's exquisite intricacy, but now we have scientists telling us it is mathematically impossible: feed the data into a computer and it just jams. Some mathematicians are saying that there was insufficient time for the number of mutations apparently needed to make an eye; the odds of this happening are 10 billion to 1 (Hitching 1982, 87). In other words, the selection of such mutations by pure chance would have taken much longer than the known age of our planet!

But the doctrine of chance and mutation is required by atheism. Married to a secular universe, big bang and evolution insist that things took shape all by themselves, that system and order happened by chance without the help of any Higher Intelligence. Bertrand Russell in *Why I Am Not a Christian* sums up the nonbeliever's view: to him, structures are merely "accidental collocations of atoms." Yet as biologist Lynn Margulis puts it, "I was taught over and over again that the accumulation of random mutations led to evolutionary change [and] new species. I believed it until I looked for evidence" (Teresi 2011, 68).

Did such precisely adapted modifications arise by pure luck? Is evolution a matter of chance? Is science (backed up by its convenient principle of uncertainty, of indeterminacy) really doing its job if it hangs its hat on nothing but accident and randomness? Are you willing to subscribe to a philosophy that says mere chance produced elaborate, almost perfect, design? This dark illogic makes me shudder. The coordination of parts into a functioning whole is characteristic of work and intelligence, not blind chance.

Well practiced at troubleshooting and skillfully turning liabilities into assets, the Darwinian spin doctors have put out all the fires threatening their preserve. To keep the theory alive and well, a continuous bloodline must be proven—from tree shrew to apes to australopiths and right on up the ladder to us. But ruining all this are those gaping holes in the hominid pantheon, like the infamous gap between 2 million and 3 million years ago, or the 200,000-year gap that separates the end, the demise, of *H. erectus* from the beginning of *H. sapiens,* his presumed successor. The fossil record is full of these improbable gaps. What's more, presumed daughter species often do not even resemble the supposed parent. Known among insiders as the black holes of evolution, these gaps are desperately smoothed over with rhetoric, abstruse rationalizations, and jargon-y catchphrases like "systemic macromutations" and "punk eek."

And when it has proven impossible to demonstrate a phyletic link (that is, a biological sequence in time) from one hominid type to the next—when the missing link failed to turn up—the problem, which is really a death blow to Darwinism, vanishes simply by invoking a "common ancestor."

Scientists have identified, living over 60 million years ago, what may be humankind's common ancestor, an animal that nourishes its young in utero through a placenta.

Fig. 2.4. Our common ancestor. Cartoon by Marvin E. Herring

Darwinism's most recent bulldog—ethologist, biologist, and author Richard Dawkins—counts up the differences between comparable molecules in two animal groups to determine the time that has elapsed since their alleged common ancestor lived (1996, 271). This is called the molecular clock (one of the latest technological boasts of genetics). The concept of a common ancestor, though, is actually a relic of eighteenth-century thinking, entertained by Darwin's own grandfather, Erasmus Darwin, as well as by Erasmus Darwin's contemporaries,

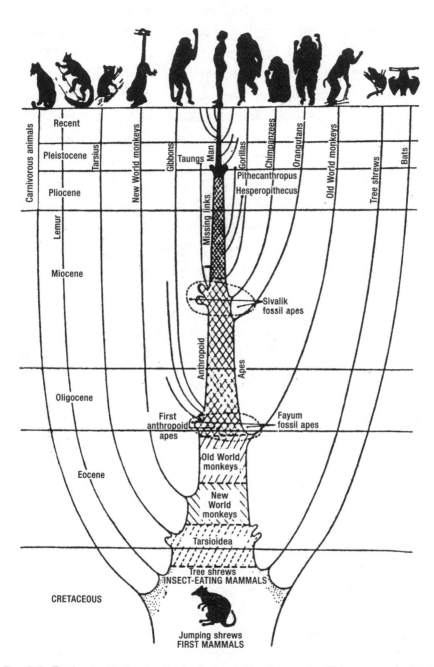

*Fig. 2.5. Textbook charts of primate descent, like this one, still leave me doubtful
that I am the granddaughter of a tree shrew or an insectivore. And no real link
has ever been found between fish and amphibians, amphibians and reptiles,
reptiles and mammals.*

Denis Diderot, George-Louis LeClerc, Compte de Buffon, and Jean-Baptiste Lamarck. Not only does this simple-minded model have all hominids branching off, forking off, from one ultimate anonymous ancestor, constituting the elusive "trunk" of the family tree, but the same setup is posited for the entire order of primates. Somewhere in the deep, dark past, humans and apes already shared a common ancestor, then went their separate ways. That ancestor has never been found, despite an ambitious search.

Do all living things descend from a single entity, which floated around in the primordial soup some 3 or 4 bya? Scientists, finding that the neurons of our cerebral cortex are similar to those inside the head of the ragworm (a lowly marine creature), then jump to the conclusion that "They were too similar to be of independent origin . . . [and] we must share a common ancestor" (Abrams 2011, 36).

Similarity, though, does not necessarily signal common historical roots: they may be merely analogous, parallel structures, just as flight in birds, insects, bats, kites, and airplanes each has a separate story. Neither does molecular unity prove that we all come from one "common ancestor," simply that this is how life is built. The same chemical elements are found throughout the cosmos, according to physical chemistry Professor Harold Urey, a Nobel Prize winner. No, the "wonderful unity in all living things" does not require a common ancestor, but it is interpreted as if it did, thus endorsing Darwinism "with a thundering voice for the validity of evolution theory" (Edey and Johanson 1990, 277). DNA has merely confirmed the basic biochemical identity of all organisms, from bacteria to man. The fact that all life forms have similar DNA patterns does not mean or even suggest we are all derived from the same batch.

The evolutionary agenda requires an unknown common ancestor, a creature that led to apes on one branch and to humans on another. But this mythical common ancestor remains a matter of promiscuous guesswork and intellectual juggling. In the context of race origins, our imagined common ancestor, a dark AMH African, supposedly branched out to Caucasians on the one hand, Asians on the other, and so on by some miraculous process of lightening or bleaching. Out-of-Africa's monogenism (the theory of the single cradle of man) also requires the different races to be formed rather late in time. But humans were cast in their mold (Negroid, Mongoloid, Australoid, etc.) from the beginning; for each was raised up in its own division of the Earth (this alternate view is called "polygenesis"). The spurious common ancestor forces us to entertain racial separation from a single (monogenetic) root. But the races never separated. In fact, they did the opposite—they came together. Nookie!

Svante Paabo of the Max Planck research team in Leipzig declared that we shared a common ancestor with the Neanderthal before diverging from them and evolving apart. But that is not how we or the Neanderthals came about. Neanderthal is simply, very simply, a retrobred race—and not a primordial human at all, but a retro stock—resulting from Cro-Magnon indiscretions. No, Neanderthal is not "premodern," but actually postmodern, offspring of the naughty backbreeding Cro-Magnons. Neanderthal did not separate from some line, some "common ancestor." He owes his existence not to any separating or splitting, but to a coming together of two different races: the Cro-Magnon type and the lowly erectoids with whom they mixed.

The "nookie factor"* actually stands Darwinism on its head. Whereas evolution has species lines branching out and separating (splitting) at some time in the distant past, the crossbreeding factor entails quite the opposite: the different stocks came together and cohabited to form new races. And we are all hybrids. Fossils taken as representing "stages" of evolution or "changes" (by mutations) really represent nothing more than the unstoppable intermixing of the Paleolithic races. The peopling of the world is about the mingling and merging of disparate types. No evolution there, just the continual confection of half-breeds, quarter-breeds, and so forth—an exchange of genes since day 1.

---

*I am in debt to my colleague, archaeologist Christopher Hardaker, author of *The First American,* for the term *nookie factor,* which Chris also calls "sex and the single species."

It is time to reevaluate evolutionism and its myth of the common ancestor. Ultimately, this doctrine is reduced to the absurdity of all life evolving from bacteria, fungi, or the single-cell eukaryotic forms, which forces us to assume that the first living creature must have had within itself the genetic potential to grow into every one of the trillions of plants and creatures that have lived since. Listen to Sir Fred Hoyle on this inane and daring claim: "No evolutionary connection has ever existed . . . between so-called prokaryotic and eukaryotic cells [or] between bacteria and yeast cells. . . . evolution from a common stock? . . . We doubt that a terrestrial* evolutionary connection ever existed between the plant and animal kingdom. . . . There is no reason why some of the categories [taxa, i.e., different species] . . . should not always have been separate" (1984, 114).

Have you noticed we don't hear about the "missing link" anymore? No. Now it is the *common ancestor;* Matt Ridley (cited above) uses the phrase no less than seven times in one short article, pointing to an optical gene that gave jellyfish and insects and apes and us all one "common origin." Yet it is equally reasonable to suppose that shared genes or molecular unity are to be expected and that they do not imply a phylogenetic relationship, no less a shared ancestor. Resemblance, after all, according to many fossil researchers, does not necessarily indicate descent. What it does tell us is that behind all life is a common blueprint: DNA confirms the basic biochemical identity of organisms, from bacteria to trilobite to man. The myth of the common ancestor, a single taproot of all plants and animals, leaves us forever wondering who that unknown, unidentified ancestor was, and why he is still missing in action. "Each species develops according to its own kind," said Lucretius two thousand years ago. I think he was right.

The complications of interbreeding, the experts now openly admit, make it impossible to draw neat branching lines of descent—Darwin's

---

*Sir Fred's use of the word *terrestrial* probably represents his idea of a nonterrestrial (that is, cosmic) source of DNA and germs.

famous "descent with modification." When straight-line evolution, say, from *H. erectus* (parent) to *H. sapiens* (daughter) didn't wash, this "ladder" model of descent was abandoned in favor of a tree diagram. But no sooner was the family tree adopted to illustrate human evolution, then new problems arose: most of the fossil humans placed on that tree turned out to be not ancestors but dead ends, blind alleys, otherwise described as irrelevant side branches or divergent cousins off the main stem and with no relevance to evolution. All these "also-rans," deemed ineligible as ancestors, were then quietly shunted from the trunk of the tree to the outer branches. The top-heavy result was that the tips were well populated, while the trunk remained shrouded in mystery. In fact, there was no trunk.

Solution? Change tree to bush. But even with bush, almost all the nodes or branching points remained unidentified, with big question marks at the point when each species first appeared.

Solution? Change bush into network, a crisscrossing affair with genes moving freely across the lines. However, this setup, this network, is not a diagram of evolution at all: It represents nothing more than a multiplicity of crossbreedings. And in that case, we no longer need Darwin's slow and plodding evolution to account for the many varieties of humans. New races (hybrids) can take shape in a generation or two. Quick work! "Suddenness" unveiled!

So let's cut to the chase. Instead of evolving, our ancestors were cohabiting! Shake that family tree and down comes the apple of amalgamation—race mixing, plain and simple. And if evolution

Refuting Darwin

Fig. 2.6. Interbreeding. A grapefruit chastising an orange for cheating on her husband and giving birth to a tangerine and an apple. Cartoon by Marvin E. Herring

claims that earlier, inferior hominids were replaced or destroyed by their betters, the fossils themselves—and their DNA—are saying this: "We were not really wiped out, we were absorbed. We melted away by mating with others outside our group. Eventually, we interbred out of existence!" "This kind of interbreeding" of AMHs, blending "with archaic species such as *H. erectus,*" announced *Scientific American* (Stix 2008, 56), left its footprint in as much as 80 percent of the modern genome.

The affairs of our ancestors, primly known as "gene exchanges," explain all the overlap, the continuity, say, from archaic Java man (the first erectus ever discovered, in 1891) to the present races of Java. One classic example is the shovel-shaped incisors of Peking man (the next erectus to be discovered, in the 1920s), which are indeed typical of today's Asians.

There is overlap, too, between Neanderthal and *H. sapiens.* As we

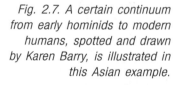

Fig. 2.7. A certain continuum from early hominids to modern humans, spotted and drawn by Karen Barry, is illustrated in this Asian example.

have seen, Neanderthal did not evolve into us; rather, he was largely absorbed into the Cro-Magnon (modern-type) population, until he finally disappeared as a distinct race. With Neanderthal's status changed more times than the miniskirt has gone in and out of fashion, as one anthropologist put it, his lineage's long-debated place on the family tree has only very recently been settled (sort of). In 2010, DNA researchers announced that some Neanderthals had red hair, pale skin, and adorable freckles. Not only do Neanderthals show Cro-Magnon traits, but Cro-Magnons show Neanderthal traits—again, reflecting genetic mixing.

What did that red hair tell us about our alleged predecessors? Actually, we learned a great deal about our Neanderthal friends in the twenty years of fieldwork between 1986 and 2006: excavations in Croatia in 1986, at Israel's Mt. Carmel (mentioned above) in 1988, in France in 1996, in Portugal in 1998, in Czechoslovakia in 2003, and finally in Romania (Peştera cu Oase) in 2006. A common thread ran through all these finds: Neanderthals turned up with perfectly modern traits, just as their AMH (Cro-Magnon) neighbors turned up with Neanderthal traits.

Both groups used similar tools; they were contemporaries. In Portugal, their affairs resulted in a hybrid race, while those recent digs in Eastern Europe came up with more blends: moderns sporting distinctive Neanderthal features, such as large noses, molars, and occipital buns (protuberance at the back of the skull), suggesting, all told, the out-and-out intermixture of Cro-Magnons and Neanderthals.

Cro-Magnon as a rule was brawny and brainy (1660 cc). His was a high forehead, a strong jaw, and a broad barrel chest. Notwithstanding his modern form, he, too, was a mixed bag, judging from his heavy brow, robust build, large teeth, beaklike nose, large rectangular orbits, wide cheekbones, broad face, and long arms—all part of the erectoid heritage, although these traits could equally reflect mixing with Neanderthals. Especially in Europe, Cro-Magnon shows an alloy of races in the proportion of limbs and in certain features of the face. Not too civilized, the more back-bred of these Cro-Magnons was an incessant warrior, with lots of weapons; he was a hunter of wild cows and horses. The great extent of Cro-Magnon retro-breeding was their undoing. Their culture did not mysteriously disappear; rather, they retro-bred with Neanderthal types, as indicated by the lapse of Lascaux-type cave art circa 12 kya.

The Ihuans [Cro-Magnon] have degenerated by marrying with the druks [erectoids]. . . . I raised up Ihuan, and I gave them certain commandments, amongst which was not to cohabit with the druks, lest they go down in darkness. But they obeyed not my words, and lo and behold, they are lost from the face of the Earth.

OAHSPE, BOOK OF FRAGAPATTI 39:1,
SYNOPSIS OF SIXTEEN CYCLES 1:23

The same picture of racial blends emerged at digs in Spain (Atapuerca), Germany (Steinheim), Czechoslovakia (Mladec), Romania (Muierii), and France (St. Cesaire and Arcy-sur-Cure). None of these results had the ghost of a chance of fitting into any sort of evolutionary sequence. With every variety of intermixture among these fossil humans, it looked instead like some serious fraternizing was going on in the Old Stone Age.

Given all this interbreeding, the whole case for human evolution is dead in the water. For this very reason man-the-mixer, man-the-hybrid has been stonewalled at every turn. Debate raged for decades before the simple fact of coexistence was established, trouncing competitive exclusion. Next to be shaken was the stubborn resistance to the actual crossbreedings between those coexisting races. Such liaisons, the hardliners say, were rare. We needn't worry about a one-night stand, they say. Cro-Magnon crossbred with Neanderthals on a very small scale, they assert. These stone age Romeos and Juliets bedded down "at least once" (Neimark 2011, 67). Kicking and screaming, the fossil keepers, sooner or later, will have to confront the facts of life; that is, transracial sex, the nookie factor. "Why not? Sex happens," concluded one of America's leading paleoanthropologists, after geneticists published findings of up to 25 percent Neanderthal DNA in our immediate ancestors' genome (Lemonick 1999, 58).

"COMMON ANCESTOR"

We are all a complete mixture.

BRYAN SYKES, GENETICIST,
THE SEVEN DAUGHTERS OF EVE

John Rennie once made the statement that "serious scientific papers" disputing evolution are "all but nonexistent," adding that one researcher who looked for such articles in the "primary literature . . . found none" (2002, 81). Of course not! Alternatives to neo-Darwinism cannot pass muster; that is, peer review, the pompous process of vetting academic works against the SM, of screening out any material that challenges Darwinism. Regarding this censorship, "archeologist Paul Bahn wrote me [John Feliks] that *Current Anthropology* published 'a lot of rubbish' while blocking good papers. . . . They are going to need dozens of attorneys defending them once the scope of this deception cracks open" (Feliks 2013, 11–12). Much is, and has been, suppressed.

I have come across one opinion, repeatedly: Darwinism, with all its flaws, stands almost by default. Why? Because, aside from creationism, no reasonable alternative has been offered. Yet that alternative has been in our midst all along—the nookie factor, crossbreeding—giving us, as far back as we can see, a continually mixed "network" of ancestors.

Read any scientific book about the races in today's world and hardly a page goes by without mention of admixtures, crossbreedings, half-breeds, and so forth. Darwin himself was witness to "the immense mongrel population" of Negroes and Portuguese in Brazil. In Chile and other parts of South America the entire population consists of Indians and Spaniards blended in various degrees, showing "the most complex crosses between Negroes, Indians, and Europeans" (Darwin 2004, 202). It was the same in ages past. As a result, we are all hybrids, without a scintilla of evolution in sight. We are a cohabiting, not an evolving, species.

Scenes of Cohabitation

The Ihins strayed and mixed with asu man. We saw a sample of such offspring in *Homo floresiensis,* the tiny Java island hobbits who came into the world when Ihins blended with the autochthonous race. In Hindu scripture it is written, "The first race asu tempted the white people," who then broke the commandment of endogamy (in-marriage), which had enjoined:

> Neither shall ye permit the Ihins to dwell with Asu, lest his seed go down in darkness.
>
> OAHSPE, FIRST BOOK OF THE FIRST LORDS 1:10

But the little people wandered out of the Garden of Paradise and began to dwell with the Asuans, and there was born into the world a new race, called "druk" (a.k.a. Cain aka *Homo erectus*).

Throughout their history, the chosen were commanded to marry among themselves, to withdraw from other peoples; this scenario in a latter-day context is repeated in Genesis 28:1. "You shall not take a wife from the daughters of Canaan [i.e., descendants of Cain]." Nevertheless, the druks (a.k.a. Cain) . . .

> burnt with desire . . . and the chosen came unto them, and they bore children to them.
>
> OAHSPE, SYNOPSIS OF SIXTEEN CYCLES 1:27*

Later (in the third wave), the sacred little people again mixed with the ground people (the druks), for the latter came to them in the winter as beggars. And they mingled and it came to pass that the Ihuans were again born into the world. And on yet a different occasion the druks fetched the root of babao (an intoxicant) and brought it to the little Ihins to eat and get drunk, saying,

---

*Producing the second wave of Ihuans, i.e., Cro-Magnons.

Lest the white and yellow people fall upon us, and our
seed perish on the earth, make us of flesh and kind, bone
and bone, blood and blood.

OAHSPE, THE LORDS' FIFTH BOOK 1:20

The seduction, reminiscent of Hebrew lore, recalls the depraved
daughters of Cain, women of a lower order who had charms and
enchantments with which they seduced the sons of God.

Competitive exclusion has fallen by the wayside, along with many
of evolution's most cherished doctrines. Man, touted as the big game
hunter, seems to have actually been more of a vegetarian, feasting on
cattails, and a scavenger; at best he was an opportunistic hunter of
tortoise, lizards, and small game (monkey brains being a great deli-
cacy). Embryology, as a map to decipher evolutionary changes over
the ages, has also been discarded. A slew of carefully choreographed
and often bizarre hypotheses have been dethroned as well. Even the
concept of progress (onward and upward) will have to bow to cycles
and the law of entropy, following the mounting evidence for fallen (or
lost) civilizations and relapses into barbarism. The waxing and waning tide of man and his wonderful accomplishments is a curve of rise
and fall, rise and fall.

Horse evolution—almost literally a dog-and-pony show, heavily
promoted as a classic example of descent with modification—has
taken a beating. Though textbooks still regale us with the evolution of
today's single-toed horse, George Gaylord Simpson, at one time the last
word in evolutionary genetics, pointed out that this was a bad example of linear evolution. Why? Because some of these different horse
types were alive at the same time! In artful diagrams, the evolutionist
selects species that look like plausible intermediates while ignoring all
the other species that don't accord with that picture. Is this natural
selection or evidence selection? Paleoanthropologist Niles Eldredge in

Reinventing Darwin objected to this habit of picking and choosing which species to include in the model. Simpson actually came out and said that the continuous transformation from its so-called ancestor to the modern horse "never happened in nature" (Eldredge 1995, 119).

One by one, the underpinnings of the standard model of human origin are washing away, like feet of clay in a flood of contrary evidence. The identity of the common ancestor hangs in the air. The linear succession of hominid types hangs in the air. The value of natural selection and fortuitous mutations (as the cause of change) hangs in the air. The origin of consciousness hangs in the air. The causes of extinction hang in the air. The "missing link" is still missing in action. Our experts have been on a hunt for something that doesn't exist, and never did, such as the link between human and ape. Are we really made-over apes?

> *[They are] looking and hunting for that which was forefather to both man and monkey. What sort of beast they expect to find I cannot imagine.*
>
> JAMES CHURCHWARD, *THE CHILDREN OF MU*

They say the genetic intimacy between humans and African apes forces us to conclude that man and ape arose from a common ancestor. Fossil "anthropoids" called dryopithecines were once spread over a wide zone of the Old World. These were thought in the twentieth century to have evolved into the ancestors of today's apes, on the one hand, and into early humans on the other. There have always been candidates like Dryopithecus, but no trace of their descendants who could conceivably link them to early humans.

It was the same with the Ramapithecus candidate. There was that yawning gap—millions of years—between this supposed "common ancestor" and his human descendants. Although they are still looking for that common apelike ancestor—a type like Ramapithecus, Kenyapithecus, Proconsul—who gave rise to the australopiths, such beasties were actually ruled out as ancestors, recognized as a false

Fig. 2.8. Three types of spider monkey (top row) and two marmosets (bottom
row). While the marmoset is tiny, the South American spider monkey (Atleles)
often grows to a height of three feet. If, as we are taught, we share an
ancestor with the apes, not the monkeys, why do we look (facially) rather
more like monkeys than apes?

alarm back in 1979. They were nothing more than anthropoid apes.
Dryopithecus was a "nasty-looking great ape" (Gamow 1948, 177),
much like today's gorilla. Ramapithecus* was too old to be a hominid
forerunner.

At the end of the day, any comparison of australopiths to apes is
overworked and frankly superficial, for homo features are dominant
even in the earliest people: cranial height, shape of occiput, poise of
head on vertebral column, structure of pelvis and limb, face and teeth.

---

*Judged to be 12–30 myr, Ramapithecus of India, as a candidate for that common ances-
tor, had molars, canines, and jaw with a decidedly human cast; the curved dental arcade
made him look quite human. But molecular dating determined that no hominid could
have existed more than 7 mya. Anyway, new evidence overturned these Ramapithecus
hopefuls. They weighed, after all, less than thirty pounds, and the mandible was finally
judged not humanlike. It turned out to be some sort of primitive orangutan.

Homo sapiens, say analysts, has hardly changed in the past 75,000 years; indeed, it is often said that the physical evolution of humanity is finished. The human brain is in stasis; actually it has been decreasing in size, shrinking, for the past 10,000 years. If it is true that we are no longer evolving, perhaps it is also true that we never evolved in the first place. All we have ever done is mix with our neighbors.

So where are the whistle-blowers? Evolution has already been refuted many times by biologists, mathematicians, lawyers, theologians, prehistorians, science writers, and even paleontologists. Today, as we speak, signatories against the obsolete theory of evolution include scientists from the National Academies of Science in Russia, the Czech Republic, Hungary, Poland, India, Nigeria, and the United States. Many are professors or researchers at major institutions, such as MIT, Cambridge University, Moscow State University, Chitose Institute of Science and Technology in Japan, and Ben-Gurion University in Israel.

How many people who think they believe in evolution have ever actually read Darwin's unwieldy book *On the Origin of Species*? Few indeed. But gospel—even secular gospel—is a hard bubble to burst. Outgunned and outmaneuvered, the dissenting voice has been ostracized and arrogantly stigmatized. This is the underbelly of Darwinism you seldom hear about. The mudslinging. The power plays. I personally know one prominent, vocal, and quite brilliant opponent of Darwinism who was accused of mental illness—pilloried and embittered for life by the House of Darwin.

Professors defend Darwinian theory as if their lives depended on it. To buck Darwinism, or neo-Darwinism, as the modern synthesis is called, would be professional suicide. There is too much to lose: grants, promotions, tenure, prestige, books and articles, lecture invitations, and so on.

"Scientists," wrote the late, great Michael Crichton in his remarkable book *State of Fear*, "are in exactly the same position as Renaissance painters, commissioned to make the portrait the patron wants done . . . [T]he system works against problem-solving. Because if you solve

Fig. 2.9. Spoofing Darwin's "transmutation" of species. Cartoon by Marvin E. Herring.

a problem, your funding ends. . . . Scientists are only too aware whom they are working for. Those who fund research . . . always have a particular outcome in mind" (2004).

But evolution, in all truth, is a "soft science." Its subjects, long gone, cannot be tested in any controlled way. The fossil record, it has often been pointed out, left no smoking gun. Therefore all must be decided by inference, by sketching in the blank spaces. But when I see a gorilla face drawn in reconstructions of our hominid ancestors of Africa, I cringe at this ersatz specimen, this glorified baboon, who allegedly evolved into us.

Let us not be so anxious to make our study a "science." (We will run into this problem again in chapter 6, where that ambition, that thirst for scientific respect, was, in my opinion, the downfall of Carl G. Jung.) As I see it, the question of human origins is history, not science per se. It is about the past—humankind's history here on Earth. Oh, you can "scientize" it, but only up to a point. Yet in doing so it is easy to dehumanize it, rendering a sterile, meaningless account with precious little that is truly human. There are some things that can never be scientized because they're too fragile, too subtle to meet the rigors of laboratory or computer. Despite great effort, scientism has not struck down the very real possibility that an Intelligent Designer created the living.

Instead of accepting the overwhelming evidence of intelligent design as proof that a great Mind planned and created the universe . . . evolution has been dressed in the language of science and deliberately passed off on a gullible public as fact. The perpetrators of this massive deception have taken millions of dollars in grants from foundations to pursue a phantom that mathematics conclusively say does not exist.

DAVE HUNT AND T. A. MCMAHON,

THE NEW SPIRITUALITY

Although Darwin's approach to scientific investigation may have revolutionized natural philosophy and righteously trumped long-standing biblical dogmas, I am afraid that evolutionism has ended up throwing out the baby with the bathwater. It is time to rescue that baby, which is none other than the utter uniqueness of humankind. Science may have found our genetic fingerprint, but it has missed altogether our spiritual fingerprint. The mystery of *Homo sapiens,* the spark with which we humans are all endowed—ingredient X—is tucked away in a corner of our own minds, ready to reveal itself, not on the stage of science, but within our own understanding. The soul-deadening materialism, the animality, to which Darwinism consigns us does not begin to tell the extraordinary story of humanity.*

Forty years ago the delightful British author Norman Macbeth wrote that Darwinian evolution was going to pieces. Another European critic thought that "one day the Darwinian myth will be ranked the greatest deceit in the history of science" (Lovtrup 1987). The interpretation of fossil evidence exclusively by convinced evolutionists should, quipped attorney Philip Johnson in *Darwin on Trial* (1991) be "scrutinized as carefully as a letter of recommendation by a job applicant's mother." If evolution were actually on trial, its mountain of circumstantial evidence, obfuscatory jargon, and factoids would be bounced out of court by an annoyed judge. Darwinism today rides on its reputation, not on its validity.

---

*That story is told in my last book, *The Mysterious Origins of Hybrid Man.*

Fig. 2.10. Spoofing genetics.
Cartoon by Marvin F. Herring.

Now let us see how Darwin almost single-handedly invented the Ice Age . . .

CHAPTER 3

ICE AGES

A Melting Myth

A great part of our geological teachings must be rewritten.
JAMES CHURCHWARD, *THE LOST CONTINENT OF MU*

Did you ever wonder how Darwin's gradual evolution, especially of tender species, could have happened amid killer ice ages? These freezes could have made quick work of all the warm-loving creatures that filled the landscape of early times.

Charles Darwin was especially puzzled by one question: How is it that herds of the same breed—presumably of the same origin*—were living separately at the very extremes of the Earth, that is, at both northern and southern high latitudes, more than nine thousand miles apart? Well, it was easy enough for naturalists to account for the migration of plants through bird droppings (deposited seeds). But they wondered at the widely separated distribution of such large plodding animals as the mammoth. How could such cold-weather creatures have traversed the equatorial zones? Or did they?

---

*As a monogenist, Darwin assumed the animals had a single ancestor, which gives creatures a single garden of Eden. The supposed migration of cold-climate animals assumes they originated in only one part of the Earth. It might just as well be supposed that if animals adapted to their environment, they did not have to travel from distant places to be found in different parts of the world. They could have originated there (through polygenesis).

Ultimately, Darwin—in the footsteps of Louis Agassiz's newly minted theory of ice ages (ca. 1840)—postulated "advancing ice" to explain the presence of these animals in different corners of the world. Agassiz, the dashing Swiss-American geologist and paleontologist, had turned to glacial observations, climbing unscaled mountains and performing other feats of daring for the sake of science. Spotting landforms in nonglaciated regions of Europe that were nonetheless similar to the Alps, he reasoned that past ages of ice could be the explanation. But the world, as one critic saw it, "was dazzled by the factitious glamour of Agassiz's rhetoric. . . . Geologists were aware of the many contradictions and seemingly impossible assumptions involved in the postulated ice-sheets" (Corliss 1980, 35–36). The acclaimed naturalist and explorer Alexander von Humboldt, for one, told Agassiz that he did not believe these "revolutions of the primitive world" (Imbrie and Palmer 1979, 31), meaning the theory that the Earth went from hot to cold to hot and back again to cold—many times.

But let us see how Darwin's solution to the mammoth puzzle went. When the hypothetical "glacial period" descended upon the northern stretches of the planet, these large animals slowly migrated southward. During the next "interglacial" warm period, the "Big Defrost" (see "Interglacials," page 118), some of those heavy-coated beasts returned North again while others remained in the equatorial regions, finding refuge from the heat in the cool mountains. (Darwin did, indeed, have a good imagination!) And when the next "ice age" came along (this is the important part), it came from the South. Some kind of alternating pattern seemed to be at work. James Croll, a nineteenth-century theorist, calculated glaciations once every 11,500 years, with ice alternating in the northern then the southern hemisphere.

Well, with the return of ice, those animals in the southern highlands presumably climbed down from the heights in search of warmer pastures. And, as Darwin reasoned, when the next "interglacial" warm period set in, those same beasts moved to higher (southern) latitudes, to beat the heat. This, then, seemed to solve the riddle of identical animals—presumably of a single and common origin—residing at the cooler latitudes of either hemisphere. Altogether, Darwin's theory of monogenesis (single origin) could not fare so well without positing great glaciers to the North and the South. He needed "advancing ice" to establish a good reason for the animals to have migrated from one end of the Earth to the other.

Interglacials

Theory says that within each ice age are warm periods called interglacials, "generally lasting only ten to twenty thousand years" (Macdougall 1996, 8). But temperate periods (like now) within or between supposed ice ages lasted much longer—100,000 years—according to professors Stanley Chernicoff, George Gamow, and others. American anthropologist and author Loren Eiseley calls them "interstadial springs" (Eiseley 1962, 83) and has them lasting about 50,000 years, in a regular waxing and waning pattern, the last ice recession having occurred about 20 kya. However, scientists say Leipzig did not defrost until 4 kya. And according to French paleontologist Georges Cuvier, the great naturalist of the early nineteenth century, the last ice receded about 6 kya. Indeed, the Canadian archipelago thawed out only a scant 1 kya, even though estimates for North American ice put the Great Defrost around 40 kya. (Charles Lyell, sometimes called the father of geology, had the Niagara Falls ice retreat 30 kya.) Then again, it is said the Wisconsin glaciers retreated only 10 or 12 kya. Not much agreement either on the present interglacial (warm period), which lasts, according to some, only 1,000 years, but to others, 40,000 years. "The Glacial Period," commented James Churchward facetiously, "is really modern and up-to-date, for it boasts of having a family of little glaciations, so that there will be some support for it in its old age. . . . [But] it is impossible that we could ever have experienced a temperature sandwich, as described by geology . . . No violent or very unusual temperature changes can possibly have taken place during the whole period of her [Earth's] cooling" (Churchward 1968, 89, 96).

Fig. 3.1. Map of North America, drawn by geologist T. C. Chamberlin in the nineteenth century. Drift deposits in North America indicate at least three ice ages.

Harvard's Louis Agassiz contended that ice ages were prompted by repeated global catastrophes, accompanied by a drastic fall in temperature. A student of Cuvier, Agassiz was clearly a catastrophist, favoring sudden, even violent, scenarios for the geological past.

He never bought uniformitarianism or Darwinism's gradualism. The norm, to Agassiz, was a cold Earth. Yet this certainly conflicts with glaciations separated by "long periods of time with much warmer climates" (Macdougall 2004, 53) and with Loren Eiseley's opinion that ice ages occur only once every 250 million years, making up a scant percentage of Earth's history. Albrecht Penck and Eduard Bruckner's ice ages were also short pulses separated by longer warm intervals. Science, in fact, declares "the presence of water, not ice, throughout most of Earth history" (Pollack 2003, 145), which conflicts with those who postulate "long deep freezes alternating with brief warm spells" (Alley 2004, 64).

The only reason we get occasional warmth, Agassiz reasoned, was owing to renewed igneous activity in the interior of the Earth. Others, favoring catastrophism, have proposed that the last ice age was abruptly ended by a collision with debris from a comet. Still others say warm periods begin when organisms decompose close to the surface of the ground or when methane is released from melting continental glaciers. All fine guesswork.

To yet others, regular decreases in the amount of sunlight have produced long sequences of ice ages separated by short, warm interglacial periods. Short? Is 2 million years short? This is the period given as the interval between the ice age that supposedly occurred 20 million years ago and the previous one (Lerner 1991, 309). And this, in turn, conflicts with the observation by glaciologists of no significant ice on Earth until the K-T boundary,* ca. 70 mya.

Indeed, Loren Eiseley throws cold water on deep-time ice ages (dated 200 mya or even 70 mya): "Suppose, just suppose for a moment, that . . . our geological estimates are mistaken" (Eiseley 1962, 100–101). Suppose, Eiseley says, the first glaciation was only 300 kya. This was micropaleontologist Cesare Emiliani's chronology, "startlingly different from orthodox estimates . . . as revealed

---

*The K-T boundary, the Cretaceous-Paleogene (K-Pg) boundary, formerly known as the Cretaceous-Tertiary boundary, defines the end of the Mesozoic Era.

in the oxygen-18 content of the minute shells from the ocean floor." These cores indicate the earliest cold period as no more than 300 kya.

And what about the next ice age? According to some, it is well overdue. Others say it's not due for another 2,000 years (Imbrie and Palmer 1979, 178). This broadly conflicts with the Serbian scientist Milutin Milankovitch's prediction that the next one to occur over North America and Europe is far in the future, roughly 23,000 years from now, though George Gamow more than doubles that to 48,000 years from now (1948, 184).

Now a few questions spring to mind: Does one "ice age" advance from the North Pole and the next one from the South Pole? The northern and southern glaciers, as Darwin saw it, had not occurred simultaneously. Rather, they took turns. So forget about ice "ages." Scientists say the Permian glaciation was confined primarily to the southern hemisphere, with vast regions—Africa, India—buried under ice. This is our first clue that ice came, not in ages, but in certain places. It has been noted, for instance, that while Europe and North America were experiencing cold conditions, there was "anomalously warm weather in the South Atlantic and Antarctica" (Alley 2004, 64). But there is nothing anomalous about it, at least not in Darwin's scheme, which had the different regions of the Earth simply taking turns being icy—one "glaciation" coming from the North, the next coming from the South. For all we know, this may turn out to be Father Darwin's biggest contribution to science; that is, the directional flow of the glaciers, marked by an alternating north-south pattern.* It might also explain why, at the height of the glacial age, only about 28 percent of the Earth's land area appears to have been covered in ice.

---

*Isaac Asimov sees it differently, though: "There is evidence that the Ice Ages take place in both hemispheres simultaneously." His supposition is that "a smaller tilt to the [Earth's] axis encourages an Ice Age in both hemispheres" (Asimov 1978, 116–23).

Evidence shows that the earth was not cooled overall during ice ages.

JOHN WHITE, *POLE SHIFT*

"Agassiz's idea . . . was incorrect. In fact, individual ice sheets had expanded from *different* spreading centers" (my italics) (Imbrie and Palmer 1979, 51). Today the boon of satellite photography has furnished us with the direction of ice flow from multiple centers. Geology Professor J. D. Macdougall, in *A Short History of Planet Earth* (Macdougall 1996), saw "many local variations in the way the ice behaved: perhaps advancing in one region while retreating in another." Indeed, geologists have found that ice came from every direction, including the equator! Rather than the expected movement of ice from the poles, glaciers have traveled northward from the tropics in Africa, Madagascar, Australia, and Tasmania (White 1980, 11). Striations indicate a "northward motion of the supposed ice sheets" (Corliss 1980, 32). How is that possible? It is possible because, as the world turned, or rolled, in a motion I call oscillation (motion #3), the ice always "approached" from whatever lands were sailing into higher latitudes. Approach and retreat of ice are figments of the imagination only.

Theorists, at best, have had to give up the idea of a single ice sheet that advanced from the poles. Yet the next question is this: How did the continents manage to wander? The most reasonable explanation is pole shift, which is the same as magnetic reversal and oscillaic motion. The theory is that the continents crept steadily through the latitudes, thus accounting for the puzzle of a once-arctic India, Brazil, and Africa. There is no escaping the signs: Almost all the continents have been through ice at one time or another. And in the case of India's and Africa's ice ages, the glaciers are known to have moved from the present site of the equator, not toward it! The ice that once covered India moved northward for a distance of more than a thousand miles. As Immanuel Velikovsky put it, "Not only are the causes of the glacial sheets unknown, but the geographical [question] . . . is also a problem. Why did the glacial sheet in the southern hemisphere move *from* the tropical regions of Africa *toward* the South Polar Region and not in the opposite direction? And why did the ice in India move *from* the equator toward the Himalaya Mountains and the higher latitudes?" Velikovsky answers this as follows:

> The change in the position of the poles carried the polar ice outside the polar circle, while *other regions* were brought into the polar circle. . . . Now we reflect: Was not the North Pole, at some time in the past, 20 degrees or more distant from the point it now occupies? In like manner, the old South Pole would have been roughly the same 20 degrees from the present pole. (my italics) (Velikovsky 1965, 23, 325–26)

It is a simple matter of observation. Professor Macdougall describes the vegetation zones of the northern hemisphere as having marched up and down the continents like so many armies surging back and forth in battle as the glaciers waxed and waned. But rather than the glaciers waxing and waning, the continents themselves did the marching! They moved into different latitudes. The globe, quite simply, rolled around, but very, very slowly.

This unknown (or barely understood) motion was indeed recognized by scientists in the nineteenth century when geologists began to discover fragments of ancient forests in the arctic lands. It then dawned on them that these forests had once occupied tropical belts

Fig. 3.2. Although we speak of the "retreat" of glacial ice, it is not the ice at all that retreated, but the landmass that "rolled" away from the pole in the inexorable planetary shift known as "magnetic reversal."

but since had moved away from equatorial space. This accounts for how warm-climate animals could have lived in North Asia, which by all accounts was once temperate, even quite mild, with plenty of vegetation.

WARM-WEATHER SPECIES BURIED IN THE FAR NORTH

PLACE	BURIED FINDS
Arctic Circle	fig palms
Spitsbergen	warm-sea animal fossils; sequoia trees (now petrified); corals and coal beds
New Siberian Islands	forests; "luxuriant vegetation"
North Greenland	magnolia trees
Alaska	corals; bones of lion, elephant, and rhino; tropical rain forest
Yorkshire, UK	cave bones of immense tiger, rhino, hippo
Northern Black Sea	river delta muds and rich soils (from core samples)

"About 375 million years ago," say geologists, "the Canadian Arctic was actually near the equator, enjoying a sub-tropical climate, which could explain the Illinois flora of 300 mya, its fossil forest a 'tropical wilderness' buried 230 feet below ground" (Gugliotta 2009, 14). The legendary northland of Hyperborea was also warm in prior ages: according to the Greeks, the interior of this fabled island, perhaps in the North Sea, was green and pleasant, possibly the seat of a lost empire. The Gobi Desert also at one time was a springtime paradise: the seat of (Churchward's) Uighur Empire, an exceedingly fertile area of land (more on Uighurs in chapter 8).

THE ENIGMA OF THE POLAR DINOSAURS

Large dinosaurs evidently once roamed the landmass that is today's South Pole, presenting a challenge to standard paleontology. How

Fig. 3.3. Left: Roy Chapman Andrews, one-time director of the American Museum of Natural History, examining the first-known dinosaur eggs during his epic expedition to the Gobi Desert (1925). Other life-forms near the Arctic Circle indicate it was perfectly ice-free and warm at an earlier time. Right: Dinosaur eggs found in Gobi sands.

could animals of temperate or tropical climes have lived in such cold regions? Science's (ridiculous) attempt at an answer: Because they were fat, very fat. This hopeful solution begs the question: Where did these fat creatures find the needed supply of food in such a cold place? There is a better explanation than the obesity hypothesis: "The polar regions have not always been polar. . . . The surface of

Fig. 3.4. Antarctica. After James Geikie, The Great Ice Age, *1894.*

the Earth has shifted dramatically in relation to its poles"* (*Atlantis Rising* Staff 2008, 11).

Writer Tom Valentine put it this way: "The polar regions themselves have been located in various spots" (Valentine 1975, 28). Apparently the landmass that is currently at the South Pole once sat at a temperate latitude. Some estimates say the continent of Antarctica arrived on the scene 50 million years ago. Before then the South Pole was an open sea. Scientists say Lake Vostok in Antarctica had been under ice 15 or 20 mya; though others say it was under ice only 10 or 5 mya, or even "no more than a few thousand years" (White 1980, 84–86). There is not much consensus here.

During explorer Robert Falcon Scott's Terra Nova Expedition of 1910–1913, at 85 degrees S, seven seams of coal were discovered. Indeed, the first leader of the IGY (International Geophysical Year) South Pole Station, explorer Paul Siple, hit the nail on the head when he explained

---

*An equally valid explanation is that the whole Earth was warmer at that time: large reptiles roamed about Antarctica 200 mya. The same species also inhabited Africa, India, and Australia. As George Gamow noted, "The conditions of milder and more uniform climate were typical of the whole period of [the] Mesozoic" (Gamow 1948, 144).

that the coal was due not to a shifting of the pole per se, but to a certain oscillation of the Earth itself.

Think of oscillation as an infinitely slow rolling of the whole Earth—not the poles and not any parts, but of the entire Earth—with every continent, in turn, inevitably reaching the polar regions in due course. And for this reason alone, ice has been almost everywhere at some time or other and coming from different directions. But not at the same time. Never ubiquitous. Never universal refrigeration or "ice age."

Snowball Earth in Pre-Cambrian Time (850–550 mya)

Professor Macdougall assumes that since glacial marks appear at very low latitudes (12 degrees S), a drastically cold era must have once frozen the entire planet. Yet he admits that this Snowball Earth hypothesis stands on "fragmentary evidence, sparseness of the fossil record, and uncertainties in dating glacial deposits" (Macdougall 2004, 150–51, 155, 208–9). At that very early time before the Cambrian, the only animals were in the seas, which were supposedly frozen over. It is natural then to ask this: How could the Earth ever have thawed out again, once it was completely ice covered? (Snow and ice cause a vicious cycle of cold, since they deflect warmth.) Indeed, how did algae survive all this with the oceans frozen from pole to pole? Did active volcanoes save them? Please! A 600-million-year-old Ice Age seems preposterous in light of early Earth's heat-loving creatures, tropical reefs, and so forth (for more on this point, see chapter 4)—clear evidence of planet-wide warmth throughout the early age of Earth.

"The concept of the ice ages has no explanation itself," averred Immanuel Velikovsky, whose book on Earth changes lay face down on Albert Einstein's desk at the time of his death in April 1955. Later that year, in September, an article titled "The Earth's Magnetism" appeared in *Scientific American,* written by Cambridge professor S. K. Runcorn, a specialist in paleomagnetics. Runcorn stated: "Reading natural compass needles in rocks at various places around the world, we find evidence of astounding changes in the Earth's main axial field. During the Tertiary the north and south geomagnetic poles reversed places several times!" (Runcorn 1955, 156). His simple, uncomplicated conclusion: "The planet has rolled about." More recently, Joseph Stone, a paleomagnetist at Oregon State University, told a meeting of the American Geophysical Union that polar reversal may be a part of a normal oscillation. Yes, of course. Normal and natural. And not catastrophic, either.

How else could we explain why magnetic fossils show opposite polarity? Hematite and magnetite crystals point like tiny compass needles in whatever direction the magnetic field flowed at the time of their formation. This discovery had been predicted in 1882:

Every portion of the earth hath been to the east, to the west, to the north and to the south. Which is proven in the rocks and boulders.
OAHSPE, BOOK OF COSMOGONY AND PROPHECY 1:11–12

In the twentieth century, the science of paleomagnetics went on to verify that landmasses have inerringly shifted from one point to another in relation to the poles. With the continents taking turns at the poles, nothing less or more than the Earth's oscillaic roll explains the changing orientation of the lands, as seen in the registry of the rocks (and even of magnetic mountains—such as those in Northern Europe at Dannemora and Taberg, Sweden, and at Utoe, Finland—whose iron has two poles acting as magnets).

At Southsea, one of Dr. Arthur Conan Doyle's patients was Major-General Alfred Wilks Drayson, an astronomer and mathematician and a professor at the Royal Military Academy. Drayson (whom Doyle befriended) had developed a theory explaining glacial effects, the history of which is written on the rocks themselves. Drayson hypothesized the second rotation of the Earth, a hitherto unknown movement which accounts for variation of climate in both hemispheres. His first book on the subject was jauntily titled *Great Britain Has Been, and Will Again,*

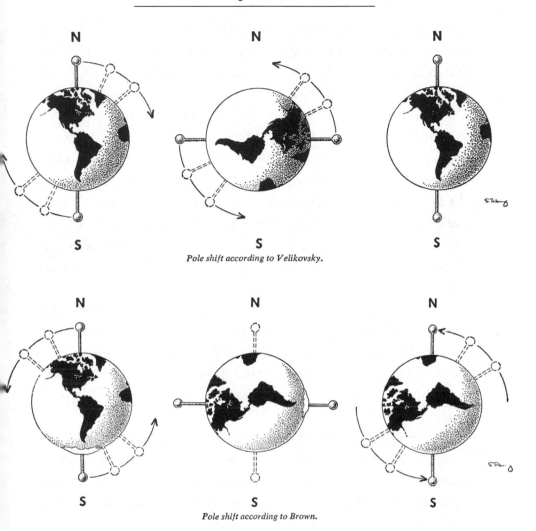

Pole shift according to Velikovsky.

Pole shift according to Brown.

Fig. 3.5. Top: Velikovsky's concept of pole shift. Over time, new regions moved into the polar circles. Bottom: H. A. Brown's concept of pole shift, in which he argues that the centrifugal force of the South Pole ice cap's eccentric rotation tends to career the globe, rolling the Earth over in a motion like a somersault. Both diagrams used with the kind permission of John White, author of Pole Shift.

*Be within the Tropics.** Was Drayson onto something, like oscillation?

The steady roll through the latitudes (think of a lazily rolling beach

---

*A more recent study of magnetism in the rocks of England showed that the British Isles were once more than two thousand miles south of their present position.

ball) is exactly the same thing as "pole shift"—the well-known magnetic reversals—a motion slow and gradual, a noncatastrophic turning about of infinite slowness. Pole shift as catastrophe (a quick somersault) is just another counterfeit crisis.

Think of it. At one time the surface of Europe was adorned with tropical vegetation. One-hundred-million-year-old amber from what was once a tropical forest in France encases fungus. But later, Europe was buried under a vast mantle of ice. Universal refrigeration? Ice age? No! Just a change of latitude. Contrapuntally, parts of the United States, Canada, and Alaska were at the equator (in the Triassic). And yes, today's equatorial lands were once under glaciation; late Paleozoic glacial deposits were found near the present equator (White 1980, 54). Thirty million years ago India, South America, and Australia were iced over.

Velikovsky pointed out that a few of the classical authors knew "that the earth had changed its position and had turned toward the south." He cites as an example the Arabian scholar Arzachel, who discovered that Babylon, which was once situated at 35 degrees N, later moved to a more southerly latitude. That information was later used by Johannes Kepler (Velikovsky 1965, 316).

It is now time to ask this: If there were ice ages, why can't scientists agree on their cause? What in heaven or earth (both are blamed) could possibly cause the world to freeze and defrost, freeze and defrost, again and again like a machine? The Earth is not a machine. The theater of ice ages has been played out in spectacular guesswork, though sometimes resorting to a simple plea of ignorance: "Periodic deep-freeze events are quite complex and not well understood." (Macdougall 1996).

The nonexistent Ice Age has been explained six ways from Sunday—nearly one theory per year. The altitude theory, for one, traces the origin of glacial periods to changing altitude: since great mountain ranges are under the sea, and seashells are found on mountaintops, you don't have to be a rocket scientist to figure out that significant changes of landscape are part of the legacy of Earth. But what caused these major displacements? And how would an altitude change here and there produce global refrigeration? Some say mountain-building forces in the Himalayas triggered the last ice age, having affected ocean circulation. The Cascadian Revolution, which is said to have built the Himalayas, supposedly brought on the Pleistocene ice age. In the same vein, Gamow has glaciations taking place "when the surface of the continents is elevated and covered with high mountains" (Gamow 1948, 145–47). Although the identical idea had been proposed by the great geologist Charles Lyell, others like the Scottish geologist James Geikie found it hard to conceive of local elevation movements affecting the whole northern hemisphere. Perhaps then the real culprit was continental drift, going back to the time when a substantial portion of Earth landmasses were presumably located at high latitudes (near the poles).

Another educated guess blames the nonexistent ice ages on an insufficiency of greenhouse gases, which trap heat, causing the temperature to drop maybe six degrees, enough to form glaciers where summer temperatures can't melt all the ice; yet Gamow pointed out that "there is no way of checking whether the extensive glaciations of the past were actually connected with a variation in the air's carbon dioxide content" (Gamow 1948, 148). In fact, the paleoceanographer John Imbrie asks: "Why should such a decrease [in CO_2] happen" in the first place? "The theory must be added to the list of plausible ideas that now seem impossible to test" (Imbrie and Palmer 1979, 64).

Or maybe, as suggested by a New Zealand scientist, an abrupt sliding of a huge chunk of Antarctic ice could have triggered an ice age. All supposition. A similar scenario has the South Pole ice cap uncouple itself from its bed and surge outward, its sun-deflecting mantle bringing on a new ice age. All supposition.

Well, then, what about volcanic disturbances? Wouldn't a great deal of volcanic dust block the sun, plunging temperatures on Earth? If so, "evidence should be preserved in ancient soils . . . or in layers of mud," notes Imbrie. But again, "it has so far proved to be impossible to make a valid test" (Imbrie and Palmer 1979, 64). And if it was volcanism that spewed out tons of ash, sulfur, aerosols, and dust, thus blocking the rays of the sun and freezing the Earth, how could such obscuring dust, even if severe and widespread, refrigerate the entire globe? This idea never caught on. Sure, cool periods have followed volcanic activity for a year or two, as we know from the 1991 eruption of Mount Pinatubo in the Philippines or from the eruption of Indonesia's Mount Tambora in 1815, which gave us the "year without a summer." The 1883 Krakatoa blast in the Sunda Strait was also bad news for farmers almost worldwide for three years running. But one or two or even three years of slightly lower temperatures does not produce an ice age.

Yet I would like to note that there *is* a record of dark times in the oral history of the world's people, when a veil came over the face of the Sun for many years. Some scholars think it was the impact of a comet or asteroid that darkened the sky, preventing the Sun's heat from penetrating. Ignatius Donnelly's *Ragnarok,* for example, argues that the Pleistocene ice age occurred when a comet collided with Earth. Or could it have been the strike of a giant meteorite? Fred Hoyle, in a sci-fi book called *Ice,* speculated that the next ice age, conceivably lasting fifty thousand years, could be caused by a meteor strike.

The guesses keep coming. It has also been suggested that changes in ocean currents might result from an increased temperature differential between the equator and the polar regions, thus increasing winds. Some speculate that the fabled continent of Atlantis (assuming there *was* an Atlantis) could have been the cause of the last ice age in Europe and America by blocking the warm Gulf Stream. Then, when it sank, it ended that ice age. As Velikovsky critiqued this idea,

> changes in the direction of warm currents in the Atlantic Ocean were brought into the discussion, and the Isthmus of Panama was theoretically removed to allow the Gulf Stream to pass into the Pacific at the time of the glacial periods. But it was proved that the two oceans were already divided. . . . Besides, a part of the Gulf Stream would have remained in the Atlantic anyway, and the periodic retreats of the ice between the glacial periods would have required periodic removal and replacement of the Isthmus of Panama.

Not a very likely scenario, in other words. Velikovsky concluded,

All such hypotheses fail if they cannot meet a most important condition: In order for ice masses to have been formed, increased pre-
cipitation must have taken place. This requires an increased evaporation from the oceans. . . . The problem is: What could have caused
the evaporation? . . . At present the cause of excessive ice-making on
the lands remains a baffling mystery. (Velikovsky 1965, 22–23)

Sir Arthur Eddington, once England's astronomer extraordinaire,
believed the ice ages were caused by a shifting of the Earth's outer crust
(lithosphere) over its interior crust (mantle). Called displacement, or
isostacy, such a mechanism would presumably affect ocean circulation,
which would, in turn, trigger ice ages. And if the Earth's entire crust
slid around the core, that could also explain how the Indian Ocean was
once located at one of the poles. But as Gamow noted: "The basaltic
bottom of the oceans is much too rigid to permit of any changes in the
relative position of the continents. . . . We can hardly expect the poles
to have changed their position by more than a few degrees" (quoted
in Eddington 1937, 146–47). Besides, isostacy offers neither periodic
nor cosmic timing in its wanton play of tectonics; for ice seems to have
played out in cycles—with great regularity—as some, like Loren Eiseley
or Australia's R. W. Fairbridge, see it. Indeed, none of the foregoing
theories accounts for the periodicity of ice.

> *The Antarctic and Arctic regions were ice-free at nearly*
> regular intervals. *(my emphasis)*
>
>   JOHN WHITE, *POLE SHIFT*

Nonetheless, versions of isostacy continue to appear in today's textbooks and scientific journals, blithely ignoring the periodicity of ice: "At
irregular time intervals," said a recent article, "the earth's plates have
moved over the eons" (my italics), causing polarity flips (Tarduno 2008,
73). How could irregular isostacy cause regular ice ages or regular pole

shifts? "The cause of the alternating warm-cold periods of the Permian-Carboniferous Ice Age is unknown; but then regularity hints at an astronomical or other external 'pacemaker'" (Macdougall 2004, 150).

Perhaps then it was some unknown fluctuation in the sun's brightness that caused ice ages, "but no evidence that the sun is such a variable star was adduced to support this hypothesis" (Velikovsky 1965, 21). Gamow also thought that "there are no indications . . . of variations of solar activity persisting for thousands of years. Here, as with the carbon dioxide hypothesis, it appears to be quite impossible to check the coincidence of past glacial ages with minima of solar activity" (1948, 148).

A century ago, the Serbian geophysicist, mathematician, and glaciologist Milutin Milankovitch had begun by considering plate tectonics the culprit; but finding this to be shaky ground, he then proposed a scheme with cosmic timing: the Earth grows cold when, on its elliptical path, it attains its farthest position from the Sun. Slight changes in the Earth's orbit, it was argued, might conceivably reduce the level of solar heat reaching our planet. If so, the periodic variations in the eccentricity of the orbital ecliptic would lead to glaciations at maximal eccentricity. But here too were problems. These changes would be too slight, the variations just too small to produce drastic changes in climate. And the timing isn't right, either.

> The biggest objection to the theory that the Milankovitch cycles caused the ice ages is that there is no statistically significant connection between the two. A careful analysis has shown that the timing of the Milankovitch cycles relative to the ice ages is no closer than what would be expected by chance. (Spencer 2010)

In fact, it seems that "the total amount of heat received by the Earth during an entire year is virtually unaffected by variations in orbital eccentricity" (Imbrie and Palmer 1979, 83). Despite this, it is still generally held that Earth's orbital cycles around the Sun "tick . . . out the rhythm of the planet's climate cycles" (Macdougall 2004, 175).

Loren Eiseley in *The Unexpected Universe* argues that intervals of ice correspond to the time it takes our Sun to make one full circle of the galactic wheel; that circuit brings recurrent ice, presumably started in the "dim pre-Cambrian"—more than 550 million years ago. But, wait, there was no ice until 70 mya! (See cooling over the ages, charted in chapter 4). Skating on thin ice, the theory for good measure tosses in axial wobble as an additional cause. Axial tilt, the theory goes, could have affected the amount of solar rays reaching the Earth, under the assumption that small changes in the axis of rotation caused great ice sheets to spread out from the poles. Nevertheless, Earth is rigid and it could not have shifted more than a few degrees: "The difference could not have been great enough to have been responsible for the glacial ages." What's more, since Siberia (above the Arctic Circle) was not ice-covered, "all hypotheses . . . [involving] solar alterations . . . cannot avoid being confronted with this problem" (Velikovsky 1965, 21–23).

> And it came to pass that great darkness covered the earth . . . and clouds came over the face of the earth; the moon shone not, and the sun was only as a red coal of fire.
>
> OAHSPE, FIRST BOOK OF THE FIRST LORDS 4:12–16
> (REFERRING TO A PERIOD ABOUT 70 KYA)

It may be fruitful to track those dark times, so often recalled in world mythology. Recognizing the importance of c'vorkum (motion #4, further discussed in chapters 4 and 5), one school of thought comes, I think, the closest to scientific truth, considering that cosmic space, through which the planets travel, consists of warmer and cooler areas as well as regions of differing light and densities. In this connection, Fred Hoyle and Ray Lyttleton put together a plausible model of astronomical agents triggering the onset of cold. When the solar system, under the rule of c'vorkum, passes through a dark cloud of gas and dust, this would automatically lower the temperature on Earth. Such opaqueness would block sunlight when the Sun's path, orbiting the Milky Way, tunnels through an interstellar cloud. Cloud cover, lowering global temperature, would usher in a cold epoch.

DARK AND COLD

Jehovih cast a veil over the face of the sun, and it did not shine brightly for many years.

OAHSPE, THE LORDS' FOURTH BOOK 4:21–24

(THOR TIME, ABOUT 14 KYA)

Indeed, some nineteenth-century theorists believed that Earth had passed alternately through cold and warm regions in its travel through space. In Oahspe, the Kii Tablet records:

An abundance of a'ji [condensed nebula]* in the firmament giveth a cold year upon the earth.

OAHSPE, BOOK OF SAPHAH: OSIRIS 56

The critical factor, though, may not be the "obscuring dust." James McGill reminds us that these interstellar dust regions, sometimes called molecular clouds, are themselves extremely cold. McGill points out that the LGM (last glacial maximum) some 24 or 25 kya ago corresponds to a time when the Earth was in a'ji for several hundred years (see Oahspe, Book of Aph 17:7). McGill also mentions that the first actual detection of cold diffuse matter in interstellar space was made by Johannes Hartmann in 1904, "twenty-three years AFTER [sic] Oahspe associated a'ji with cold." Published in 1882, Oahspe had predicted a rain of a'ji starting in 1914. McGill took a look at the statistics for that period: heavy snow "of unusual size" over England; it was one of the severest winters of the twentieth century, indeed a major problem in the Great War for all parties concerned, suffering bitterly freezing conditions on the front (www.studyofoahspe.com).

---

*Chapter 5 picks up on this a'jian cold.

"The Lord cast down great stones from heaven," reads the Holy Bible's Book of Joshua (10:11). That too is a'ji. A'ji marks off cold, dark periods on Earth, which may be accompanied by stone showers and other strange skyfalls (nebular detritus). In 3700 BP, for example, came a "trail of dust and clouds"—thought to be from Venus or a comet's tail—but was it simply a'ji?* Finnish epochs as well as the Mexican *Annals of Cuauhtitlan* say stones fell from the sky, while the Chinese say stars fell like rain and vermin and darkened the land for twenty-five years (see the link between pestilence and a'ji in chapter 5).

A'ji rains stuff on Earth. This is why the amount of dust in ice cores has proven to be higher during cold a'jian periods. Indeed, ice cores from the cold time, called "younger dryas," contain much more dust than usual, confirming the association of a'jian cold and extraneous matter (cosmic junk). Fossil pollen (taken from lake sediment cores) during the younger dryas came from plants that thrived in cold climates. The climate of the younger dryas (ca. 12,000 to 9500 BP) is noted in Canaanite tablets, which specify that it was so cold 11 kya that "they never felt certain that winter would be followed by spring" (Kolosimo 1975, 122). The Polish Academy of Science says that 10,800 years ago the weather turned cold for several hundred years. Known as the Big Freeze, this roughly corresponds to a period noted in Oahspe's Book of Fragapatti (35:8) that records a spell of a'ji lasting more than a thousand years, beginning around 10,500 years ago.

A'jian cold may start rather suddenly and persist for centuries.† For example, Oahspe reports that right before the time of Brahma 6 kya, a four-hundred-year period of dim light occurred. Examining

---

*See Velikovsky (1965, 58–60) for a fuller account of that "Egyptian darkness."

†Darkness due to cometary impact, though, might only last a matter of days. One example from Oahspe describes an event from about 6 kya:

> A comet came within the earth's vortex and was drawn in just like floating debris is drawn within a whirlpool in a river. The substance of the comet was condensed and fell on the earth in mist and dust and ashes. Consequently, the earth and its heavens were in darkness for twelve days, and the darkness was so great that a man could not see his hand before him.
>
> OAHSPE, BOOK OF DIVINITY 17:2

ice cores in the Near East, scientists have found particularly low temperatures for this time, that is, between 6400 and 6000 BP. It was then that

> there fell perpetual atmospherean substance on the earth, similar in all particulars to the composition of earth things, but atmospherean. … The belt of meteors was itself nearer the earth by thousands of miles. The stones fell like a rain shower.
>
> OAHSPE, BOOK OF DIVINITY 9:1–4

At that precise time, "a series of catastrophes occurred. . . . It seemed to the people that the end of their world had come" (Trench 1974, 144–45).

Later, after Moses, approximately 1000 BCE (see Oahspe, 572, Remarks on the Book of Ben), there came partially interrupted darkness, again for four hundred years.* This a'jian forest (atmospheric densities) darkened the Earth

> on every side … and the sun shone not …
>
> OAHSPE, GOD'S BOOK OF ESKRA 9:3, 9:13, AND 11:1

Those dim years match the Iron Age cold epoch of the North Atlantic, which the sixteenth-century Incas called "the night of mankind," for the world, says legend, turned dark and sunless. Around Lake Titicaca (Tiahuanaco), the sun was thought to have hidden itself, just as Mexican oral history recalls a time when the world was left in darkness. In the North, too, Wyandot oral history speaks of spells of darkness, while a California myth (Wintu) states that there was a kind of dim light all the time.

Returning to why ice ages actually begin, Macdougall frankly reports

---

*On the significance of four hundred years, see appendix D.

There are hypotheses, but none have yet attained the status of an accepted scientific theory. Charles Hapgood, though, thought significant shifts of the planet's poles are one important cause of ice, with regions moving from temperate areas to cold ones and vice versa. However, in his scheme this happened quickly and violently. This matches the catastrophic bent of Louis Agassiz who built his ice theory on the popular notion of outright crises in Earth history, assuming that temperatures had plunged suddenly. Later research, however, found Agassiz's ice age not as rapid in onset as he proposed. (Macdougall 2004, 7)

Nor is pole shift itself a calamity, occurring, at least according to Lord Kelvin, "without any perceptible sudden disturbances of either land or water" (White 1980, 56). In other words, the earth has not and will not capsize or tumble over! as suggested by the "HAB" theory of Hugh A. Brown. Brown's theory (the "somersault") transforms pole shift into a grand catastrophe, wherein accumulation of ice at the poles upsets the equilibrium of the Earth, causing it to tumble over like an overloaded canoe, the weight of the caps presumably causing such a disaster every eight thousand years. But, on the contrary, it was precisely the slow and steady (not sudden) oscillation that *prevented* this very thing from happening! Oscillaic (motion #3) of the Earth brought balance, simply causing it to roll, untraumatically, a steady continental creep taking perhaps half a million years for North and South to change places.

Catastrophists have made much of pole shift (see the box on page 15), but for the Earth as a whole there never has been a backward step or a sudden jump forward. Is there any real evidence of a planet changing its progress (or temperature) midcareer? To answer this question let us have a look at ice age statistics. If ice ages are a fact, why are there such wild discrepancies in their chronology? I am going to briefly summarize what I found in the literature. Just skip this section if you don't want to get confused. "Inconsistencies in scientific theory, and poor correlation [with] observational data,

sometimes are indicative of . . . a fundamental error in assumptions"
(Corliss 1980, 49). Just check out these ice age dates for an idea of the
chaos that lies just below the surface of this accepted science:

When was the first ice age? Some say it was the Gunz glaciation,
1 mya. Others say only 300 kya (as per Cesare Emiliani's work in
1956), though some put that at 300 *million* years ago! (Gamow
1948, 144). Some say even earlier, in the pre-Cambrian, before
the dawn of the Paleozoic, as discussed in the box on Snowball
Earth (page 127) (Macdougall 1996, 42; and Macdougall 2004,
141, 143, 158). It purportedly involved four "glacial episodes"
between 850 and 550 mya. Others posit sporadic ice ages for the
past 3 *billion* years—but no real ice age until 300 mya—and it
lasted 80 million years. (Wouldn't that ice have killed off the
dinosaurs or their "evolutionary" predecessors?)

How often do ice ages occur? Once every 250,000 years (Field 1955);
or every 100,000 years (according to sea core readings; there were
twenty cycles of cold in the past 2 million years, Arctic ice presumably expanding every 100,000 years); or every 150,000 years
(judging from gravel terraces along the Alpine river valleys); or
once every 150 *million* years (four ice ages in the last 600 million years, according to a student textbook; or four ice epochs in
the past billion years; or with intervals of a quarter-billion years
(White 1980, 10, 13); or perhaps as little as 40,000 years apart,
being an estimate of thirty glacial advances during the Quaternary.

How long do ice ages last? More than 100,000 years. This is the
consensus for the duration of the most recent ice age, the Wisconsin (Wurm) glaciation, said to have begun around 130 kya,
retreating about 20 kya. But according to Professor Macdougall,
the glacial epoch that supposedly started 850 mya lasted till
600 mya, that is, for 250 million years (Macdougall 1996, 42).
Permian glaciations, on the other hand, lasted only 25–30 million years. Thus does Professor Chernicoff (1995) call for tens of

millions of years' duration, though others say only 200,000 years!

How many ice ages have there been? Twelve of them (Chernicoff 1995); five of them (Jane Goodall); thirty of them in the last 3 myr (Pollack 2003, 55); seven of them in the last 300,000 years (according to Caribbean cores studied by Emiliani). Or ten of them in the last million years (Milanikovitch's figure), as against only four of them in the past million years (Field 1955, Eiseley 1962). Or four of them in the last 600 million years (Chernicoff 1995). Five or six glacial periods are mentioned by Velikovsky, though he adds: "Neither the cause of the ice ages nor the cause of the retreat of the icy desert is known; the time of these retreats is a matter of speculation" (Velikovsky 1965, 21).

When did the last ice age end? That figure is somewhere around 10,000 or 11,000 years ago, according to most scientists. But others say as recently as 6,000 years ago; or make that 40,000 years ago for North America—although the ice retreat in Ohio and Minnesota was not until 15,000 years ago. Nevertheless, James Croll had the most recent ice age ending 80,000 years ago (having begun 250 kya). Others say thawing is noted for the Arctic, in the area of the Canadian Archipelago, only 1,000 years ago.

Had enough of ice age "statistics"?

The Little Ice Age (LIA)

Also known as the Maunder Minimum or the mini–ice age, the period from about 1350 to 1850 was colder than usual (at least in Europe). Does the end of the LIA around 1850 correspond to the "heat" generated by the industrial revolution? Or is the timing just a coincidence? I think the hidden factor again is a'ji, which brings both dark and cold: "In times of great darkness, [a'ji] which shall come upon Earth and these heavens, lo, I will bring the Earth into danha" (Book of Jehovih). Otherwise put:

A'ji cometh near a dawn of dan.

BOOK OF DIVINITY 14:1

The most recent dawn was the year 1848. Dawn (or danha) refers to the beginning of a three-thousand-year cosmic arc (see chapter 8 and appendix B, on the three-thousand-year cycle). A'ji falls at the end of that cycle, rather like the proverbial darkness before the dawn. (The last several hundred years of the three-thousand-year etherean arc are dark where etherean arcs as seen in fig. 2 in appendix B are shown as vertical white lines). In this scheme, a'ji represents the non-terrestrial flotsam and jetsam (nebulous material) typically released prior to dawn times. The phenomenon may be thought of as a kind of "housecleaning" that ushers in a new cycle of light (dan).

Brian Fagan in *The Little Ice Age* (2001) aptly regards extreme weather fluctuations as part of natural global cycles. Although the LIA cold rippled through Europe, Fagan says there "was no deep freeze." It was cold enough, though, to shrink the forests, kill the cod, and drive the inhabitants from Greenland, and, in England, to freeze the Thames regularly, also covering English farmlands in snow until late spring. "Frost fairs" (carnivals) were held on the frozen Thames until 1814. But according to H. N. Pollack, the LIA may only have fallen on the North Atlantic. How can we call a regional event an "age"? (Pollack 2003, 56–57).

The deepest chill of the LIA fell from 1650 to 1715. Very few sunspots were seen in this span, which, I noticed, was a 65–66-year period (the prophetic number $66 = 11 \times 6$; see appendix B). Now the question was asked: Does the sun brighten and give off more heat when sunspots appear? And, conversely: Does it cool when they disappear? The sunspot cycle runs approximately 11 years. But there are also multiples of 11, such as the 33-year and the 66-year cycle. According to McGill, "A'ji shows up in multiples of 11 years." Wasn't the 66-year (33×2) deep chill and sunspotless phase of the LIA a factor of a'ji, the veil of rarified matter that intercedes between Earth and the heavens?

In the passage of the earth in its own roadway, it goeth amongst these etherean and atmospherean worlds regularly, so that the periods of . . . darkness are not haphazard.

OAHSPE, BOOK OF COSMOGONY AND PROPHECY 7:15

Scientists cannot explain why there were no sunspots visible during the deepest chill of the LIA. I think the view of the sun was dimmed by dense a'ji. Looking in the other direction, away from the sun, was the view of other celestial bodies also obstructed? Apparently so. Astronomers register the "disappearance" of the red spot on Jupiter during that same period. Maybe sunspots have nothing to do with extra heat, after all, but are merely easier to see when there is no (cold) space junk (a'ji) in the way.

———

Neither does the actual behavior of ice convince us of ice ages. The ice age theory was originally put forth to explain the presence of clay, sand, and gravel on igneous and sedimentary rock. Was the depositing mechanism really ice or was it flood water? Or something else? Until the middle of the nineteenth century, natural philosophy had long embraced the biblical dogma that the geological features we see on the land today are remnants of the Great Deluge: Noah's Flood. Defending the diluvial theory, geologist Sir Henry Howorth opposed the new glacial theory, arguing that its assumptions were contrary to physical laws and the actual properties of ice.

The Ice Age, like so many of the theories examined in this book, came as a victory for the secular stronghold. It was during the first half of the nineteenth century that "certain influential shapers of scientific opinion wished to sever geology from the biblical tradition of the Great Flood" (Heinberg 1989, 175), and by midcentury, glaciologists took over where biblicists left off, the former attributing all kinds of geophysical phenomena—from rock streams to buried forests to animal migrations and mammoth graveyards—to the action of ice. Even human-made megaliths have been blamed on glaciers! The diluvial theory, in any case, was put to rest. But was its replacement valid?

The most important geological remains in question were called "drift," meaning all the loose materials that glaciers supposedly eroded, ground up, and carried, then later deposited. Drift layers, once ascribed to the Great Deluge, were now considered tokens, indeed proof, of ice ages. These layers were noticeable because they were composed of different ("erratic") material than the bedrock in situ, and presumably were transported by moving ice.

Nevertheless, it has been truthfully argued that "intense opposition to biblical teachings . . . hindered a really objective attitude to the evidence" (Corliss 1980, 35). Critics, for example, have pointed out that there is no trace of ice flow in places where we would most expect it: Alaska's interior, northern Canada, northern Greenland, even northeast Siberia, one of the coldest places on Earth. Instead, drift is often present where we would least expect it: vast deposits of clay, boulders, sands, and gravel are found in America and Europe, as well as in tropical lands like subequatorial Africa, Madagascar, India, and the Amazon. Yet there is no till in Siberia (till comprises randomly mixed boulders, pebbles, and fine soil, presumably deposited by ice). We also find mammalian remains, which signal a milder climate than what we have now. Clearly, Siberia was warmer when elsewhere it was cooler (White 1980, 25). Doesn't this suggest oscillation (taking turns)?

DRIFT AROUND THE WORLD

The land turns round as does a potter's wheel.

IPUWER PAPYRUS, EGYPT

Whereas drift is held up as the main "proof" of glacialism—which amounts to a circular argument—the ice sheets of today's Antarctica and Greenland do not seem to be forming any drift like that attributed to past ice. Rather, their rocks are angular (unlike rounded boulders) and their deposits are more like "heterogeneous muck." Another problem: although icebergs and "erratic boulders" are said to be millions of years old in places like Arkansas, most of the drift in the western hemisphere is "remarkably fresh," with minimal erosion, so it was apparently formed recently. In fact, "structures composed of drift *around the world* are all very well preserved . . . [and] actually must be quite recent . . . if the degree of erosion is considered as an indicator of age" (my italics) (Corliss 1980, 34, 37).

Is there, then, something other than super-ancient ice that could have formed drift and till? Like mud flow or even magnetic cataclysm à la Churchward? In Mexico, for example, Churchward found strata of sand, gravel, and boulders at ancient city sites, which are more than a thousand feet above sea level, with mountain ranges several thousand feet high intervening between them and the sea. Sarcastically, he asks: "Did the ocean raise waves . . . more than 5,000 feet in height? Did Mexico borrow a glacier for the occasion in order that these boulders might be deposited where they now lie? Nothing of this sort happened. . . . These cities were built before the mountains were raised." Scratches and boulders, Churchward points out, are also seen on the top of Mt. Mansfield in the Green Mountains, more than four thousand feet above the sea, just as there are boulders on the White Mountains of New Hampshire six thousand feet above sea level: "certainly some jumpers! The geological Glacial Period," Churchward had to conclude, "is one of those bizarre, fantastic, mythical theories that has taken strong root in the minds of . . . scientists . . . surrounded with a corollary of impossibilities" (Churchward 1968, 89–93).

I discussed these problems with Canadian engineer Gary Nicholls, who wrote

> The accepted geological history of southern Vancouver Island makes no sense regarding the formation of rivers, streams, creeks, pebbles, cobbles, boulders, and the erosion, cracking, and shattering of the granite and basalt bedrock . . .

What Nicholls sees are

telltale signs of electrical vortices having cut out the rivers and their

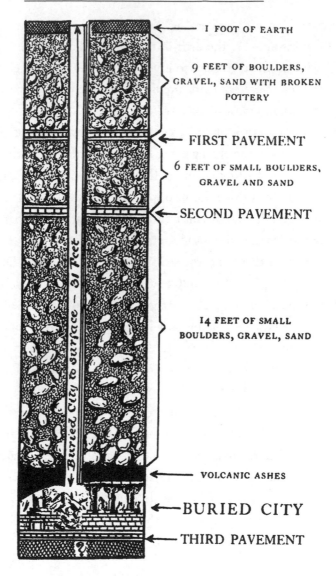

1 FOOT OF EARTH

9 FEET OF BOULDERS, GRAVEL, SAND WITH BROKEN POTTERY

FIRST PAVEMENT

6 FEET OF SMALL BOULDERS, GRAVEL AND SAND

SECOND PAVEMENT

14 FEET OF SMALL BOULDERS, GRAVEL, SAND

Buried City to surface — 31 Feet

VOLCANIC ASHES

BURIED CITY

THIRD PAVEMENT

Fig. 3.6. Niven's Mexican buried cities. From Churchward (1968). These cities are now 7,000 feet above the level of the sea, mountains of 5,000 feet higher intervening.

valleys: unmistakable to an electrical engineer anyway, but not to a geologist. This activity, judging by the erosion . . . would indicate a very young age, as little wear of the rock is evident. . . . The surface features are not the result of millions of years of slow glacial erosion.

Nicholls adds,

I've been looking into the local geology [Vancouver], photographing examples of what I see as electrical/plasma modification of the land, as opposed to the slow, long-term glacial erosion model. There is no doubt in my mind that most of the shaping was by very rapid, high-intensity events. . . . The standard [uniformitarian] explanations based on tectonics, glaciers, and erosion are physical impossibilities. (Nicholls 2013)

If the glaciologists trumped the biblicists 150 years ago, will the magnetic cataclysmists (Churchward, Nicholls) soon trump the glaciologists?

Features of ice disintegration also tend to melt the Ice Age myth. Glaciologists have assumed that ice disintegrated in isolated blocks (rather than at its perimeter), if only to account for the various mounds, hollows, and plateaux along Canada's Great Plains. However, this assumption is not confirmed by present-day ice: "Melting ice influences the temperatures of the environment in such a way that isolation of various blocks would be most unlikely. The glaciers melt at their perimeters" (Corliss 1980, 48–49).

The ice theory further claims that telltale scratches and striations were gouged into bedrock by moving ice sheets. Flutings (streamlined landforms) are also explained as created by onrushing ice. But could ice have formed these flutings, some of which cover thousands of square miles? Glaciers do not move very rapidly, normally only a few inches or yards per day. From the air, America's largest glacier appears to be quite still, almost frozen in place. Ice moves slowly. How could rocks harder than ice have been fluted or eroded by slow-moving glaciers? And why are there drift deposits and striations well south of the charted ice sheets, some of their striations indicating a northward motion?

If it was, in fact, ice that left its mark on land and rock, fluting and striating them in Europe, Russia, and America, why is Northern

Asia free of such remains? The Ice Age, as noted, never touched northern Siberia, and "no satisfactory solution to this question has been proposed" (Velikovsky 1965, 23). Eskers (long gravelly ridges occurring in formerly glaciated areas) are found in Britain and North America but have no counterpart in Siberia. Are eskers really ice-made? Geologists have found it difficult to account for eskers by glacial theory, for their courses go up and down hills. "Eskers are actually enigmas in the glacial theory" (Corliss 1980, 48).

Concerning the problems presented by uplands: Why, we might ask, are there striations on mountaintops? Drift material and boulders sit high up in many parts of the world. Indeed, such "erratic boulders" are offered as proof of advancing ice. Yet Velikovsky put into words the wonderfully obvious fact that "The glaciers of the Alps push the stones down, not up the slope" (Velikovsky 1965, 76). Agassiz would have had us believe that ice slid up the Labrador, Adirondack, and Catskill Mountains, proceeding merrily along to the fortieth parallel! But even the Mohawk and Hudson Valleys would have stopped them, anchored them.

Something else must have transported the great boulders—perhaps, à la Nicholls, "high-intensity events." Was it volcanic action? Mountain raising? When the mountains were thrown up, the boulders could have gone right along with them to their present resting places. I find Nicholls's explanation intriguing.

> Energetic events such as sputtering, ion etching, plasma sculpting, and pulsed electric fields are well recognized by electrical and plasma scientists, but not applied to earth sciences.
> . . . The Sooke River Valley would appear to have been formed in minutes, initiated perhaps by a very steep vertical electrical gradient, the flow being from ocean shore to mountaintop. (Nicholls 2013)

But such quick work remains unacceptable to the keepers of the SM. Rather, the doctrine of uniformitarianism (gradualism and unchangeable mechanism) is the standard that upholds both evolution and ice ages: "The appeal to ice removed the glacial period from the position of a catastrophic phenomenon. It placed the ice sheet . . . at the disposal of the geological uniformitarian. That was the real basis and inspiration of the theory . . ." (Corliss 1980, 36).

This good argument notwithstanding, don't you think ice itself would be a catastrophe for the living? Here we find that the glacial theory suffers from incongruent types of fossils occurring in the drift. We would expect these to be fossils of cold-weather creatures, right? But some turn out to be more tropical (cave lions, for example). Those large cats, ice age apologists explain away, probably came from interglacial deposits. To others, however, "The list of mammals cannot confirm the idea that the Quaternary was really a time of cold climate. . . . Unsolved problems, contradictions, and mystery surround the question of the fossils of the [so-called] glacial period. Is it time for a new explanation?" (Corliss 1980, 49).

How could all those species, especially those in tropical and temperate climes, have survived at all in frigid ice ages? Keep in mind that along the Alpine River terraces, some layers of gravel contained fossil mollusks that are found today only in warm climates. "How could sediments containing such fossils have been deposited during an ice age?" Most European geologists, John Imbrie candidly admits, "chose to ignore the problem" (Imbrie and Palmer 1979, 119). When those gravel terraces of the Alps turned out to contain inconsistent fossils, theorists escaped embarrassment by claiming that "The terraces had not been formed by glaciations, but had another origin altogether" (Macdougall 2004, 132). Such as?

Meanwhile, North America presented its own stumbling block—this one at glacial Lake Algonquin in east central North America—where human-made flakes, cores, and points were discovered beneath the gravels, the artifacts older than the "glacial" till. "I had been taught," said one miffed geologist, "in the great and infallible universities, that glaciers destroy all trace of man in their advance" (Corliss 1980, 390–91). According to Churchward, though, "The whole of western North America was peopled by highly civilized races . . .

Fig. 3.7. The last magnetic cataclysm. The biblical Flood and the geological myth: the glacial period. After Churchward (2011, 208).

before the [so-called] geological Glacial Period" (Churchward 2011, 204).

THE CARBON FACTOR

Human remains and human artifacts of bone, polished stone, or pottery are found under great deposits of till and gravel, sometimes under as much as a hundred feet.

IMMANUEL VELIKOVSKY, *WORLDS IN COLLISION*

The carbon factor throws yet another wrench in glacial theory. During most of geologic history, there was much more CO_2 in the atmosphere than there is today. CO_2, as we will see in the next chapter, indicates warmth. Another chemical clue to past temperatures is heavy hydrogen (in ancient wood), which has been measured with the "surprising result: winter temperatures in North America . . . during the height of the last ice age may actually have been higher than they are today!" (White 1980, 12).

In fact, pre-Neolithic fossils (i.e., before the end of the last ice age) show the Earth as warm and paradisal, with a great variety of species on every continent. Western America was filled with elephants, camels, antelope, horses. The Papago Indian (Arizona) creation legend recounts that "all the seasons were warm, and no one wore clothing" (Heinberg 1989, 106). Other legends of native peoples recall an early era when animals were more abundant and wild plants more plentiful. Concerning those animals and plants, Charles Hapgood, in *The Path of the Poles* (1999) asked, "If the temperature of the whole Earth fell enough to permit ice sheets a mile thick to develop on the equator, just where did the fauna and flora go for refuge?" Indeed, if ice extinguished "all organic life" (as indicated by Imbrie and Palmer 1979, 33)—just as the Flood supposedly did—how then did it all spring back to life in the next interglacial period?

Finally, we are led to ask how all this ice and snow (the current ice age beginning supposedly 3 mya) computes with *Homo sapiens*'s naked advent (evolution) on this Earth. "What are we to think," pondered Loren Eisely, "of the story of man? Into what foreshortened and cramped circumstances is the human drama to be reduced?" The ice theory as it stands "would involve a complete reexamination of our thinking upon the subject of human evolution" (Eiseley 1962, 81). Think of it: the last ice event is claimed to have run from about 130 kya to 12 kya. Says Macdougall, "For most of the time since then [since 125 kya] it has been much, much colder" (Macdougall 2004, 7). Nonsense. This was the very period of humanity's beginning, which took place in a tropical to temperate clime. Most of that era was the time of Neanderthal man, who occupied the valley of the Ilm in Germany, enjoying "a warmer climate than now" (Keith 1929, 190). In Croatia as well, "the famous

Neanderthals of Krapina . . . enjoyed a pleasant climate" (Tattersall and Schwartz 2000, 176, 205), as did their cousins in the Near East and North Africa.

Past eras on our planet were warmer and wetter, a major theme of the next chapter. Until the mid-Pleistocene, all humans lived in mild climates. Hippos roamed Europe and North America, the giant size of many fossils another indicator of warm temperatures.

Man's origin in the tropics is not really controversial (or shouldn't be):

- "The spread of humans [was] out of the tropics" (Fagan 2010, 43).
- "Until 50 kya, all humans lived in tropical and mild climates" (Diamond 1992, 218).
- "Homo erectus and early man came forth in a time when 'the climate was warmer'" (Cannell 2010, 14).
- "The primordial climate was mild; it was a sort of Eden," recalls the Chinese text *Huai-nan-tzu*.

If the most recent retreat of ice from Europe began about 25 kya (as per textbook gospel), why do layers below the Aurignacian (more than 40 kya) reveal more tropical fauna, including rhino, elephant, hippo, and so forth? In France, the grotto of Vallonet Cave at Roquebrune-Cap-Martin is a veritable cemetery for exotic animals of the early Quaternary: elephants, rhinos, lions, monkeys. Later, during the Magdalenian, according to French prehistorian Robert Charroux, other areas of France were "not particularly cold," such as Glozel, Lascaux, Lussac-les-Chateaux. He contends that the cave drawings of that era, painted by our Cro-Magnon ancestors, "are a great embarrassment to the pundits. . . . They supposedly lived in an 'ice age' . . . [yet] the men often walked around half naked with their penises ostentatiously protruding from their trousers. . . . We have serious doubts about the polar climate decreed by academic authorities!" (Charroux 1974, 83). Charroux is here referring to the famous Paleolithic cave artists of France and Spain, and indeed it has elsewhere been noted that, despite the supposed Ice Age, "our ancestors recorded little on their cave walls about such cataclysmic events" (Corliss 1980, 1).

Nor did the aborigines of Australia ever wear clothing; they have gone naked for 60,000 years. Although it is said that dozens of climate sequences occurred over the "millions of years" when humans were evolving, we seriously wonder how they could have survived at all. Eiseley reasoned that, over those millions of years, "powerful selective forces must have been at work, as ice sheets ground their way across vast areas of the temperate zones." That means that humans were sorely challenged, our genes somehow adapting to the cold, under "natural selection" (Eiseley 1962, 81).

Ever turning a liability into an asset, the Evolution Club trades on good old natural selection and "adaptation" to avoid the implausibility of early humans surviving an ice age. Indeed, the Darwinists have gone so far as to suggest that our evolution from hominids was actually spurred by ice. Using lots of qualifiers and hypotheticals, they invoke the SM, wherein the African forests shrank due to aridity. It then (3 mya) became drier and cooler, which forced our ancestors down from the trees to become bipedal! Such "adaptations" on the open savanna (after our ancestors straightened up and learned to walk upright) were allegedly followed by learning to hunt, developing cooperative society, language, and bigger brains. Gamow, for example, asserts:

> *Homo sapiens* . . . developed from . . . anthropoid apes some time during the Pleistocene glacial epoch. . . . During one of the successive advances of ice, some group of these animals . . . was forced to get accustomed to the new, more severe conditions of life. The hardships that such an existence . . . involved might easily have given the first impulse for the development of the primitive brains of these creatures, sending them along the path of discoveries and inventions characterizing the entire progress of the human race. (1948, 178)

The same argument appears in the most recent studies: "The emergence of human culture . . . had been instigated by the life-or-death challenges confronting humanity. . . . Glacial conditions compelled social cooperation, the advancement of language skills, and communication" (Frank Joseph 2013, 66). This, in a nutshell, is how the conjectured Ice Age and the conjectured evolution of humanity become mutual apologists! There was even a belief among some Nazis that their "Nordic ancestors grew strong in ice and snow" (DeCamp, 1975, 92).

It was my goal in this chapter to show that there were no ice ages—only ice in lots of places—in fact, almost everywhere, but never ubiquitous, never at the same time. The different regions of the Earth merely took turns being icy, probably due to oscillation.

> In the early times of earth's formation, the earth was longer north and south than east and west. But the . . . earth then assumed the globular form, which was afterward attenuated east and west. Then the earth turned again, to adapt itself to the north and south polarity of the vortex. In these various turnings of the earth . . . every portion of the earth has been to the east, to the west, to the north, and to the south.
>
> OAHSPE, BOOK OF COSMOGONY AND PROPHECY 1:11–12

His knowledge of the Ice Age didn't extend much beyond the glass.

Fig. 3.8. Scientists say that when a significant chunk of the Greenland ice cap slides into the ocean, it will raise sea levels instantaneously, as do ice cubes dropped into a glass of water. Cartoon by Marvin E. Herring.

It was the combined clout of a Darwin and an Agassiz that sufficed to establish and enshrine the ice ages. But we have come a long way since Victorian science. Ice Age now stands as nothing more than scientific fundamentalism. However, the world will end in ice. And this is the perpetual cooling of the Earth to which the next chapter is devoted.

CHAPTER 4

GLOBAL WARMING?

Shift Happens

Magnetic reversal . . . is most economically explained by moving the whole earth. Likewise, ice ages.

JOHN WHITE, *POLE SHIFT*

As we saw in chapter 3, oscillaic motion allowed us to do away with ice ages. We saw through the telescope of time that tropical species (see table "Warm-Weather Species Buried in the Far North" p. 124) once thrived in today's most frigid latitudes, leaving traces of nothing less than fig palms in the Arctic Circle, magnolia trees in North Greenland, lion bones in Alaska, hippos in Yorkshire, UK, and coal beds in frozen Antarctica. We asked: Why would northern terrains have entombed the bones of elephants, crocodiles, hippos, and other warm-weather creatures? The simple answer: These creatures lived there. Only not at that latitude. Over the long haul, those equatorial lands sailed poleward. Several paleomagnetic studies suggest that large fragments in the western USA, Canada, and in Alaska were located near the equator perhaps at Triassic times. (Childress, 240).

Siberia, we saw, was ice-free at a time when North America was glacier-bound down to Missouri. The remnants of "luxurious vegetation" in Siberia, as well as corals in Spitsbergen and Tibet and rhino bones in Yorkshire, UK, are not "anomalous" or "misplaced"; for we find a perfect counterpoint in the torrid zones: Brazil, Africa, Australia,

156

and India were once under ice. No, there is nothing anomalous about it. Things are not really "out of place." The world seems topsy-turvy only because of a change of orientation for every landmass, which was, in the long term, as natural as day and night. Oscillaic motion. Magnetic reversal. Pole shift. All the same thing.

Just as ice is now depressing (scooping out) the Antarctic continent by virtue of sheer weight, the "bowls" of the Sudan basin, the Caspian Sea, and the Hudson Bay basin were formed in the same way when they were located at the pole. Not only have great shifts in climate been noted for all parts of the globe, but no part of Earth's landmass—when you come right down to it—is where it once was! Magnetic observatories report that none of the continents seem to be in their old places.

Which brings us to the twin fallacies of global warming and ice ages. Yes, they are intrinsically related. Apparent warming as well as alleged "ice ages" turn out to be nothing more than regional events, not global, never worldwide. When developing countries met in New Delhi, India, in 1989, they acknowledged that warming will not be globally uniform but will differ significantly between geographical regions. Over the past thirty years, while winter temps in parts of the North have risen noticeably, the worldwide average increase has hardly been 1°F (Laird 2002, 77). John Christy, Alabama's climate chief, reports that satellite data give a slight rise in U.S. temperature, but a lower one in the southern hemisphere.

Of course, Dame Science is well aware of magnetic reversal and the changing orientation of the continents, variously explaining it as the result of nutation (axial wobble), continental shift, migration of the poles, isostacy, and so forth (see the box, page 15). However, plate tectonics is unable to explain the uniform "migration of the poles," which is actually a misnomer, for it is the continents themselves (not the poles that changed location. Just where the continents happened to be at a given time, with respect to the poles, tells their story.

Lava layers and magnetic fossils with "inverted polarity," which zig when they should zag, have long since proven the rolling or oscillation of the Earth. The North and South Poles reversed places several times.

No controversy here. Position is everything in life. Determine the shifting position of each climate zone with respect to the magnetic pole, and you are on the way to solving the climatological puzzles of the past that baffle even the experts.

More than a hundred years ago, Giovanni Schiaparelli, the great Italian astronomer, cautioned the science world against pseudosolutions, predicting the answer to these terrestrial mysteries: "The permanence of the . . . poles in the very same regions of the Earth cannot be incontestably established" (Schiaparelli 1889, 31).

And in the long term the Earth, rather than warming, is actually cooling, but very, very slowly. René Descartes and many other scientists of the past recognized the cooling state of the Earth. There is, as nineteenth-century sages of science noted, a constant heat loss from a once fiery Earth. Indeed, some saw the cause of the so-called ice ages as the decrease of the original heat of the planet. After all, a proto-world begins as a seething ball of liquid fire. Boiling and roiling, whirling and swirling, embryonic Earth was a twisting and turbulent vortex of friction, her gas clouds in rapid rotation, condensing particles in solution. The molten Earth, before we could go for a walk in the park, must turn down the lights, slow down, cool down, and solidify.

The very slow process of cooling, as James Churchward pointed out, occurs in three stages: "from a gaseous to a molten state; from a molten to a hot solid; from a hot solid to atmospheric temperature" (Churchward 1968, 10). Or as George Gamow, Russain-born theoretical physicist and cosmologist, put it: "The temperature of the surface of the Earth had dropped more than 1800 degrees F since the Earth first began to cool down. . . . We [can] follow the successive stages of the Earth's evolution from its early molten state . . . to its final thermal death in the far-distant future" (Gamow 1948, 99).

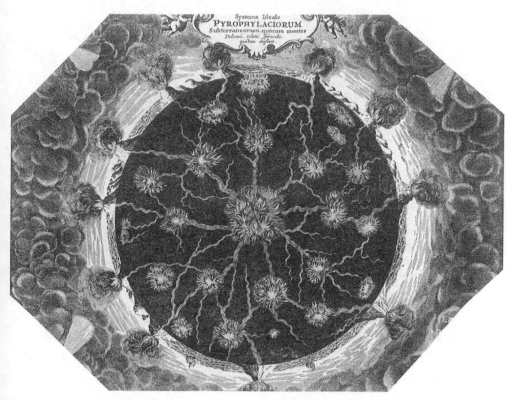

Fig. 4.1. Jesuit scholar Athanasius Kircher's depiction of the Earth's fiery beginnings. Amazingly, geologists still do not agree that Infant Earth was hot and molten.

The travelogue for cooling Earth could take us through rockfields, the dinosaur extinction, the planet Mars, and other sites of interest, each a stopping point on the way to the truth about climate change.

What's the deal with Mars? What do we hope to discover (or terraform!) on a nonagenarian, used-up, desertified, and near-extinct world? Simply put, all these questions can be resolved under a single axiom, which is to say, the planets—Earth and Mars included—experience aging as a cooling and drying affair. As we saw in chapter 1, age and aging are critical but egregiously overlooked factors. Tell me, what planet is known to reverse this trend midcareer? Youth means hot and wet, as was our own early world. Cold and dry will be her aspect (like Mars) as she approaches old age and death.

GLOBAL COOLING

Not only has Mars been wet; it has also been warm in the past. . . . Mars was slowed down when the planet's water supply diminished and the temperature dropped.

ROBERT JASTROW, *UNTIL THE SUN DIES*

Wherefore, then, global warming? Or the propped up mystique of Mars, which, as the senior citizen of our solar system, offers little more than the opportunity to study planetary extinction. In her frigid dotage—windblown and pockmarked, dark and desiccated—Mars has lost most of its atmosphere and gravity. Its orbit is markedly elliptical, another sign of slackening. There are no earthquakes. The volcanoes of Mars have been extinct for hundreds of millions of years. It's history. . . .

Yet the death of Mars offers a unique object lesson in climate change and temperature loss. Once wonderfully wet, warm, and wild, the red planet betrays her senescence now in tawdry erosional features, dry channels, and arroyos. More to the point, her summertime temperature registers a paltry −40°F.*

Planet Earth, too, is losing moisture, radiance, and heat. Past middle age, Mother Earth is no youngster, and she is not getting any warmer. Worlds do not warm up. Rather, they undergo a steady loss of heat. According to satellite data, our globe has been cooling for at least 90 million years, but very, very slowly. How does this compare with bugbears that have today's warming "outstripping anything the earth has seen in the past 100 million years" (Lemonick 2001, 26), which is sheer "climate porn." There wasn't even ice at the poles 100 mya. Besides, the Medieval Warm Period (see page 196) was warmer than today; even the 1930s were a bit warmer than today!

Notwithstanding hypothetical asteroid or bolide hits and the like,

---

*Some of our sages would contend that Mars's moribund frigidity was the result of being in the grip of an ice age, while her tomblike drought is presumably due to some unknown change in the planet's atmosphere, causing water to vaporize. It is here that the simple fact of aging has been unscientifically ignored.

it is more soberly thought that it took the dinosaurs a long time—perhaps more than a million years—to die off. Scientists Comte de Buffon, Richard Owen, T. H. Huxley, author Isaac Asimov, and others agree on this much: cooling was the cause of the dinosaur extinction, not some giant asteroid. The loss of two or three degrees of blood heat was their probable downfall.* Ice, around that time, had begun to accumulate at the poles.

Much later—toward the end of the Pleistocene—the mastodon, the saber-tooth tiger, and the giant ground sloth also met their end, most likely because of a critical drop in temperature. The point is, while exotic theories and sudden, catastrophic causes, like extraterrestrial bombardments, may titillate the popular imagination, they have little to do with animal extinctions or climate change, both of which reflect perfectly natural and exquisitely gradual change.

Predictions of warming abound: George Gamow forecasted a day, five thousand years from now, when Boston will have the present climate of Washington, D.C., and, in ten thousand years, of the West Indies. The average summer temperature in Boston, others go on to predict, could increase 14°F by 2100 (Mone 2011, 46), while a UN-sponsored study sees a rise of from 2.5°F to 10.4°F by the year 2100. Other climatologists foresee a jump of fully 9°F by 2050. But how is any of this possible when geologists say, "If the past is any guide, our present warm period will soon end?" (Macdougall 1996).

With the global warming craze in full swing, you just don't hear much about the long-term cooling of our planet. In the year 1900, Florida's panhandle was the center of citrus growing, just as Georgia and Louisiana were major producers. But devastating freezes have driven growers as far as three hundred miles south. Most recording stations in the Southeast, including those in Georgia, Alabama, and Texas, register marked cooling since the end of the nineteenth century (Jones and Smith 2007). The marine air temperature worldwide, at least according

---

*But don't be surprised when you hear from alarmist paleontologists like Peter D. Ward (*Under a Green Sky*) who blame the mass extinctions of the past on rapid global warming!

to the British Meteorological Office, was reduced slightly, 0.24°C between 1888 and 1988.

In the late 1970s, scientists were wondering if a decline of the Sun's fusion reactions was responsible for a "long-term cooling trend" (White 1980, 359). More recently, scientists at the National Solar Observatory also predict a period of lessened solar output; while atmospheric data reported by the George C. Marshall Institute of Washington show slight cooling of the Earth. "Are Sunspots Causing Global Cooling?" asks one investigator (*Atlantis Rising* Staff 2010, 10), citing NASA reports that indicate our planet has not warmed for fifteen years* and "world temperatures may be heading downward."

In 2012, a solar scientist named Henrik Svensmark stated publicly that the warmists' conclusions based on computer models "have already been discredited by the failure of world temperatures to rise since 1997" (*Atlantis Rising* Staff, 2010, 10). The question of computer models came up again when two Oregon chemists announced that "global weather is so complicated that current data and computer methods are insufficient to make such predictions. . . . Computer climate models are very unreliable. . . . The weatherman still has difficulty predicting local weather even for a few days" (Robinson and Robinson 1997).

Enter Roy W. Spencer, Ph.D., an atmospheric scientist, author of the eye-opening *The Great Global Warming Blunder*, who contends that "one can get just about anything one wants with computer models. . . . The output of computers is no better than the information that the programmers put in." It is also Spencer's contention that "you don't need fancy climate models or supercomputers to do some very good experiments . . . [say,] in a spreadsheet program on your home computer. In contrast to the IPCC's magical mystery megamodels running on supercomputers, I have used a simple climate model to demonstrate these concepts" (Spencer 2010, xxi, 120, 121, 160). Another writer,

---

*Even IPCC (the UN's Intergovernmental Panel on Climate Change) admits the twenty-first century, so far, shows a cooling trend. We will see why when we get to PDO (pages 164 and 172).

this one at www.Climate-Skeptic.com, calls these agenda-driven computer models "scientific money-laundering. . . . The models are built on the assumption that anthropogenic [human-caused] effects drive the climate, and so they therefore spit out the results that anthropogenic effects drive the climate."

These models, other critics add, have left out the complex interaction between warm southerly winds, variations in cloud cover, and sunlight reflection from open water.

Clouds and water vapor, we learn, are the main greenhouse gases (GHGs), acting as a "thermal blanket" and accounting for most of the Earth's natural greenhouse effect. Carbon dioxide actually contributes relatively little—less than 4 percent—and methane even less. Why have researchers ignored the natural variability in clouds and instead aggressively pushed "anthropogenic forcing" on us? Spencer has written extensively about not human-made, but *natural, internally generated cloud variability* [that] is responsible for most of the climate change we have seen . . . and will likely see in the future" (my italics) (Spencer 2010, xxvi).

Purveyors of global warming have been peddling bugbears and drop-dead scenarios. Chief among them is Al Gore, whose film on the subject, *An Inconvenient Truth,* grossed $24 million, and who holds the threat of warming as "second only to nuclear war." Britain's Sir David King says he's more scared of global warming than of terrorism. These knights in burning armor would have us all out there "fighting climate change" tooth and nail. Gore was awarded the Nobel Peace Prize in 2007 for his politically correct campaign, spotlighting human (that is, anthropogenic) causes of warming. His lovely improvement scheme would only cost the world's taxpayers $46 trillion.

"Former US Vice President Al Gore," comments a South African physicist/mathematician, "is being totally simplistic in his movie . . . most of which exhibits the absence of genuine science and rather presents itself as part of an election campaign" (Kelvin Kemm, quoted in www. studyofoahspe.com). "Al Gore's apocalyptic vision of a global-warming Armageddon," Spencer points out, has been supported by a "$300 million

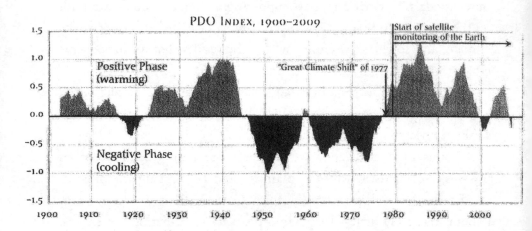

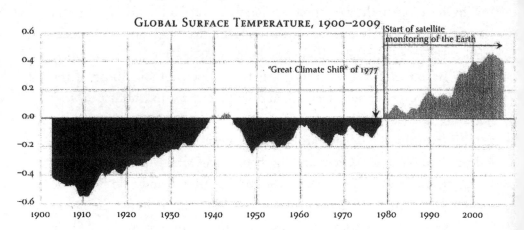

COULD NATURAL CLOUD VARIATIONS CAUSE GLOBAL WARMING?

Fig. 4.2. Pacific Decadal Oscillation (PDO).
The top panel shows five-year averages of the PDO index. The bottom panel shows globally averaged surface temperature variations during the twentieth century.
Spencer asks, "Is it just a coincidence that these features correspond to global temperature trends that also changed in these three periods, warming up to the 1940s, slight cooling until the 1970s, and warming since then? I don't think so . . . PDO might cause a small fluctuation in cloud cover resulting from those circulation changes." Quoted from and courtesy of Spencer 2010, 110, fig. 23.

advertising blitz . . . [this] disinformation campaign based on a litany of scientific half-truths, exaggerations, and inaccuracies" (2010, xxv, 26, 124). And according to Timothy Bell, Canada's first Ph.D. in climatology, Gore's film is "an error-filled propaganda piece." Bell has actually received death threats for his contrarian view (Horner 2008, 52).

In this astonishingly quixotic campaign, with its inflammatory tactics, thousand stooges, and formidable arsenal of factoids, we are spoon-fed choice bits and pieces to help us understand greenhouse gases (GHGs) and rising sea levels. Don't believe it. All the industry in the world—all the secreta, waste products, and heat islands of our bustling civilization—could not warm up an aging planet. "Something doesn't add up," twits investigative reporter John Stossel, who notes that "half the global warming of the past century happened from 1900 to 1945. If man is responsible, why wasn't there much more warming in the *second* half of the century? We burned much more fuel during that time. What about that? Huh?" (Stossel 2006, 202).

Anthropogenic, as we've seen, is a term referring to human-caused phenomena. But some scientists say warming is caused not by humans but by changes in the Sun. Mathematician and science writer James D. Stein tells us that "the Sun is going to start to heat up" (Stein 2012, 56). That's a far cry from predictions of "lessened solar output" that we read about a few pages back. Who's right? Astrophysicist Sallie L. Baliunas of the Harvard-Smithsonian Center has suggested that the Sun is responsible for up to 71 percent of the Earth's temperature shifts; other experts say that percentage is way too high. So who's right?

One of global warming's most hysterical sideshows features dangerously high sea levels. In coastal areas, we are warned, flooding will become routine, eventually wiping out New York, Tokyo, and many of the world's largest coastal cities. Doomsters say global sea levels could be raised more than 200 feet (Bell 2008, 60); IPCC has put that figure at between 46 and 150 inches by the end of this century. Some scientists say sea levels are increasing 0.3 inches a year. That's about the height of a pea; but some say sea levels in the past 100 years have risen 4 inches, which would be more like twelve "peas" per century or roughly a pea per decade (not per year). Why don't these figures match? One Florida science writer says that the twentieth-century rise amounted to one foot; another analyst says that by 2100 we will see a rise of 23 to 24 inches, while another calls it several feet, at which time "Faneuil Hall [Boston] would be inundated by six feet of water" (Mone 2011, 46). Still another alarmist has the global warming scenario of melted ice raising sea levels from one to three feet by 2100 (Helvarg 1999, 38), which sounds pretty good compared to yet another prediction that warns us of a rise of "up to 21 feet worldwide" by 2050, thanks to "a catastrophic melt" of Arctic ice (Laird 2002, 77). On the other hand, that figure for midcentury is otherwise predicted to be only 2.5 feet (Walsh 2013, 10). Still others (if you're still with me) declare that "sea levels might eventually rise five meters [16 feet], submerging the world's low-lying land . . . " (Lemley 2005, 30). Eventually?

How good are all these assorted and inconsistent (computer-generated) predictions when the most recent announcement has a 60 percent increase in ice-covered ocean water since 2012, leading some scientists to believe that the planet is actually undergoing global cooling.

An aging planet is a drying planet. Just think of Mars's dry-as-dust canals. Will our planet really reverse the long-term trend toward lower ocean volume? Shall we just ignore geologists who say "the past few million years have been a time of steadily *decreasing* ocean volume and temperatures?" (my italics) (Macdougall 1996, 216). Rivers like northwest India's Sarasvati (once five miles wide) have disappeared. Spain's Guadelquivir is shrinking. The Jordan River is nearly dry. The height of Lake Titicaca is going down year by year as is Africa's Lake Chad. Lakes Huron and Michigan are now at their lowest ebb since monitoring began in 1918, and a continuing drop in the Great Lakes levels is anticipated (Pollack 2003, 232).

The Aral Sea of Central Asia has shrunk to 10 percent of its original

Fig. 4.3. A dried lake at the shorelines of ancient Lake Bonneville, Utah, a vast body of freshwater, of which the Great Salt Lake (now salty) is the only remnant. (U.S. Geological Survey, 1890).

size. Three hundred thousand years ago, the Mediterranean was eighty-five feet higher than it is today; in fact, it is thought that part of North Africa once lay under the Mediterranean Sea. Fossil marine shells in the Nile Valley indicate those lands were once covered by the sea. In Northern Europe, land is sinking into the North Sea, but in Spain and France it is rising. Biarritz gains more than an inch of altitude each year, Cadiz two inches (Charroux 1974, 118). There are high-water marks on plateaus and raised beaches throughout the world. Former beaches, as in South America at Paracas, Peru, and Valparaiso, Chile, have been raised many hundreds of feet.

Sea-level expert Nils Axel-Morner was amazed to find that he had never heard the names of the so-called world's leading experts informing the IPCC on sea-level rise. A little probing seems to indicate that some of the real experts have simply backed off the IPCC agenda, which, as critics warn, produced "doctored data," tailored to report dangerous levels of GHGs due to humanity's emissions. What's more, Axel-Morner objected to the Hong Kong tide gauge used by the IPCC, because "every geologist knows that that is a subsiding area. It's the compaction of sediment . . . [giving] a falsification of the data set" (quoted by Horner 2008, 323). And what about Greenland, which was once (5 kya) three separate islands? On Greenland's coast some beaches have been raised 1,700 feet!

Glaciologists usually explain sea levels as a result of ice ages. With the coming of ice, levels decrease, but when glaciers melt, sea levels increase. This, they say, would explain why the seas fell dramatically around 100,000 BP, around the time that the last ice age began.

But aren't sea levels receding overall? Why are there dead coral reefs in Indonesia on dry land, well above sea level, and coral atolls all across Oceania? Today there are scientists who believe the Pacific island of Tuvalu is disappearing beneath the waves not because of rising sea levels but as a result of sinking coral. Nevertheless, in 2002, that tiny Pacific nation (exploiting the global warming craze) threatened to sue Australia over their greenhouse gas emissions, which they argued are bound to drown the island as warmer temperatures send sea levels rising. But how do these rising waters square with the actual trend of drought? (See the section on drought, appendix C, pages 400–401.) And speaking of

Is global warming myth or reality?

Fig. 4.4. Global warming. Cartoon by Marvin E. Herring.

GHGs, predictions that double the current amount of CO_2 in the atmosphere claim that its increase would mean four or five degrees Fahrenheit of warming for the world as a whole (Lemley 2005, 31). Nonsense. No CO_2 footprint is about to trigger the Chicken Little meltdown any time soon—or ever. A waste product of burning fossil fuels (coal, oil, wood, gas), poor old carbon dioxide has been made the villain of global warming, accused of the heinous act of melting the glaciers! And whereas we are taught to fear the devastating "heat waves" CO_2 will trigger, a sober look at this minor trace gas quickly reveals that

1. CO_2 makes up less than 1 percent of the air we breathe, and Mother Nature produces a lot more CO_2 than does humanity.
2. In the past half-century, "during the twenty years with the highest carbon dioxide levels, temperatures have decreased" (Robinson and Robinson 1997).
3. CO_2 is not guilty! Consider the folly of demonizing "heat-trapping" carbon dioxide, which makes up Mars's polar caps (dry ice)! If rising temperatures are really due to increased levels of CO_2, how come frigid Mars is 95 percent carbon dioxide? And if Uranus and Neptune have so much methane, another "heat-trapping" GHG, why are their waters frozen? And why are there "methane oceans" on Saturn's cold (–300°F) moon Titan? Methane (CH_4) which, on Earth, comes from swamps and wetlands, is abundant on all the "cold" giant planets—Jupiter, Saturn, Uranus, Neptune—but (to circumvent the obvious contradiction) this is ascribed to a nonorganic source.

The simple fact that CO_2 is emitted naturally from the Earth each spring makes that gas far more likely a result of warming, not a cause. Have we, once again, mistaken symptoms (or correlation) for cause, or even effect for cause? Atmospheric CO_2, after all, has been seen to increase during past periods of "deglaciation," when temperatures increased. "That correlation alone does not tell us that GHG changes *cause* temperature changes. . . . Perhaps the temperature increased first and was closely *followed* by the greenhouse gases . . ." (my italics) (Pollack 2003, 168). In fact, changes in GHGs correlated with global temperature long before humans were around. Such increases in carbon dioxide may be "a consequence of . . . warming rather than a cause" (Schoch 2012, 119). When oceans cool, they absorb CO_2. When oceans warm, they release CO_2. Like the other GHGs, water vapor follows temperature changes rather than causing them. Huge clouds of methane hover above forested areas: "Turning up the heat seems to increase the rate at which the plants produce methane" (Kruglinski 2006, 15).

Nor can we blame humanity for these gas levels, when the amount of carbon dioxide has plummeted "with each ice age" and shot up with each interglacial. These, of course, were times when no humans were around.

While Al Gore's science man, James Hansen (of the NASA Goddard Institute for Space Studies), argues that CO_2 has already reached dangerous levels, chemist Arthur Robinson, on the contrary, points out that the only evidence of increased CO_2 comes from computer models (simulations), which have not taken into account the effects of cloud cover, El Niño, variations in solar activity, snow and sea ice, or urban heat islands.

But even if CO_2 *is* holding in more heat,* that's good news for a cooling planet. John Christy, for one, finds little warming in the lower troposphere (the bottom five miles of atmosphere) since satellite records began to be collected in 1979. Christy, in fact, thinks surface temps will cool down in the future. When I spoke with Christy, the gist of his comments was this: The actual science has little impact on the government's energy agenda. In his opinion, all the GHGs in the world can only "tweak the planet's temperature up and down." For this blasphemy, one NASA warmist threatened him: "I'm paying people to come at you with bricks and bats." In addition, Gore's man Hansen tried to boot

---

*The only reason, says one warmist, that carbon dioxide hasn't raised temperature to predicted levels "is that the atmosphere is full of crap (dust and aerosols) that act as a global coolant" (Begley 2009, 30). Clouds also act as a coolant: when things warm, more water evaporates from oceans and forms clouds.

Christy from the Senate panel he was advising. But that's how this thing works. Many of the "cooler-headed" analysts, as Chris Horner points out (punningly), "are afraid of employment repercussions"; one distinguished Israeli atmospheric scientist talked about this kind of intimidation: "Many of my colleagues . . . report on their inability to publish their skepticism in the scientific or public media" (Horner 2008, 72).

"CO_2 is a natural and necessary component of life on Earth . . . and Earth loves it. . . . Maybe we are doing nature a favor by adding more of this essential nutrient to the atmosphere. . . . Atmospheric carbon dioxide is just as necessary for life on Earth as oxygen" (Spencer 2010, xiv, 132–34). Extra carbon dioxide is actually a boon for the environment. CO_2, for starters, is the best airborne fertilizer there is. Arthur and Zachary Robinson, that heretical pair of chemists at the Oregon Institute of Science and Medicine, burst a major bubble when they announced that atmospheric carbon dioxide enhances the growth of plants and is especially beneficial in dry regions. With increases in this gas, "our children will enjoy an earth with twice as much plant life. . . ." These scientists stress that hydrocarbons are needed to feed and lift from poverty vast numbers of people across the globe. "Global warming is a myth," they say. "The reality is that global poverty and death would be the result of Kyoto's rationing of hydrocarbons" (Robinson and Robinson 1997).

"Fundamentalist doom mongers," adds John Stossel, "have been sidelining scientists like Harvard's Sallie Baliunas," who points out that additional carbon dioxide in the atmosphere promises to benefit the world not only because it encourages plant growth, but because warmer winters would give farmers a longer growing and harvest season. The "most scary" predictions of heat waves, says Stossel, are based on "computer models [which] are lousy because water vapor and cloud effects cause changes that they fail to predict" (Stossel 2006).

Not long before the global warming craze took off (in the late 1980s), a cooling trend was evident in the North Atlantic, Britain, northeast Canada, and southwest Greenland. In the mid-1970s, computer models told us to prepare for global *cooling;* major crop failures (due to colder temperatures) was the prediction of many climate experts crying "nuclear winter." Worldwide, the atmosphere had been growing gradually colder for the past three decades, according to meteorologists (but the trend was just a function of the 30-plus-year PDO cycle).

Hugh A. Brown, in his book *Cataclysms of the Earth,* thought (as we saw earlier) "The growing ice cap . . . will annihilate all of us. . . . Let us attack the ice cap" (with atomic bombs!) (quoted by White 1980, 81). The extremely cold winter of 1976–1977 had experts alarmed. As early as 1963, the UN Food & Agriculture Organization convened a conference in Rome to assess the threat of global cooling. The cooling craze continued until the late 1970s, with documentaries on the subject running in America and England and a slew of popular books as well. "Snow even fell in Miami. . . . Some feel a new ice age is imminent"[*] (Jeffrey Goodman 1978, 137–39).

The research-based books of science fiction writer Kim Stanley Robinson are considered scientifically sound. In his eco-saga *Fifty Degrees Below,* we are back in Gamow's blistering Washington, D.C., but this time it is gripped by a polar winter! At –50°F, ice skaters, skiers, and snowfighters glide along the Potomac, with its two-foot-thick ice. Everyone is dressed like Russians. A fictionalized warning, the story introduces the possibility of a Gulf Stream stall that could throw the North Atlantic into frosty conditions. Scientists are indeed apprehensive about an abrupt change that could alter the thermoha-

---

[*]According to Macdougall, "Today's climate is just a geologically short warm spell in this continuing ice age. . . . If history is any guide . . . the ice will return, and quite soon . . . perhaps New York and Chicago will be buried deep in glacial ice . . ." (Macdougall 2004, 7–8).

line circulation of the Atlantic Ocean, ending the transport of heat to Northern Europe. With ice caps diluting the salt content of the sea, the Gulf Stream could slow down. The increase of freshwater entering the North Atlantic would then cool Western Europe considerably; England would become like Greenland over the course of a few years.

As a result of ancient farming, suggests paleoclimatologist William Ruddiman, our forebears kept the planet warmer and may have even averted the start of a new ice age. Today's temperatures, according to Ruddiman, might well be on the way toward glacial, if not for the greenhouse gases established by early agriculture. Without it, our climate would "cool enough to produce the long-overdue glaciations" (2005, 48, 52–53). Contradicting this rather far-fetched theory, researchers now say the "biochar" from ancient fertilizer enhanced the soil's ability to seize carbon, thereby trapping GHGs.

Today, wannabe do-gooders build CO_2 extraction machines, the device designed to remove CO_2 from the air. More than 250,000 such "synthetic trees" would be needed to soak up the excess of this so-called nasty GHG (Lemley 2005, 31). Hmm. Would this costly effort be a boon, or a huge, even deadly, mistake? And if we set up treatment plants along the coasts to absorb CO_2 into seawater, wouldn't this raise its alkalinity, killing off marine life near these treatment plants?

What is the Big Picture? Why don't we hear about scientists (like England's J. Falk) who are saying that some zones of the earth may be cooling while others are warming. We are, I believe, top heavy in data from our own northern latitudes, neglecting possible countertrends in the southern hemisphere. Not surprisingly, one researcher was warned that his funding would be cut off if he pursued his study of figures "for the *tropics*. . . . In fact, it [funding] has been cut" (my italics) (Horner 2008, 73). Temperatures in the equatorial zone should be searched out. As John Christy has pointed out, though average temperature in the United States has recently been a bit higher than usual, the average temperature of the southern hemisphere has been lower. They sort of average out.

International data must be pursued in order to draw a climate map of the world, suggested H. N. Pollack, who identifies one source of uncertainty in establishing global temperature—where it was measured. "There are many more weather stations in Europe and the USA . . . than in the Amazon . . . [or] the Sahara," giving us a skewed result, "regionally biased" and effectively misrepresenting the temperature "for the entire Earth" (Pollack 2003, 80). Underrepresented is climate data not only for the Southern Hemisphere but also for remote deserts, thinly inhabited regions, oceans, and rainforest areas.

In the late 1800s meteorological observatories were set up in the major cities of the world. Urbanization, of course, means heat-absorbing pavements and buildings. The development of towns and cities led to an increase of 1°F in the twentieth century. Human-made surfaces just sit in the sun and bake, giving us urban heat islands. Since so many of the world's weather stations are near cities, they record temps warmer than the global average. Although the IPCC has dismissed this differential, the temperature of even small towns is typically increased, thanks to the addition of concrete and steel. Rural readings, on the other hand, tend to show cooling. One shameful exposé of the warmist agenda uncovered official thermometers sited in all the wrong places—near chimneys, burn barrels, exhaust vents, and in asphalt parking lots!

While textbooks speak of a warming trend in the North (Norway, for example, being 2°F warmer since 1930), waters of the tropical Pacific have shown cooler temperatures in recent years. In other words, for every area with rising temperatures, there may be an equal and opposite area with a cooling trend. Even warmists wonder if changes in the North Atlantic truly constitute a "global" climate shift. Perhaps they are merely "regional climatic events that have been misinterpreted as global." The same scientist poses the question of "how climate change in one part of the globe is linked to changes elsewhere . . . [indeed] we must be cautious in making global interpretations from geographically limited observations" (Pollack 2003, 56–57).

Our friends in Australia, for instance, reminded us of this differential a few years back. While the Northern Hemisphere was experiencing a warm winter in 2006–2007, Australia was hit with record lows; the snow was back earlier than ever. Eastern Australia, the journals were reporting, had not seen that kind of cold for more than one hundred years. Queensland "went frosty," and on June 24, 2007, Brisbane saw the coldest morning on record.

Alaskans, too, say the last three winters were colder than ever, now that the PDO warm phase has ended. "When the PDO flipped sign in 1977," Spencer explains, "Alaska suddenly got warmer . . . characteristic of the warm phase of the PDO" (Spencer 2010, 111). Then, when it flipped again at the end of the twentieth century, Alaska got colder. Winter 2010–2011 saw a record-setting number of snowstorms in North America. In South America, the Andean snowpack has increased. Most recently, a rare snowstorm paralyzed Jerusalem (January 10, 2013). Two months later, massive snowfalls covered Japan (March 4, 2013). The year 2014 saw 91 percent of America's Great Lakes frozen by March and November temperatures of −25°F in Casper, Wyoming.

Nevertheless, spokesmen keep warning us in dark and ominous tones that all the world's glaciers are melting—in the Alps, the Himalayas, the Andes, New Zealand, East Africa, Greenland, and the Rockies. Others, however, point out that the glaciers recede naturally as the Earth continues its three-hundred-year warming phase, following the Little Ice Age (see chapter 3). Not to worry: Swiss glaciologists inform us that natural shifts in ocean currents are largely responsible for the melting of the Alpine glaciers in the past decade. In fact, the IPCC had to retract its scaremongering scenario predicting a Himalayan meltdown by 2035.

As for Greenland, it is only her southern glaciers that are

melting; Russian and Norwegian scientists reported in 2005 that most of Greenland's ice was actually thickening. Ninety-five percent of Greenland is ice covered, the glaciers 3,300 feet thick; only the margins are calving off (Macdougall 1996, 1). The facts have been grotesquely misportrayed. "Surging glaciers in Greenland . . . and huge ice shelves in Antarctica breaking off . . . are things that have always happened and always will happen." But those Greenland surges "have stopped . . . [which is] a natural fluctuation in glacier behavior" (Spencer 2010, 62).

Warmists managed to spread panic in the 1990s when the Larsen A ice sheet disintegrated; other ice shelves have disappeared since the 1950s as temps along parts of the Antarctic Peninsula have climbed. At the same time that ice is melting away at the continent's edge, ice at the center of Antarctica is growing thicker. Despite grim reports on Antarctica, Patrick Michaels, editor of *World Climate Report,* having pooled data on the whole of Antarctica, found no overall warming. Shouldn't we factor in the cooling of Antarctica's desert valleys? Sure, parts of the vast southern continent are warming and melting, but other parts (shelves on the eastern side and elsewhere) are cooling, and here the ice is thickening. One side of Antarctica is warmer than usual and the other side is cooler than usual. Back in the 1970s glaciologists were saying that the total ice mass of the frozen continent was increasing. Even in 2007 the Associated Press announced, "Antarctica cooling as Earth warms." A cooling trend in the deserts and other parts of the white continent was reported by researchers from eleven American universities and government laboratories.

From these facts it is clear why critics complain about manipulated data, tailored to produce evidence of record-high temps (see www .junkscience.com/jan05/breaking_the_hockey_stick.html). "We have been duped," protested John Coleman, founder of the Weather Channel (quoted in Horner 2008, 41); "global warming is a lie."

Spencer also frankly states that "the IPCC was formed for largely political reasons, not scientific . . . [and it] has systematically ignored the 800-pound gorilla in the room: natural, internally generated climate variability . . . [opting] instead for the protection and dissemination of the IPCC party line" (Spencer 2010, xiv, xvi, 107).

Global warming is not science. It is gospel, and its master is politics. Incestuous relations and conflict of interest be damned—everyone is in on it, from the American Bar Association to the news media, from do-gooder celebrities to multinational giants, PR organizations, the nuclear industry, venture capitalists, the National Academy of Sciences, the White House, lobbyists, NASA, the American Academy of Pediatrics, the Public Utilities Commission, the state of California, and the EPA. One dissenting member of the American Meteorological Society warns that it will be their downfall. Gagged and bagged, scientists who refuse to cooperate have had to shut up, stand down, or ship out. Chris Horner's astounding exposé, *Red Hot Lies,* has all the dirt, a thousand scandals still unknown to the (largely indifferent) public. We have been gulled by what is likely to go down as the boldest climatological hype of all time. The only crisis, as Buckminster Fuller once warned, is a crisis of ignorance. To Spencer, this "astoundingly naive, if not scientifically corrupt" phenomenon, this "huge house of cards . . . [is] the greatest scientific blunder in history . . . the worst case of mass hysteria the world has ever known" (Spencer 2010, xxvii, 105).

Speaking uneasily about the green agenda, Hans Oeschger, a Swiss atmospheric physicist, saw the handwriting on the wall a few decades ago, when the warming scare first began: "I'm very much afraid of these things getting into the hands of people who do not understand it, but make a big fuss about it" (quoted in Weiner 1989, 86).

In vortexian science, the temperature of a planet ultimately depends on its axial velocity and not on the behavior—for good or ill—of its inhabitants. Let's look at axial velocity for a moment. In chapter 1, under the discussion of redshift, the neglected question of planetary age came into play. So let us look at the relationship between axial velocity and planetary age, and the effect of these factors on temperature. The relationship, as I understand it, is direct: an aging planet is a slowing planet. And a slowing planet is a cooling planet.

As outlined in chapter 1, densities, consistencies, and motions are a gauge of planetary age, just as age is a factor in determining the physical properties of a star. When we looked at redshift, we saw that it could simply be a function of the gravitational field (rather than of recession speed or distance). After all, magnetic fields around fast-rotating stars are particularly strong. I stand with those astronomers who take redshift as indicating the age (not distance) of the body in question. And it is here that the study of field—that is, planetary vortex and its speed—becomes crucial. High redshift (greater brightness) would then indicate the "fiery" beginning of a brand-new planet, a new, rapid, vortex (gravitational field) in formation. If the power to predict is one key to scientific merit, the vortex is the thing to study. The vortexian calculus promises to deliver the underlying mechanism responsible for a tremendous range of phenomena. The long-sought theory of everything will find its home in vortexian science.

We know appallingly little about magnetic fields.
DAVID RAUP, *THE NEMESIS AFFAIR*

VORTEX

Eric Lerner talks about "galaxies being vortices [and] electrons are vortices too" (1991, 257). In a nutshell: just as a nut kernel is encased by a shell, so is the Earth the precious kernel of its atmospheric "shell," which we are calling its vortex. It is here in our own magnetic field (the vortex) that we will discover the long-sought unifying mechanism, for the field is the dynamo, the incessant whirlpool of energy containing all the ingredients for a living world. To Kepler, the Force was a vortex, a raging current.

Stocked up in our vortex's galleries and chambers are all the winds and elements a world could need to spin its fate. Most notably, it is the age of the field, the vortex, that determines its amount of moisture, light, velocity, strength, and temperature. When young and still forming—nothing more than a molten ball of liquid fire—her temperatures are extremely high. Low in density are her gases before solidifying, and rapid, very rapid, is the rotation of those gases.

PLANETARY AGING

The ultimate source . . . of a quasar's immense power is rotational energy.

ERIC LERNER, *THE BIG BANG NEVER HAPPENED*

The faster the [rotation] period of a pulsar, the younger it was likely to be.

ISAAC ASIMOV, *QUASAR, QUASAR, BURNING BRIGHT*

How is it, then, that the youth and age of a planet has been all but forgotten in the calculus of astronomy and earth science itself?* Yet nothing could be more central to the question of climate change. As a vortex grows older and slower, disturbances and heat decrease. The galaxies, we realize, are populated by cosmic bodies of all different ages: young, middle-aged, and old. And our own planet is in the middle category, already showing signs of aging. In the future, the aging Earth will

rapidly give off its life force and its moisture, rapidly grow old.

OAHSPE, BOOK OF COSMOGONY AND PROPHECY 8:6

As geologist Paul Sylvester, Ph.D., has poignantly observed, "Earth is fueling far fewer major tectonic events. . . . New continental crust is piling up at a geriatric pace. Like Mars, earth is on its way to becoming a dead planet; the heyday of its continents is long gone. . . . It's kind of sad." (Sylvester 2001).

Scientists are well aware of the loss of atmosphere on deathlike worlds such as Mars, whose atmosphere is two hundred times thinner than our own. As we saw at the beginning of this chapter, Mars's water

---

*Rather than accept aging as the cause of slower rotation, theorists come up with scientific-sounding guesses like "lunar and solar gravitational effects" (Gribbin and Plagemann 1974, 47).

is gone; flagging, too, are its light, heat, gravity, plasticity, zoology, velocity, and magnetism. Yet the same scientists who report these facts are loath to comment on Mars's (obviously old) age. Some even propose to "terraform" that old rock for future use!

As our own planet's axial speed decreases, the Earth is also declining in magnetism (weaker EMF), atmosphere, gravity, plasticity (plate tectonics), moisture, radiance, and heat. As verified by space satellites, our electromagnetic field (EMF) is weakening and thinning, declining noticeably,* some say by as much as 15 percent since the seventeenth century, or, as others have it, 10 percent in the past 150 years. Another guesstimate says 50 percent in the past 2,500 years. Whatever the correct figure, we have already seen that young Earth registered a fabulously turbocharged atmosphere, recorded in ancient rocks, some of which give readings of one hundred times today's magnetism. These are the oldest rocks on Earth and have dutifully recorded the enormous amplitude of our planet's original force field, its vortex. The subsequent decay of EMF can be measured against these old rocks and boulders, silent record keepers of our planet's early field—a supercharged dynamo, bristling with energy, electrical storms, monumental surges, tumultuous winds.

As the great gestalt of planetary aging unfolds, abating field strength testifies to a slowing pace. But none of these decrements is an isolated— or even unexpected—event. Rather, slowing, weakening, and thinning, along with cooling and drying, are the normal and natural conditions of an aging orb. As rotation decelerates, the overall heat of the planet also begins to subside. Oxygen-isotope readings and carbon dating, as well as radioactive chronometers and instruments for measuring atomic decay, have all permitted us to probe the depths of time, gauging temperatures long vanished. In unison they speak of global cooling over the vast stretch of eons.

How simple and consistent is this new science! It is not complicated

---

*Scientists don't agree on what weakening EMF means. Some say it is the precursor of a magnetic reversal—meaning that the EMF disappears as a prelude to a pole flip. Alas, the weakening is only a sign of aging, and has little, if anything, to do with pole shift.

or beyond our reach. The vortex, quite plainly, weakens with age. Most critically for climate study, our planetary spin (rotation) is slackening. Last time I checked, heat was a measure of motion, and Mother Earth is losing axial speed—some say to the tune of .00073 seconds per century; others "two milliseconds" per century (Corliss 1980, 650). Yet another source says 3/340th of a second annually (representing a total of 100 minutes in the past 80,000 years). Although other estimates give only 1 second per 120,000 or 600,000 years, most recently that figure has been put at .000016 seconds per year, according to astronomy's cesium clock.

Spin a top, and it will eventually spin out. Start a planet, and it too will decline in swiftness and power over time: lots of time. With age comes slowing, and with slowing comes cooling. And this process is completely natural. This correlation of speed and heat was dramatized in October 2006 when astronomers revealed the fastest-known planet, named SWEEPS-10, with a "year" just ten hours long and a surface temperature of about 3,000°F. Two years later, the hottest exoplanet, WASP-12b, was discovered—gaseous and with a surface temperature above 4000°F. Significantly, "It's also the *speediest* exoplanet . . . circling its star once every 1.09 Earth days" (my italics) (M. M. 2009, 26).

A younger Earth spun around more jauntily on her axis—giving it a much shorter day— only four hours long 2 bya, according to George Gamow (Gamow 1948, 187). Of greatest relevance is the relationship of that axial velocity to Earth temperature: the quicker the spin, the warmer the planet. When it ages and that motion gets geriatric, well, everything begins to cool down. A planet warming up? Please. That simply does not happen. An aging, exhausted world is a limp, under-powered, slowing, and cooling affair. The Book of Cosmogony and Prophecy 5.20 and 3.6 indicates a loss of vortexian temperature at the rate of 1°F per 18,000 years: "There is ever a trifling loss toward perpetual coldness." Life on earth, according to pre-Agassiz philosophy, was once a "perpetual summer," cooling down ever since its creation.*

---

*On fiery ("seething") beginnings and very gradual cooling, see Churchward's incisive account in *The Second Book of the Cosmic Forces of Mu* (Churchward 1968, 24, 62–66, 93, 96).

I have said that Planet Earth came forth as a molten mass of hot, undisciplined gases, nothing more than a flaming ball of fire. It solidified perhaps at 600°F. Temperature loss continued without stint, for it is an intrinsic part of planetary life and most likely the factor behind large-animal extinctions. With the globe steadily cooling, loss of blood heat was the probable downfall of the dinosaurs.* Ice was just forming at the poles, and the oceans were cooling, too. Most of the marsupials died off during this time. Survivors, after all, are found only in hot Australia and tropical South America.

Much later the Pleistocene mastodon, saber-tooth tiger, giant ground sloth, and other large beasts met their end, in all likelihood, due to a critical drop in temperature. The woolly mammoth, according to new research using ancient DNA, was not killed off by the excess of hunters, but by a cooling trend in which drier conditions also prevailed. It seems that temperature affects the ability to generate viable sperm. This route to infertility could have finished off the mastodon, the ground sloth, and the saber-toothed tiger.

Heat loss could also kill off the flora on which certain animal species depend. "The drop in pollen abundances seems to be . . . the signature of a cooling climate" (Macdougall 1996, 174). Each new epoch of flora and fauna appears to have emerged in concert with the temperature of the times. Adam Sedgwick, Charles Darwin's geology mentor at Cambridge University, believed that in each succeeding period new tribes of beings were called into existence. Earth history, in this view, was a series of creations, one outstanding example being the spontaneous appearance of new creatures at the Triassic-Jurassic boundary. Even before then, most animals made their debut abruptly in the fossil record. Nor do we have a clue as to why or how the Cambrian Explosion (of new life) occurred when it did. The answer could be quite simple: the Earth was ready and the temperature was right.

---

*Dinosaurs are beginning to look like warm-blooded animals, particularly based on recent alligator research.

The earliest life-forms, we know, were hyperthermophiles ("heat-lovers"), organisms that live at temperatures above 176°F. The simple forms of life, says molecular biologist Michael Denton in *Evolution: A Theory in Crisis,* appeared as soon as surface waters were cool enough to support them. Cellular life itself could not exist until hot Earth began to cool down.

And for humans to live on it, Earth temperature had to come down to 98°F. Tertiary temperatures were still too high. Accordingly, when humans were first quickened into life, the Earth must have been 98°F for at least nine months of the year—the period of gestation. "The earth upon which we live," expounded geologist G. F. Wright, "is but a cooling planetary mass and has become fit for the habitation of man only during recent geological ages. A few million years ago, the heat upon the surface of the earth was so great that it would have been impossible for man to have endured" (Wright 1912, 1).

We see continuous cooling not only in the succession of life-forms (from naked hyperthermophiles to fur-clad mammals) but also in satellite data, the geological record, the history of ocean temperatures, and cosmology. One astronomer, in a context having nothing to do with the global warming debate, said "the earth is unquestionably colder today than it was in the past. . . . The temperature of the earth has fallen over the past six hundred million years" (Lerner 1991, 394). Let's take a look at this overall trend:

- 2 billion years ago: According to Gamow, after molten Earth solidified a crust, rocks began to form, though Earth was still a "molten spheroid." 1 bya: still "extremely hot . . . [and] boiling" (1948, 28, 6, 4).
- 540 million years ago: Heat loss marks the pre-Cambrian kills, where lush super-tropical fossils give way to new forms of life on the cooling planet. Until 500–300 mya (Paleozoic), the Earth was warm everywhere, according to climate scientists. Then, 300 mya, much of plant life began to dwindle; the only species that survived were hardier varieties adapted to a cooler clime.

*Fig. 4.5. Mid-Paleozoic landscape—mostly marshlands,
covered with giant horsetails and ferns.*

- 285–245 million years ago: CO_2 is at ten times its current levels. Entering the Permian, 90 percent species-kill goes hand in glove with climate cooling. Theorists, pulling the catastrophe card, like to say it was some unknown cataclysm, perhaps a bolide, that wiped out so many species. But could it simply have been heat loss?
- 100–65 million years ago (end of Cretaceous): A cooling trend, traced back at least 90 million years, is based on marine fossils and mud cores. Eighty million years ago, ocean temperatures of 70°F began to drop. Before the dinosaur extinction came (at the end of the Mesozoic), it was 18°F warmer than today, with no continental glaciations. In the Cretaceous that followed, winters appeared for the first time; the character of vegetable life also indicates a significantly cooler environment. At the end of the Cretaceous, trees begin to show wintering rings. Tropical reefs are also hard hit, zoo plankton die off, especially in tropical waters. All waters are now cooling. The best survivors were deep-sea creatures, who, we might surmise, were less subject to surface chill. Thus, with ice caps and continental ice appearing at the end of the Cretaceous, the great herds of dinosaurs began to decline. (Smaller animals survived by burrowing.) Besides cooling, it seems to have also been drying* that finished off the dinosaurs. The landscape of

*Cold and dry go together. One example would be the 8.2 kiloyear event of 8,200 years ago, when an abrupt decrease in temperature occurred together with prolonged drought.

this time saw inland seas and swamps receding, most of them drying up. Indeed, it has been suggested that "drought was the killer," for now the huge beasts were congregating "in parched riverbeds, where they perished as food and water disappeared" (Rogers 2007, 50).

- 55 million years ago: In the Cenozoic "global climates were significantly warmer" than today (Palmer 2010, 174). Until 35 mya, says Professor Macdougall, the world was warm; the Earth, he says, "has been progressively cooling for the past 50 or 60 million years . . . [entailing] a persistent, gradual decrease . . . in ocean water temperatures" (Macdougall 2004, 7–8, 172). The cooling in this period has a name: the Cenozoic climate decline.

- 50 million years ago: Between 50 and 20 mya, the temperature dropped 9°F. By 30 mya, ice domes have begun to form on Antarctica. After the Eocene extinctions, the Oligocene saw a colder and drier climate, inhabited by new mammals adapted to the cooling clime: cats, dogs, pigs, bears. "Ever since the Eocene, the earth has experienced a gradual, irregular cooling" (Howells 1993, 116).

- 20 million years ago: Cooling climate marks the Miocene; temperature now drops another 10°F. Ten mya, glaciers begin to form on high latitude mountains. Five million years ago, with worldwide drying and cooling, older animal species fade, replaced by many new (cold-adapted) terrestrial species. According to Macdougall, the North Pole ice cap formed only 4 mya.

- 2–3 million years ago: With the end of the Pliocene and the coming of the Pleistocene, earlier warm-weather mammals ebb away. Continental ice sheets first form in the Northern Hemisphere, and permanent glaciers, like that of Kilimanjaro, now appear in abundance in the Northern Hemisphere. "Hundreds of thousands of years ago . . . global temperatures were balmy" (Kruglinski 2006, 15).

- 80,000 years ago: Earth temperature drops to 98°F.

- 15,000 years ago: Shifting weather patterns bring more snow, mosses, and sedges. Snow cover affects grazing animals; the mammoths die off.

- 10,000 years ago: The Arctic was warmer than today, with less ice cover. "The cooling trend that began 7,000 years ago *will continue into the future*. . . . Vegetation belts have . . . moved steadily southward" (my italics) (Imbrie and Palmer 1979, 178, 181).

- Today: 10 percent of planet Earth is glacier-covered. Tomorrow's forecast: cooling and drying. Thus, when today's warming alarmists say things like "1998 was the warmest year on record . . . eight of the ten hottest years in history have occurred in the 1990s" (Helvarg 1999, 38), we have to laugh. "On record" sometimes means since 1972 in the northern latitudes only! Indeed, it was hotter in the 1930s than today. The year 1998 was simply a particularly warm El Niño year.

Should we all go out to fight this nonexistent warming? Such a battle royale would stand in absurd opposition to H. A. Brown's suggestion (during the previous PDO) to fight the icebergs!*

Finally, I must ask: Why has global warming been so slickly promoted? The agriculture commissioner in North Dakota has publicly stated that human-made climate change does not exist, noting, "I think an agenda is being pushed." The manager of a North Dakota wheat mill in Minot also doesn't believe in "man-made climate change." In his view it is "a ruse to justify greater government regulation" (quoted in Hertsgaard 2012, 34). After all, schemes to reduce emissions involve more taxes, fines, regulations, and energy controls. "Once CO_2 regulations are implemented," notes Spencer, "the price of virtually everything will increase, because all goods and services require some input of energy." This has already happened in England, with little effect on GHG levels. As a result, the British, as Spencer observes, view global

---

*See John White's best-selling and extremely informative book *Pole Shift* (1980), chapter 4, which thoroughly covers this HAB (Hugh Auchincloss Brown) theory of our Earth frightfully endangered by the accumulation of polar ice.

warming "as one more excuse for the government to get its hands on the people's money . . ." (Spencer 2010, xix, 67).

The policy of controlling pollutants with economic incentives is known as cap and trade. It shifts jobs overseas (businesses moving to countries that have no tax or cap on carbon). Concomitantly, government bureaucracy grows larger and more intrusive. The price tag would be huge, raising the cost of gas, coal, and oil. Pleasantly dubbed "sky trust," cap and trade takes a disproportionately bigger bite out of the paychecks of the poor. Investment bankers, though, do very well. Worse yet, "the whole cap-and-trade process is a breeding ground for new sources of cheating and corruption that haven't even been invented yet." Neither does it "help to reduce CO_2 emissions . . . Europe and the UK have discovered just how damaging a cap-and-trade approach is to their economies." (Spencer 2010, 148–49)

These are all costly "improvements." In 2005, a European Space Agency probe, called CryoSat, designed to study the effects of climate change on Arctic ice, plunged into the ocean, the crash costing taxpayers $136 million. In other quixotic plans to cool the globe, five hundred £1 million "spraying yachts" have been proposed: their droplets into clouds would slow down the warming that is supposedly caused by carbon dioxide. In the United States, "Many corporations believe they can make a killing in the burgeoning carbon trading system," mirroring the time when the big producers of CFCs,* another GHG, decided that a CFC ban could generate a new and lucrative market in substitutes "that they were well placed to exploit" (*New Scientist* Staff 2005, 5).

---

*CFCs—industrial gases, called "chlorofluorocarbons"—come from aerosol cans, air conditioners, refrigerators, and so forth. Industrial sources (and volcanoes) give rise to CFCs, tiny suspended droplets of sulfur oxide and water. Such aerosol particles create an atmospheric haze that deflects sunlight and therefore cools the Earth. Well, then, which is it? Do CFCs cool the Earth or warm it?

The official pro-warming position of the American Physical Society, like many other mainstream scientific organizations, holds that human activities are a cause of climate change; that dire effects can be expected; and, yes, that more research funds are in order. Such declarations, muses geologist Robert Schoch, "become virtually equivalent to papal encyclicals. They form the basis and rationale for investing major financial resources in . . . [combating] global warming." Schoch goes on to remind us of the late University of California physicist Harold Lewis, who, after sixty-seven years of membership, resigned from the American Physical Society in disgust. "The money flood," the veteran physicist fulminated, "has become the raison d'etre of much of physics research. . . . The global-warming scam, with the trillions of dollars driving it, has corrupted so many scientists. . . . It is the greatest and most successful pseudoscientific fraud I have seen in my long life as a physicist."

Lewis then referred whoever is listening to the ClimateGate documents. Academic misconduct on the part of official warmists made the news in 2009 with the scandal dubbed ClimateGate. More than a thousand e-mails and documents exchanged by leading IPCC scientists had been hacked from their computers. Allegations against these scientists included schemes to hide or destroy data not supporting the global warming craze, as well as "exaggerations, . . . suppression of dissenting opinions, and plotting to remove from influential positions . . . people who disagreed with climate change dogma" (quoted in Schoch 2012, 279–80).

Bottom line: much of today's research on warming is funded by government money. "Is it any surprise if they find what they were paid to find?" asks gadfly Spencer. "Most government contracts and grants go toward those investigators who support the party line on global warming. . . . For those who think government-funded research is impartial, I can tell you from firsthand experience that it is not. . . . If you pay scientists enough money to find evidence of something, they will be happy to discover it for you" (Spencer 2010, 18, xii, 66, 157).

Today, more money is thrown at global warming than at cancer research. If the global warming gravy train is lining the pockets of some, it comes at the expense of billions of tax dollars wasted on fruitless research. What bothers me the most is the misplaced pessimism: pseudo-issues, alarms, and "threats" distract from, even cloak, the actual assaults on the environment and on the people of today's world. This is not a harmless mistake. Moving industry (and its toxins) to poor countries, which do not abide by emissions rules, solves nothing (and only raises our own unemployment stats). In reality, carbon dioxide may have no ill effect at all compared to depleted uranium, bombs, nuclear fallout, poisons, pesticides, contaminants, and the pollution of air, water, and ground by chemicals, like the deadly waste products of mining operations. These have sickened, irradiated, and killed thousands of people in the western United States, mostly in the vicinity of Indian reservations (causing cancer, miscarriages, etc.). We are talking about all-out ruination of our waterways, fisheries, and soils.

Insidious initiatives and draconian rules have been put in place by advocates of global warming, who pose as friends of the environment. Among these are zoning restrictions and the "no-till" rule (decaying organic material in the soil is rich in carbon; hence limited tilling keeps the carbon from mixing with oxygen to form CO_2). With all of this, poor people in developing countries are now under the gun: disastrous controls and regulations restrict rice paddies in hopes of reducing marsh gas (i.e., methane), which is produced by microorganisms. Rice paddies flooded by irrigation generate carbon in the form of methane—natural gas—because of decomposing vegetation in stagnant, standing water.

These farmers are the very real victims of the West's global warming hysteria: "Green energy policies will push many of the world's poor . . . into starvation. . . . [We also] run the risk of killing literally millions of people" if Gore and Hansen's call for halting electric power plants (a major source of CO_2 emissions) is implemented. Cutting off the most inexpensive energy source "will lead to starvation for many of the world's malnourished. . . . Forcing expensive alternative forms of energy on people . . . is nothing less than a war on the poor" (Spencer 2010, xviii, 27, 136).

Members of the news media have blithely, inanely, hyped the alleged danger of mammoth dung exposed in melting permafrost, which allegedly releases more deadly greenhouse gas. Deadly? Instead of promoting such silly myths, why don't we exploit this methane as a source of energy? The methane trapped in the Arctic, Siberia, and Alaska (billions of tons of this natural gas) could be a huge source of energy (if the permafrost melts). The possibility has already attracted venture capitalists. If sea ice is eliminated along the Arctic Coast, it will expose vast fossil fuel reserves. The Arctic, it is estimated, holds more than 20 percent of the world's oil and gas resources.

Costly programs to bury carbon are as foolhardy as the threat of mammoth turd. I think I see Mother Nature shaking her head and rolling her eyes. The IPCC proposes to bury one-third of the world's CO_2 by 2050, and if the "technology" is there, maybe dump it in the seas. Inject it deep in the ocean? Then it would react with seawater and increase its acidity, affecting all marine life and the food chain; this, at the same time that warmists are hollering that perilous acidification of our waters is driven by carbon dioxide emissions! Perhaps, then, we should inject the nasty stuff in abandoned oil and gas wells or pump it into rocks! But underground "sequestration" of carbon dioxide, some say, is a disaster waiting to happen. Just think of the day when an earthquake unleashes all that pumped-in gas, suffocating the nearby plants and animals . . . and people.

In 2009, after flooding displaced 600,000 Filipinos, the peasant movement rallied in front of the UN Climate Change Center in Bangkok, claiming that climate change "is being used by multinational corporations, who are the main contributors to global warming, to rake in more profit from our misery. Corporate schemes forcibly drive us from our land" ("Bangkok Activists" 2009, 6). Vast tracts of agricultural lands around the world, protesters declared, are being controlled and converted by these plunderers into cash crop plantations for biofuels.

Concerning these biofuels: the drive for "green energy" is having the perverse effect of destroying tropical rain forests. From Borneo to Brazil, virgin forests are being denuded to make room for soybean and palm oil plantations, which will fuel cars and power stations in the developed world. European law actually requires countries to produce a certain percentage of such biofuels in order to meet Kyoto targets for reducing GHGs. But the consequences are troubling. Palm oil production in Southeast Asia and soybeans in the Amazon are a leading cause of rainforest destruction. "We are trying to solve our environmental problems by dumping them in developing countries where they have devastating effects on local people" (Pearce 2005, 19). Meanwhile, converting corn, sugarcane, and soybeans into fuels will ultimately hike food prices, and may release more carbon dioxide than they use.

The developing world is paying the price in yet another way. A few years back, a faction in the Sierra Club tried to pass a motion to reduce the number of immigrants allowed into the United States, based on the contention that Americans pollute more (five times as much as Mexicans). Therefore, we should stop Mexicans from becoming Americans! Though the motion was defeated in the United States, Australian leaders are floating similar ideas. Opponents rightfully ask: Should we hold immigrants hostage to our own environmental crimes?

And now, with "global warming" as the popular wisdom, developing countries can shift the blame for wildlife extinction away from their own handiwork (deforestation). No, they say, it is not the destruction of breeding habitat and woodlands that have decimated certain species (particularly state birds and favorite songbirds); it is the long-term effects of global warming! Welcome to the big-time business of blaming global warming for everything from megadroughts to wildfires, floods, extreme rainfalls, La Niña, tornadoes, infectious diseases (malaria, dengue fever), retreat of wildlife, extinction of species, high school dropout rates, ethnic cleansing in Darfur, the 2007 bridge collapse in Minnesota, not to mention the heart-wrenching reports of ski resorts in peril.

The blame game has blinded us all to disruptions of the climate system due to other causes. "We can't write down a precise equation between carbon dioxide and storms," admit the experts, but "rising levels of green-house gases have already been accompanied by weird rainfall patterns" (Begley 2012, 87). Should we just assume, then, that global warming is responsible for "weird rainfalls" and all these other climate extremes? Even the explosive growth of coral-eating starfish has been blamed on warming ocean waters. Earthquakes and hurricanes as well: "The only development that could cause a megacatastrophe of volcanoes, hurricanes, and earthquakes . . . is global warming . . . [which] in the Gulf of Mexico has undoubtedly energized hurricanes. . . . Global warming may lead to earthquakes," guesses author Lawrence Joseph (Joseph 2007, 81).

Nevertheless, hurricane scientist William Gray has cast doubt on the claim that warming is to blame for stronger storms. He attributes them instead to the natural Atlantic multidecadal oscillation, which was at its peak in 2005, around the time of those particularly destructive hurricanes, earthquakes, and tsunamis. Other studies (Vecchi and Soden 2007, 4316) revealing no link between hurricanes and warming have fallen on deaf ears. Are warming seas the real cause of hurricanes? I would like to recommend a few pages (328–29) from Chris Horner's 2008 book *Red Hot Lies,* describing the brazen tactics of IPCC to muzzle the truth as told by the real hurricane experts. (See also appendix A.)

LONG-TERM VARIATIONS

What an observer perceives as a trend may only be one phase of a cycle.

JOHN IMBRIE AND KATHERINE PALMER, *ICE AGES*

What if the warming trend could be reduced to mere fluctuations, part of a natural cycle? As we have seen, only a few decades ago, in the 1960s and 1970s, the great ice mass of Antarctica was expanding, each year "winter sea-ice build-up more than doubl[ing] the total expanse of ice in the Antarctic" (White 1980, 360). The North Atlantic was also in a cold snap, with sea ice returning to Iceland's coast. So authors, naturally, were writing about "our cooling climate . . . the Arctic expanding.

. . . Some experts thought the ice was on the move again" (Jeffrey Good-man 1978, 138). But it was not a long-term trend, after all. "The last time the PDO changed phase was in 1977 [see fig. 4.6, page 195], an event that some have called the 'Great Climate Shift of 1977.' This event brought an end to the slight global cooling that started in the 1940s, which was then replaced with a warming trend, from the late 1970s through the 1990s" (Spencer 2010, 19). Today, John Christy thinks the warming that so alarms scientists is well within the range of natural variation, or even measurement error. So the IPCC carefully threw in a disclaimer: questions remain about how much of the warming stems from long-term variations.

The icon of global warming—the endangered polar bears—are actu-ally doing quite well, as of 2007, increasing in number in the Arctic. But even when this statistic was openly acknowledged in *New Scientist,* it was downplayed by the scientific community and accused of being propped up by the oil industry "and part of an orchestrated campaign to undermine the environmental movement. . . ." But as Chris Horner realistically and humorously puts it, "The greatest threat to polar bears is computer models" (Horner 2008, 133, 45).

Speaking of computers: new technology (RAPID, which measures ocean currents) has shown that computer models are unreliable if only because fluctuations are extremely variable. The weather is fickle; North Atlantic Ocean currents fluctuate wildly. One science writer admits, "It's tough to pick out overall changes in climate in the face of natural fluctuations. . . . Even three warm years in a row don't necessarily signal a general trend" (Lemonick 2001, 24).

We need to factor in this "natural variability of Earth's climate," advises H. N. Pollack, who then asks whether the slight temperature increase in the twentieth century is really important. After all, "short-term fluctuations usually can be accommodated without serious dam-age to living things" (Pollack 2003, 57).* This sober view flies in the

---

*Studying the ancient tropical forests of South America, for example, it seems that a periodic warm spell only spurred biodiversity, rather than rapid extinction. Many new species appeared. No harm, no foul.

face of panicky alerts that say, for example, "Climate has suddenly flip-flopped in the past and will surely do so again. . . . Weather patterns in Europe could shift in a matter of years, making that area's climate more like Siberia's" (Alley 2004, 64).

One such "climate jump" was the Dust Bowl on the American Plains in the 1930s. But here is where the cycles come in (appendix B outlines these cycles). Much of nature goes by cycles. Raymond H. Wheeler's 100-year weather cycle has a cold/wet phase alternating with a warm/dry phase every century, as noted in the alternating wet and dry swings of South Africa during the Quaternary. That study was based on gravels, sands, and silts in the Vaal River basin. Wheeler's 100-year scheme reminds me of the Egyptian 99-year "wave," a multiple of the basic 11-year cycle. The weather cycles discerned by the ancient Egyptian system were based on the 11-year "ode" (a Phoenician word meaning "short measure"; see more on the ode in chapter 5) and its multiples: 22, 33, 66, 99. As an example: To those Dust Bowl years of the Midwest in the 1930s, add 33 (11 x 3) = 1961–1966, giving us the Great Northeast Drought, the most prolonged ever. Then add an ode to that year (1966 + 11 = 1977): California was dangerously parched in 1977 (see appendix C).

Drying—not warming—seems to be the primary cause of ice loss on Africa's famed Mt. Kilimanjaro. No longer the poster child for doomster warmists, the fabled mountain has been stripped of the myth, but also of moisture. Deforestation at the base of the mountain—caused by extensive farming—is the culprit, according to researchers in Africa, Brunei, the United States, England, and Austria. Without those woodlands at the foothills, humidity evaporates into space. The problem, say scientists, is trees, or rather, the lack of them. The iconic snows of Kilimanjaro, it is further suggested, "have come and gone . . . the fluctuations are nothing new." (www.foxnews.com/story/02933.282033.00.html)

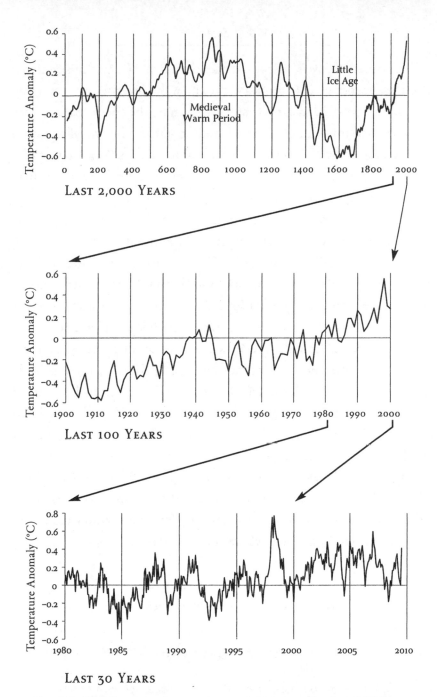

Fig. 4.6. Global average temperature variations over the last 2,000 years. In the top panel are the Medieval Warm Period (MWP) and the Little Ice Age (LIA), both showing considerable but natural change in climate. Data in the bottom two panels shows twentieth-century warming, a pattern typical of the last two millennia and unlikely to be caused by humans. Reprinted with the kind permission of Roy W. Spencer.

The "times of eleven" affect weather on Earth. What is misleadingly called the sunspot cycle (running approximately eleven years) is believed to influence crops, geomagnetic storms (at max, the thermosphere bristling with electrons), rainfall, and flooding. At the height of the cycle there are more cyclones over oceans and more earthquakes. Varve cycles (fossil mud) and tree rings also go by the eleven-year ode, which at max gives warmer and wetter seasons.*

The climate of the past millennium has, in fact, been quite variable. For example, what do we make of the MWP, Medieval Warm Period, running from about 800 to 1200 CE, the time when the Vikings established themselves in Iceland and Greenland? In the ninth century, Norse voyagers set their courses for ice-free parts of the sea, which are today frozen. When Eric the Red happened upon Greenland in the 980s, the place was actually green, warm, and bucolic; the inhabitants were raising cattle and cooking outdoors. Old Scottish tales, as well, reminisce about an earlier, warmer clime, as do other recollections of North European peoples. I guess this is why one IPCC author said to a colleague: "We have to get rid of the Medieval Warm Period" (Spencer 2010, 9), for it contradicts gospel.

Approximately every four hundred years, solar activity shifts (White 1980, 359). Just so, the MWP lasted four hundred years, a key prophetic number, not only in the ancient Egyptian winter tables but in the Mayan system as well, where the all-important *baktun* lasts four hundred years.

"The key unit to look at is the baktun cycle" says Jose Arguelles (1987). On the Four Hundred Years of the Ancients, see appendix D.

VORTEXIAN LENS

Electromagnetic waves are intimately connected with cycles.
CYCLES RESEARCH INSTITUTE

*The association of warm and wet was also evident in the Holocene Wet Phase (sixth millennium BCE), which saw abundant rainfall along with higher temperatures than today.

Fig. 4.7. Ancient sign of Earth's lens. From Oahspe

The regularity of these prophetic numbers suggests that the vortexian currents ("electromagnetic waves") manifest in a periodic manner. But what exactly is the mechanism underlying those currents? The answer lies in the magnetosphere and the Van Allen Belt. This region of the Earth's vortex is critical, for it is both the lens and the ceiling in the sky that throws back to Earth our radio waves as well as the Earth's heat and light. Its great hemispherical lens allows us to see the Moon, the Sun, and the stars.

> The earth's vortex is concave to the earth.... Because the vortex follows the same curvature, the total effect is that of a convex lens.
>
> OAHSPE, BOOK OF COSMOGONY AND PROPHECY 3:2

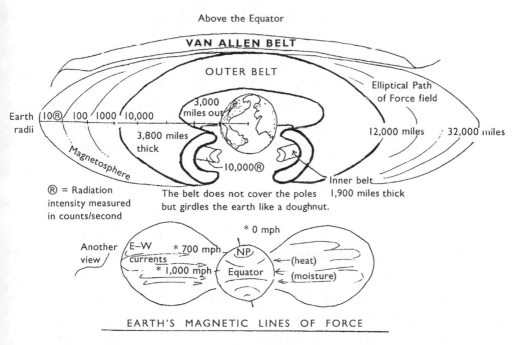

Fig. 4.8. A drawing by the author showing the dynamo of Earth's vortex. The Van Allen Belt is actually in the shape of a doughnut.

The visibility of the stars, the planets, and even the Sun would be impossible if not for the Earth's ionosphere. Understanding this "could turn our modern beliefs on their head," says Gary Nicholls (quoted in www.studyofoahspe.com/id10.html). The vortexian lens acts just like a telescope. But it is in a state of flux, or shape change. Nevertheless, there is rhythm to that flux.

> The lens of the earth's vortex varies constantly. . . . The lens power loseth by flattening and increaseth by rounding. . . . In periods of thirty-three years, tables can be constructed expressing very nearly the variations of vortexya for every day in the year, and to prophesy correctly as to the winters and summers. . . . This flattening and rounding of the vortexian lens is one cause of the wonderful differences between the heat of one summer compared with another, and of the difference in the coldness of winters. . . .
>
> OAHSPE, BOOK OF COSMOGONY AND PROPHECY 2:16

It is, then, the shape-changing lens that, when compressed (that is, rounded), triggers the magnetic storms that are associated with sunspot maxima every eleven years. Compressed, it also tends to result in warmer and wetter weather. Sunspot minima, on the other hand, harbor cold and drought (this conclusion arrived at by studying tree rings).

Moisture and warmth are the principal ingredients of the magnetic envelope, the Earth's vortex. When that envelope is in its phase of enhanced power, compressed at maxima, it charges the Earth abundantly with its products: rain, wind, warmth, electricity, and so forth. Compressed (ESA, European Space Agency, uses this very term) as well as *deformed,* during maxima, it incites radio-wave disturbances and power-grid shutdowns on Earth. By studying the changes in the magnetosphere, the flattening and rounding of the Earth's magnetic lens, we come very close to understanding climate change, whose current models, as we will come to see, are not so hot!

There are plenty of human-made disasters to worry about, but global warming is not one of them. It is a work of fiction. Any resemblance to actual science is purely coincidental.

> *The first place one should look to find explanations for climate change is in nature, not in the tailpipe of an SUV.*
>
> Roy W. Spencer,
> *The Great Global Warming Blunder*

PART TWO

SELF

CHAPTER 5

ASTROLOGY

Fortune Cookies in the Sky

Scientists today will incorrectly blame exotic viruses and drought on global warming. How does this compare with the enlightened ancients who, foretelling seasons of fever and pestilence, set about draining the marshlands or applying themselves to irrigation projects when the prophets foresaw a time of drought? Today, humans are discovering how to dissipate dangerous pathogens. An article titled "Weather, Climate, Dark Matter, Airborne Pathogens," talks about new devices, such as the soft X-ray electrostatic precipitator (SXC-ESP), that trap and zap airborne microbes. The unit, using charged particles, completely inactivates biological particles by irradiating them (see www.studyofoahspe.com/id20.html).

But where do airborne pathogens come from? We have already touched on the space dust called a'ji (chapters 3 and 4), which is felt when the Earth passes through a nebulous region in space. Its substance enters the atmosphere and humans absorb it.

According to the corpor solutions in the firmament and their precipitations to the earth . . . so will man be affected and inclined to manifest. . . . The earth goeth amongst these etherean and atmospherean worlds regularly; so that the periods of inspiration and periods of darkness are not haphazard.

OAHSPE, BOOK OF COSMOGONY AND PROPHECY 7:11 AND 15

This esoteric knowledge was part of ancient science. The sages of India and China computed the dark (a'jian) periods of a cycle and used them as an index of prophecy. The seasons of a'ji, they knew, brought not only cold, pathogens, and dimness, but also darkness of the soul. This condition spiraled into lawlessness, drunkenness, brawling, and warfare.

In the years of a'ji, mortals become warriors.

OAHSPE, BOOK OF SAPHAH: TABLET OF SE'MOIN 56

Ancient astronomical (not astrological) texts of the Near East could prophesy everything from inundations to moods ("The land will be extremely happy") to warfare ("The army will fall to battle-swords") (Lindsay 1971, 175).

In the two World Wars, astrologers on each side, citing Nostradamus, predicted victory for their respective sides.

L. SPRAGUE DECAMP, *TIME AND CHANCE*

Astrologers, fools, riff raff, stay away . . .

NOSTRADAMUS, CENTURIES VI: 100

I have come to wonder whether Nostradamus learned how to prophesy through a secret (undisclosed) knowledge of the Egyptian or Mayan system, whose cyclical *baktun* equals precisely the "400 years of the ancients" (144,000 days; see appendix D). This periodic unit of time, the 400 years that rule over the seasons of the Earth, was given in the Winter Tables of the Egyptians 12,000 years ago and was called "Sky Time." Such was the theme of my book *Time of the Quickening* in which these and other units of time, starting with the basic "ode" of 11 years, comprised the "sacred numbers" capable of predicting everything from wars and riots to abundance of grasshoppers, economic cycles, tornados, witch hunts, plagues, treaties, scandals, fads, blood pressure extremes, coal-mining accidents, and the price of wheat. More on these prophetic numbers shortly.

But, for now, what mechanism is behind the prophetic numbers? It is here we turn again to the Earth's own energy field (its vortex) and its periodic rhythms. No, not the Sun and the Moon or the stars and planets (astrology), but the Earth's own electromagnetic envelope—its dynamic vortex—generates these cycles.

> Planetary disturbances are not caused by any power or effect of one planet on another; the cause of the disturbance lieth in the vortices wherein they float.
>
> OAHSPE, BOOK OF COSMOGONY AND PROPHECY 3:18

We are all sensitive, suggested French researcher Michel Gauquelin in *The Cosmic Clocks,* to the "gamut of electromagnetic waves. . . .

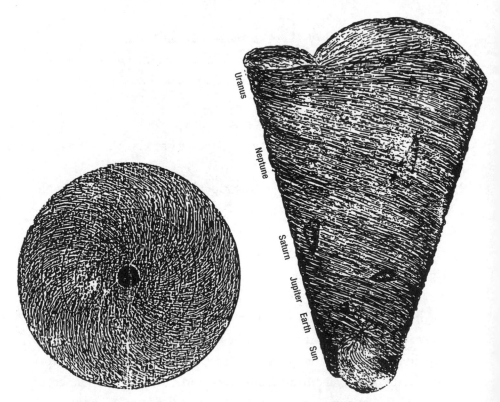

Fig. 5.1 Left: Vortex, as seen from above. Earth is the black center. Right: The master vortex, enclosing the entire solar system.

Atmospheric conditions and human physiology are very closely linked" (1967). Although science has placed a "prohibition against investigating any form of electromagnetic field associated with living organisms . . . there is actually very substantial physical evidence for such a field" (Milton 1996, 84). After all, the structure of water is demonstrably sensitive to the electromagnetic field (EMF) and our bodies are 75 percent water. As an example of electromagnetic influence, when the vortexian "needles" are direct, the day is bright and we feel energetic; when they're indirect (transverse), the day is gray and dull, and we tend to be gloomy, sluggish, arthritic.

The field of force has been depicted as an infinite array of "arrows," that is, needlelike, simply because, in space, solutions of corpor take the shape of needles. Indeed, scientists have found the Higgs boson, the so-called God particle, to be needle-shaped. Matter in its most infinitesimal size is needlelike because the currents in a vortex are shaped as thin, vibrating lines (or "strings"). Corpor, then, assumes the shape of the currents.

As the lines of vortexya are in currents from the outer toward the interior, so do the solutions of corpor take the shape of needles . . . pointing toward the center, which condition of things is called LIGHT.

OAHSPE, BOOK OF COSMOGONY AND PROPHECY 1:34

That said, we can now step into the subject of this chapter—astrology. Why did astronomers of the ancient world study the skies? Was it to find out about their own behavior or destiny through astrology? Were they fortune-tellers? Or were they astronomers, such as the Chaldeans of Mesopotamia, who left cuneiform tablets recording eclipses, the heliacal risings and settings of Venus, as well as the positions of the "seven planets," which were then imitated in the seven stories of their magnificent ziggurats? Obviously, they did not build those lofty towers "just to plot horoscopes, but for some great purpose whose secret we have lost" (Drake 1968, 207).

But it is not really lost, or even a secret. The Chaldeans, long before astrology or personal horoscopes came into vogue, were engaged in predicting the weather, the outlook for crops, water levels, and other conditions directly affecting their general welfare. The Archives of Nineveh, for example, record that "the heliacal rising of the planet Mercury is near; there will be hard rain and the storm god Adad will thunder."

Unlike today's "ego-astrology," the first skywatchers were concerned with profoundly practical aspects of the heavenly cycles. "Man," observed Lawrence Jerome, "was making careful observations of the night sky long before astrology entered the picture; he was putting those astronomical observations to a practical use: keeping track of the seasonal comings and goings of the plants and animals that were important to him" (Bok and Jerome 1975, 40).

> Babylonian men of science discovered that the rhythm of cycle recurrence . . . was discernible on a vaster scale in the motions of the planets. . . . The never broken and never varying order . . . of the stellar cosmos was now assumed to govern the Universe as a whole: material and spiritual. . . . Was it not reasonable to assume that human affairs were just as rigidly fixed and just as accurately calculable? Was it unreasonable to assume that the newly revealed patterns of the movements of the stars were a key to the riddle of human fortunes, so that the observer who held this astronomical clue in his hands would be able to forecast his neighbor's destinies if once he knew the date and moment of his birth? Reasonable or not, these assumptions were eagerly made; and thus a sensational scientific discovery gave birth to a fallacious philosophy of determinism, which has captivated the imagination of one society after another, and is not quite discredited yet after a run of nearly 2,700 years. The seductiveness of astrology lies in its pretension to combine a theory, which explains the whole *machine mundi,*

with a practice that will enable Tom, Dick and Harry to spot the Derby winner here and now.

ARNOLD TOYNBEE, *A STUDY OF HISTORY*

Babylonia developed astrology around omen-reading for kings and kingdoms. The focus was the royal horoscope, which essentially served imperial interests. Some of the prophecies were nothing more than propaganda for the regime. But when Babylonian astrology was planted in Greece, the populace as a whole—with its democratic spirit—began to crave knowledge of their personal destinies. This, I would stress, was in contrast to earliest Mesopotamian astrology, which had been concerned with events "affecting whole nations . . . not with the lives of individuals." Only with the Greeks came the popularity of the individual horoscope, "which has dominated the art ever since" (Cavendish 1967, 183).

And when the torch was passed to Rome, practitioners (known as *Chaldaei*) were now called "astrologers of the circus," since most of their craft involved predicting winners at the chariot races. "Of little service are . . . the astrologers of the throng'd circus . . . [who] would teach others, to whom they promise boundless wealth, and beg a penny in return, paid in advance," railed Ennius (quoted in Cicero 1997, 194). It was a gambler's thing (and, often enough, still is).* But let's keep in mind that astrology's popularity in Rome corresponds to the downfall of that great empire, immediately following its most decadent period.

The same applies to Greece: "In the spiritual bankruptcy following the Macedonian conquest, astrology fell upon the Hellenistic mind as a new disease† falls upon some remote island people. The phenomenon repeated itself after the collapse of the Roman Empire. The medieval

---

*The Romans were also rather obsessed with their prophesied death date. So were some moderns: Hitler's astrologer, Karl Krafft, held that the constellations still play their part at the moment of one's passing.

†"Astrology is a disease, not a science," the medieval sage Maimonides (1135–1204) is quoted as saying.

landscape is grown over with the weeds of astrology and alchemy" (Koestler 1959, 111).

Arthur Koestler, further noting the popularity of astrology in periods of trouble and decay, observed: "As always in times of crisis, belief in astrology was again on the increase in the sixteenth century, not only among the ignorant, but among eminent scholars. . . . Kepler's attitude to it was typical of the contradictions in his character, and of an age of transition. . . . Court Astrologer to the duke of Wallenstein, he did it for a living, with his tongue in his cheek.

. . . His publication of an annual calendar of astrological forecasts . . . brought an additional remuneration of twenty florins per calendar, which Kepler direly needed as his miserable salary was a hundred and fifty florins *per annum*." As Koestler sees it, the chapters of Kepler's book *Mysterium,* which deal with astrology, are "medieval, aprioristic, and mystical," while those that expound on astronomy are "modern and empirical" (Koestler 1959, 242–43, 255). Split mind.

Among the Romans, men like Caesar, Lucretius, and Petrarch remained skeptics to the end. Hippolytos, Basil, and Eusebios also scorned the chimeras of astrological divination, as did Cato, who warned against consulting the *Chaldaei,* while Plutarch wrote that these speculations attract readers by their novelty and extravagance. Tertullian was a bit more scathing in his criticism, not unlike Origen who ascribed to angels the influence that astrologers gave to the stars. Finally, the great orator Cicero stated, "When these astrologers maintain . . . that all children that are born on the earth under the same planet and constellation, having the same signs of nativity, must experience the same destinies, they make an assertion, which evinces the greatest ignorance of astronomy" (Cicero 1997, 238). Earlier, under Greece's (fifth century BCE) heresy trials, Anaxagoras was fined and banished for expressing similar sentiments, the pressure for prosecution coming from the class of professional diviners, who saw in his rationalism a threat to their livelihood.

By the second century CE in Greece, Ptolemy codified the principles of astrology, which are still used today, on the premise that the position of the heavenly bodies at the time of one's birth influences personality and temperament, even destiny. However, in Ptolemy's time, classical astronomy was "sadly deficient," regarding the Earth as the center of the universe. In Ptolemy's view, the planets and the stars surely governed the Earth, for they presumably surrounded it, orbited it. And although we are no longer "geocentric" in outlook, this is nonetheless the same fatalistic astrology that is practiced today. Ptolemy also rejected the axial motion of the Earth. "In the long centuries that followed [Ptolemy], no essential feature was added or subtracted" (Gauquelin 1967, 36). Most of the rules of analysis and interpretation of horoscopes go back to the work of Ptolemy, and although "no astronomer would think of referring to it today . . . it is still the standard reference guide for the astrologer. . . . Astrology has essentially stood still since the days of Ptolemy . . . [even though] modern concepts of astronomy and space physics give no support—better said, negative support—to the tenets of astrology" (Bok and Jerome 1975, 28, 22).

> *There must be very few people who, deep down, do not see the anachronisms of astrology.*
> MICHEL GUAQUELIN, *THE COSMIC CLOCKS*

Since Ptolemy's time, the wobble of the Earth's axis has changed the positions of the stars relative to Earth. Most astrologers today, however, continue to use the Chaldean star positions that are more than 2,500 years old. "It's hard to apply the term 'science' to a field of knowledge that makes no advances . . . astrology [is] totally static" (Stein 2012, 54, 58). Mathematician James Stein also objects to decision making based on horoscopes:

I have nothing against reading and enjoying your horoscope. . . . But if you start using it to make decisions for others (as

Nancy Reagan did when scheduling presidential appointments for her husband) . . . you may be causing someone else to suffer in consequence. . . . Astrology and numerology are basically fortune-telling devices and have no causal connection to any of the principles known to govern the Universe." (Stein 2012, 54, 58)

When the star groups (zodiac) were first given through Osiris, they were of interest to astronomy, not astrology. Like the earliest Chaldeans, their interest was environmental, practical, not occult. They sought to trace the seasons and to avert potential calamities like drought or famine. It was only later that the Chaldeans—with their haruspex, who examined the liver of animals for omens—looked to the stars for a personal oracle, betraying a "strong vein of fantasy . . . in their concept of causation" (Lindsay 1971, 27).

The Egyptians, after Osiris, were inspired to build temples of observation to verify the cycles of time: "The four hundred years of the ancients and the half-times [200 years]; the base of prophecy; the variations of 33 years and the times of 11* [ode] . . . so that the seasons might be foretold, and famines averted on the earth" (Oahspe, Book of Osiris 12:3).

Nothing could be more pragmatic than the stability of the food supply. The Egyptians, as Diodorus reported, "not infrequently foretell destruction of the crops, or on the other hand, abundant yields, and pestilences . . . and as a result of their long observations they have prior knowledge of earthquakes and floods, of the risings of comets, and of all things which the ordinary man regards as quite beyond finding out" (Lindsay 1971, 143).

The pharaoh sent his wisest mathematicians into the far-off lands of the earth to observe and enumerate the winds and droughts and seasons and fertility and famine and pestilence and all manner of occurrences in the different regions. And when they returned to Egypt, he had them compare their collected data with their own

---

*See appendices B and D for details on these sacred numbers.

accumulated records, comparing one year with another, 11 with 11, 33 with 33, and so on for thousands of years.

OAHSPE, BOOK OF WARS 50:19

The Egyptian's "times of eleven" have today been rediscovered and labeled "the sunspot cycle." But we do not need the Sun to understand the "times of eleven," for the "ode" originates right here in the Earth's own energy field (EMF), its vortex, whose thermosphere follows an eleven-year rhythm. Extremely long atmospheric waves recur like clockwork at the height of the vortex's eleven-year cycle. Science writer Richard Milton talks about how chemical activity (specifically, the rate at which bismuth oxychloride forms a precipitate) varies with these changes. Fluctuations in the rate of the reaction are related to "changes in the Earth's magnetic field . . . the reaction time varies regularly with the eleven-year cycle of sunspot activity" (Milton 1996, 40).

Yet I doubt that pseudosunspots or solar flares 93 million miles away are hurling supercharged particles at the Earth's magnetosphere. These "spots" or blemishes that appear to obscure the Sun's face may be an optical illusion, created by dead (nonradiating) planets in the solar system—there are millions of them. Alternatively, they may be caused by dark nebulae in belts not very far above the Earth itself, or by planets orbiting the photosphere of the Sun. Within the Sun's photosphere are planets, some of irregular form, which have been mistakenly called "sunspots" because when they present their negative surface toward the Earth they seem black.

No, the correlation between sunspots and earthly events is better understood as the rhythmic ode, the eleven-year vortexian cycle, which, at max,* is known to produce, for example, an abundance of insects, simply due to the extra warmth and moisture generated at the height of the cycle (see page 213). (The vortexian lens, when compressed at max, gives a warmer/wetter season.) And it has nothing to do with the Sun—

---

*It may simply be the heightened effectiveness of the vortexian lens at max that allows us to see the "sunspots."

but everything to do with our own EMF—just as Richard Milton concludes in his remarks concerning those bismuth experiments: "It is the water, not the other chemicals involved that is sensitive to electromagnetic fields. . . . All living organisms—whose bodies consist of chemical reactions taking place in water—are in some currently unknown ways capable of being affected by electromagnetism" (Milton 1996, 40).

The vortex's eleven-year cycle has been found to correlate with the French and Russian revolutions, both world wars, business cycles, the ups and downs of the stock market, car accidents, increased births, and sudden deaths. Even the level of lymphocytes in blood may be affected by it. When EMF magnetism is heightened, atmospheric atoms are ionized, affecting the body's electrical system and blood pressure.* In the very same way, magnetic storms, resulting from the compressed shape of the vortexian lens at max, tend to coincide with more frequent heart attacks, accidents, suicides, increased admissions to psychiatric hospitals, and so forth. All because human behavior is susceptible to ions in the Earth's own atmosphere (not the Sun or the stars). *Ionized* means electrically charged,[†] with vortexian energy. In other words, these are not "astrological" but bio-meteorological influences, akin to the atmospheric cycles that affect plants and animals.

The central nervous system is sensitive not only to electromagnetic changes but also to those atmospheric densities called a'ji. As noted elsewhere, the Earth is constantly traveling through the galaxy (c'vorkum, motion #4); in so doing, our planet's particular location at any given time can affect the speed of chemical reactions inside a beaker[‡] (Gauquelin 1967, 218). The ancients studied these regions of

---

*Medical researchers now realize that the brain "reads" the fluctuating field strength of the Earth's magnetosphere. (As the hippocampus is stimulated with negative-ion discharge, the brain's alpha frequency deepens and the mind may experience time-space alterations, apparations, or other psychospiritual phenomena.)

†Ion therapy, popularized in the 1990s, is known to produce euphoric states. An increase of positive ions tends to trigger human tension and irritability, as observed when the Alpine foehn winds blow.

‡The radiant energy in spectral regions—ultraviolet, for example—can cause chemical changes, which, in turn, relate to the speed of photochemical reactions.

atmospheric densities (a'ji and ji'ay), which recur along the pathway of the solar phalanx; the fall of a'jian dust from interstellar space was known to bring darkness, cold, and pathogens. Researcher James McGill has told me that by checking the year you were born on the worldwide historical temperature charts, you get a more accurate assessment of your personality than by checking your horoscope because "periods of warm and cold are based on cosmic (C'vorkum) cycles of light (dan) and darkness (a'ji), which are unseen causes of man's inspiration and behavior" (see his website, www.studyofoahspe.com).

Nevertheless, the actual effect of a'ji on human behavior depends very much on our own spiritual light, measured by our grade, which is rated lowest at one and highest at ninety-nine, reflecting basically the selfishness (grade 1) versus the selflessness (grade 99) of a person or even of a people, as described in Oahspe:

> The prophet is enabled to determine, by the vortexian currents [of the ionosphere], the rise and fall of nations, and to comprehend how differently even the same [a'jian] showers and shadows . . . will affect different people. And the same rules apply in the manifestation of Dan [Light]; according to the *grade* [my italics] of a people, so will they receive its light. If below thirty-three, they will become magicians and prophets without virtue; if above thirty-three, but below sixty-six, they will become self-opinionated malefactors. . . . But if above grade sixty-six, they will become true prophets. . . . As man calculates motions of the corporeal earth and foretells an eclipse, so shall you calculate the [grade] of man and nations, and the vortices of the unseen worlds, and foretell coming events, and cause the hidden things of the earth to deliver up their secrets.
>
> OAHSPE, BOOK OF COSMOGONY AND PROPHECY 8:10
> AND BOOK OF KNOWLEDGE 3:95

Our forebears did not exactly figure all this out by themselves. In days of old, this knowledge of atmospherean currents was given to humans through oracles.

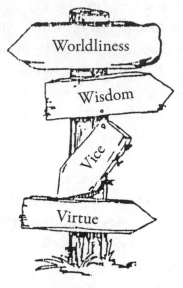

Jehovih created man to go as readily down the mountain as up it.

Fig. 5.2. We make our own grade in the here and now, according to our choices.

THE FALL OF A'JI

Algonquin: Let man build consecrated chambers that My spirits may come and explain a'ji, and they shall be provided against famine and pestilence.... Save your prophets understand a'ji, they cannot tell what the next year will be.

OAHSPE, BOOK OF SAPHAH: TABLET OF SE'MOIN 56

Thus, in the New World, sitting in their sacred round huts, originally called *hoogadoah,* later the Navajo *hogan,* the Algonquins made estimates based on the fall of a'ji (space dust). These were the "weather forecasts" of the day, predicting seasons of drought and seasons of disease, the latter referring to airborne viruses that do not spread from person to person but fall simultaneously as microorganisms from space. Fred Hoyle believed that complex organic chemicals found in meteorites exist in deep-space clouds, and he suggested that such cometary fallout may cause epidemics, just as this fallout was the harbinger of sickness, famine, and war in times past.

Times of war were part of the calculus of a'ji:* "In the years of a'ji, man's minds and hearts lose sight of heavenly things . . . and mortals become warriors" (Book of Saphah: Tablet of Se'moin 56). To the ancients, meteors and comets, which release a'ji, were omens of war and disease. Manilius ascribed the plague at Athens during the Peloponnesian War to a comet. In Rome, the meteors that struck in the Octavian War were considered harbingers of great calamities. Their sages were keen to note the comets that heralded the Battles of Philippi and Actium. Later, in the seventeenth century, comets were feared as evil omens, and when three were seen in 1618 they were thought to signal the onset of the Thirty Years' War, which indeed broke out later that year.

Much earlier, literal darkness had loomed during the backsliding of the followers of Moses and Capilya. Four hundred years after the time of these prophets, the chosen of those countries lost faith. This period was contemporaneous with a long time of partially interrupted darkness that was "cosmological as well as spiritual."

And God foresaw that the travel of the earth would cause her to pass through an a'jian forest for four hundred years, and that darkness would be upon the lower heavens. . . . [Therefore] compute the regions of the earth where it will fall most; and having determined, go to those mortal prophets who are in su'is [clairvoyant], and cause them to prophesy to the inhabitants of the earth.

OAHSPE, GOD'S BOOK OF ESKRA 9:3, 7

As predicted,

When the second shower [of a'ji] of a dozen years had fallen, mortals in many nations of the earth rushed into war.†

OAHSPE, GOD'S BOOK OF ESKRA 11:5

*This reminds me of Wheeler's analysis wherein periods of cold and drought (i.e., times of a'ji) correlate with revolutions, strife, and unrest (migrations). "The periods of deficient rainfall . . . during the nineteenth century were marked in various parts of the world by rebellions, wars, and shiftings of populations" (Haddon 1911, 14).

†Chronicles from various periods mention the coincidence of dim light and war, such as the medieval battles of Liegnitz and Muhlbach, during which the Sun was obscured.

But for another period of seven hundred years, when there was no a'ji, war ceased completely on Earth, people were gentle and killed not any living thing.

Even the peasants of the Dark Ages knew about this poisonous dust that brought strife, bad weather, and pestilence. Was the Black Death of the fourteenth century (as astrology might say) caused by a conjunction of Saturn, Jupiter, and Mars in Aquarius? I don't think so. Rather, it broke out during a darkened time, just as ancient Mexican chronicles report that pestilence took the lives of a great many people during a time of dim light.

To sum up: a'ji brings three primary atmospheric conditions: darkness, cold, and density (space junk, i.e., potential pathogens). Raymond Wheeler (whom we met in chapter 4), after studying two thousand years of historical records, postulated a hundred-year climate cycle that affects human behavior, the cold phase of which makes us more aggressive, fanatical, extravagant, decadent, superficial, and skeptical. But the warm phase of the Wheeler cycle brings better self-control, more energy and enthusiasm, as well as better judgment.

This warm phase (more light) seems to correspond to James McGill's "up spikes," in his study of the most generous, compassionate people in America. Eight out of eleven of the philanthropists he surveyed were born in an up-spike year. Out of the box, McGill calls this "a new form of zodiac." He explains that a human baby is protected from a'ji while in the fluids of the womb and is not exposed to a'ji until the moment it is born, at which time the DNA in nebula (a'ji) makes its impact. Among down-spike births, occurring in years of aji'an cold and darkness, McGill found more violence and destruction. Out of twenty randomly selected serial killers, he found that fifteen of them had been down-spike births. Looking also at twentieth-century race riots in America, McGill found more than 83 percent occurred during down-spike years.

In Roman times, according to Heraclides of Pontus, the skywatchers of Cea used similar guidelines, carefully observing the Dog Star, for if it rose "with an obscure and dim appearance, it proved that the atmosphere was gross and foggy . . . heavy and unwholesome." But if bright and clear, "then that was a sign that the air was light and pure, and therefore healthful" (Cicero 1997, 198).

Philostratus the Elder wrote of the prophetic powers of the little people (the Ihin wise men and astronomers; see chapter 8), who could herald the coming of the great floods of the Nile. Aristotle also believed in the existence of this sacred race who inhabited the marshes of Upper Egypt. After they migrated out of Egypt to become the Picts* of Scotland, they became Europe's legendary fées (little and fairylike) who could correctly predict years of famine. Among the Swiss, these clever little people predicted plague and war; to the Bretons, they gave science and prophecy.

They led not only Europe but all the world. The branch of these little people who inhabited North America (the moundbuilders) is remembered on petroglyphs in caves near Bighorn Wheel in Wyoming. The medicine wheels had to do with astronomical turning points and were essentially their calendar for sowing, harvesting, and so forth. Later, the Indian calendar itself went by the *moons*, not in any astrological sense, but environmentally; that is, how it would affect the people as a whole—the Snow Moon (January), the Hunger Moon, the Awaking Moon, the Grass Moon, the Planting Moon, the Buck Moon, the Heat Moon, the Thunder Moon, the Hunting Moon, the Falling Leaf Moon, the Beaver Moon, and the Long Night Moon (December).

Likewise, the South American calendar was of practical—not astrological or personal—interest. The Amazonian Tukanos, for example, used their knowledge of the Sun, Moon, and stars to find the times and seasons of the Earth. One early observatory in Peru's Chillon Valley served as a kind of farmers' almanac, tagging key annual events to the movement of the sun and stars. Here the landmarks point to the rising Sun of the summer solstice, at which time the Rio Chillon floods mark the start of the growing season. When the floods subside at the

---

*The story of that migration is told in my book *The Lost History of the Little People.*

spring equinox, landmarks point to a certain constellation associated with rainfall and agriculture. Even the designs of Peru's famous Nazca lines* were "laid in order to show the paths of the stars, so that the agricultural work of the year might be organized in due season" (Charroux 1974, 33–34). In Peru, "the movements of the heavenly bodies were so important in relation to planting and harvesting of crops. . . . The fertility rites of the Incas, the planting of seeds, and the time of ploughing were all tied in with the calendar. . . . The heavenly bodies were very important in their daily lives" (Wilson 1975, 120–21).

In short, all of man's earliest surveys of the skies were profoundly pragmatic, focused largely on the agricultural calendar. In the same way, the Alexandrian book *Calendar* dealt with droughts, rains, winds, and storms, according to the rising and setting of the stars. So, too, were the stone circles of Western Europe's gigantic calendars. At Stonehenge, for example, priests announced the coming of the seasons and the eclipses of the Sun and the Moon. By the same token, stone spheres found in the jungles of Central America appear to have once been arranged to represent the solar system.

The Greeks, of course, were no less concerned with food supply; their myth of Demeter describes how famine was prevented. Told as a family drama, the story of Demeter's mourning for Persephone is a well-disguised allegory clothing the secret knowledge of their astronomer-priests, whose predictions, based on sacred time-counts, could forecast the vagaries of climate, averting not only crop shortfalls but also the pestilential pathogens of a'jian or ji'ayan skyfalls. The Eleusinian Mysteries of ancient Greece gave the town of Eleusis, northwest of Athens, its name. *Elefsis* literally means "devised of eleven." Eleven years (the "ode") was and is the base of prophecy (appendix B) and, as we have seen, it records the Earth's, not the outer planets', rhythms. At this celebrated shrine,

*"The significance of the Nazca lines" lies in their "elaborate astrological cults. . . . The astronomer-priests probably found that the more complex was their astronomical knowledge . . . the more they could impress the populace with their mysterious supernatural powers. Thereby they strengthened their privileged position. . . . The attempt to obtain the aid of the heavenly bodies to rule the world continued as an important endeavor of the state-directed priesthood" (Corliss 1978, 500–1).

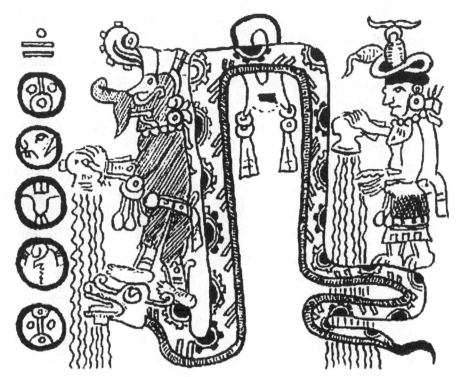

Fig. 5.3. Return of the rainy season, from the Troano manuscript. The Mayan drawing depicts astronomy's role in predicting the rainy season.

the Earth goddess Demeter inspired rites and initiations of the highest order. Their pagan mysteries were respected and observed even in the Christian world, for the Eleusinian Order retained the last traces of the wisdom-keepers, the teachers of old who knew the prophetic numbers.

Demeter's rites were secret; initiates were given a password. Demeter not only brought the blessings of agriculture (cultigens), but, more significantly, knew the mysteries of cycles of growth, the seasons of the Earth, a recondite calendar. She gave not merely seeds (Outer Court), but the agrarian secrets (Inner Court) of her forefathers. She gave prophecy. She gave sky-time. For the Greeks had borrowed Demeter from the Egyptians, who were the cleverest nation in the world. There along the Nile they built star temples* wherein

---

*One such temple was rediscovered at Merowe, an old Nile city, during the Garstang-Sayce excavation of 1910, which found a building that proved to be an astronomical observatory.

astronomers plied their formulas, producing estimates that predicted the heat and cold of seasons, the wet and dry. And they put the people to work, accordingly, on canals, irrigation, horticultural schemes, granaries, as well as on strategies for disease prevention—all things for the benefit of the nation, the group, the people as a whole. Not the individual. Applied astronomy, not personal astrology (see fig. 8.6 showing a star temple, page 344).

And this was all based on revealed cycles of time: the prophetic numbers, starting with the ode. Although the Greeks tended to use the words *astronomia* and *astrologia* almost interchangeably, let us not confuse astronomy with astrology, the latter belonging to the star-worshippers and cults that took hold more than 7 kya.

To the Persian Zoroaster is ascribed a table calculating the Moon's rising and setting, making astronomy at least 8 kyr. It was at that time that King Asha spoke to the prophet Zarathustra (a.k.a. Zartusht and Zoroaster), saying: "I am old now and have observed thousands of men, yes, even kings and queens, as to whether the stars rule over them. . . . Behold, I have cast the horoscope a hundred times, a thousand times. I have proved all the stars in heaven and named them, and made maps of them. . . . Hear me, then, O Zarathustra. . . . In all the stars there is nothing but lies; nor does it matter if a man is born under this star or that star!" Then Zarathustra answered, saying:

> I declare to you, O king, that the corporeal man is . . . only half his own master; only half the controller of his place and behavior in the mortal world; nevertheless, he is the first half. . . . Do not think that spirits and Gods rule men as if they were slaves or toys; for another power also lies over man, which is neither spirits nor Gods nor stars, nor moon nor sun; but the corporeal surroundings that feed his earthly desires.
>
> OAHSPE, BOOK OF GOD'S WORD

The Stars Never Lie!

Lucky you, Pisces
February 19–March 20

Your charisma kicks into high gear at the beginning of the week. A romantic interlude can deepen love, if you give your heart free rein. Your sexy side comes out to play midweek. Look for romance during a healthful activity. Open the door to love and passion.

Fig. 5.4. Cartoon by Marvin E. Herring.

Zarathustra, in other words, was the first sociologist, instructing the king to consider the foremost influence of one's immediate surroundings.

Yet it came upon the minds of humans that the heavenly bodies were divinities of a sort, with power over them. Could something be done to influence those powers?

The priests of Babylonia, Egypt, India, and Mexico and then the medieval sorcerers practiced an astrological magic in which the angels of planets and stars were invoked [while] burning specific incenses. . . . These astral worshipers maintained that the twelve zodiacal houses were thrones of celestial hierarchies, or superior intelligences. (Tomas 1973, 119)

WORSHIPPING THE STARS

As the culture regresses, the stars become gods.
CHARLES BERLITZ, *MYSTERIES OF FORGOTTEN WORLDS*

Astronomy was perverted to astrology
GUS CAHILL, *DARKNESS, DAWN AND DESTINY*

Cuneiform texts show the stars were worshipped as gods. These star-worshippers burned incense to the Sun, the Moon, and all the planets, as if they were gods. The official star cult in biblical Palestine worshipped Mars, Jupiter, Venus, Mercury, and Saturn, just as the Assyrians had worshipped Saturn (*Nisroch*) and the Greeks worshipped the Moon as the goddess Selene and the Egyptians worshipped Jupiter as the god Ammon. This idolatry went hand in hand with the despotic kingdoms of the Neolithic.

> By calling the planet Mercury the Star of Hermes, instead of by its older name of *Stilbon,* "twinkling star," Plato showed that the Babylonian custom of naming the heavenly bodies after the gods (a custom that gave rise to astrology in the first place) had already reached Greece in his time.* Plato's contemporary, the astronomer Eudoxos of Knidos, warned his pupils against the Babylonian superstition of astrology that was beginning to corrupt astronomy (de Camp 1970, 245–46).

This Eudoxus, as Cicero later recounted, had been

> in the judgment of the greatest men, the first astronomer of his time. [He] formed the opinion that no credence should be given to the predictions of the Chaldeans in their calculation of a man's life from the day of his nativity. . . . Scylax of Halicarnassus, a first-rate astronomer, and chief magistrate of his own city, likewise rejected all the predictions of the Chaldeans. (Cicero 1997, 236–37)

Despite this, Michel Gauquelin says, "It is no exaggeration to say . . . that astrology was . . . the first religion and the first science man

---

*Plato lampooned the astrologer: Looking aloft, he falls down a well.

developed" (Gauquelin 1967, 5). But this *is* an exaggeration; in fact, it's a flaming falsehood. Modern writers, so influenced by the idea of Darwinian progress, tend to portray bumbling astrology as eventually replaced by the more scientific discipline of astronomy. But quite the reverse: astrology appeared as a popular corruption of astronomy; this happened when the Persians, then the Chaldeans (and then the Greeks) began to attribute to the zodiacal signs a mystical power over our personal destiny. This idolatry was remembered in the Holy Bible as the Tower of Babel (Babylon's ancient name). Isaiah 47:13–15 declares:

> Let your astrologers come forward, those stargazers who make predictions month by month, let them save you from what is coming upon you. Surely they are like stubble; the fire will burn them up. They cannot even save themselves from the power of the flame. . . . Each of them goes on in his error; there is not one that can save you.

Astronomy, as Cyrus Gordon so aptly pointed out in his classic *Before Columbus,* fed the astral superstition that we call *astrology,* which retains its popularity, unabated to this day in an age that we like to call "truly scientific" . . . (lest we scorn the ancients for their superstitions, let us remember that our newspapers to this day print more astrology than astronomy). No, it was astronomy, not astrology, that was the first science of humanity, even older than twenty thousand years. For example, all of Egyptian life revolved around the annual flooding of the Nile, which watered Egypt's croplands. Their skywatchers followed the rising of the Dog Star, Sirius, from year to year; it appeared on the horizon at the same time as the flooding of the great river. Did Sirius cause the Nile to rise? Of course not. The return of the same pattern of the stars simply signaled the recurrence of the season.

But they, too, regressed: after Osiris (ten thousand years ago), knowledge of the stars fell to idolatry. Here, as elsewhere, false prophets made the Sun and the Moon into divinities who guided the affairs of nations and men, marking out the destiny of things according to the dates of corporeal births. According to Hebrew lore, these false prophets were under the influence of fallen angels who

> set themselves as masters over the heavenly spheres, and forced the sun, moon, and the stars to be subservient to themselves instead of the Lord. . . . Armaros taught them how to raise spells; Barakel, divination from the stars; Kavkabel, astrology; Ezekeel, augery from the clouds . . . Samsaweel, the signs of the sun, and Seriel the signs of the moon. (Ginzberg 1961, 124)

> Attributing the highest central cause to the sun and stars . . . [they] reasoned that if the sun made winter and summer . . . so it ruled over men [as well]. And the temples built to observe the stars . . . became the places of decrees to horrid deaths of all who taught of or believed in spirit.
>
> OAHSPE, BOOK OF FRAGAPATTI 2:12–13

"The gloomy Mayan religion," the brilliant polymath L. Sprague de Camp reminds us, "entailed a good deal of human sacrifice . . . and was closely connected with astrology, a well-developed pseudo-science that made much of the revolutions of the planet Venus" (de Camp 1970, 120). For their successors, the Aztecs, it was necessary to placate the Sun with human sacrifices, lest the world turn dark. In the North, the Pawnee Indians, like their Mayan confreres, sacrificed a young girl to the Morning Star, Venus, to "appease" it for its imagined destructions. The Babylonians also pleaded with Venus, the Queen of Heaven, to leave the Earth in peace, and as among other peoples of the world, this appeasement of that goddess/planet led to human sacrifices and bloody rituals. Nefarious astrologisms persist even among modern cults: "If a magician wants to turn the current of destructive energy associated with Mars against an enemy . . . he fills his mind with images of blood and torment . . . unleash[ing] all his own . . . hatred and violence . . . and hurl[ing] it against his victim" (Cavendish 1967, 7).

Thus did the Egyptians develop an elaborate Sun cult, casting aside the Great Spirit and instead becoming worshippers of the Sun, distorting not only the good science of spirit* but also losing the system of prophetic numbers (the Winter Tables). Before it was lost, though, the early Egyptians had studied the heavens for the rhythms that foretold weather, the best seasons for planting, the right moment for sowing, for harvest, for husbandry, the times of drought, warm and cold. They applied themselves to useful knowledge, the many aspects of timekeeping that affect the welfare of the people as a whole—meteor showers, cycles of growth, navigational factors, the cyclical alternation of wet and dry, atmospheric conditions, tidal currents, temperature fluctuations, floral and faunal cycles. All these were based on multiples of the eleven-year ode (sky-time). That cadence, affecting, for example, the dispersion of insects and the reproductive rates of fish, was known to them, just as today we have rediscovered the eleven-year cycle and its effects on plants, animals, and human beings. All this is bio-meteorology, not astrology, for it has nothing to do with the constellations, but everything to do with the Earth's own EMF and its phases. Nor is it "sunspots" that, as some believe, "influence the growing season," (Gribbin and Plagemann 1974, 126), but only the atmospheric changes of the vortex.

"MODERN" ASTROLOGY

Superstition is to religion what astrology is to astronomy:
the mad daughter of a wise mother.

FRANÇOIS MARIE AROUET VOLTAIRE

Astrology as today's leading school of occultism offers no apparent mechanism for stellar influence. Yet-to-be-discovered "vibrations" or radiations leave us hanging. Even writers who believe in astrology, like Paul Von Ward, admit that "no one understands how the energetic interactions between planetary . . . positions and human behavioral

---

*These events are addressed in chapter 7.

predispositions actually work" (Von Ward 2011, 170, 178). Yet he speaks authoritatively of "the effects of subtle energies on consciousness and behavior." But how is that possible when "the immense majority of cosmic particles do not attain the speed . . . which would enable them to penetrate the magnetic field of the earth?" (Charroux 1971, 151). The gravitational effects of the planets on a newborn child are so mild, says writer Malcolm Dean, "they don't even move one cell the length of its diameter" (Dean 1980, 61). The configuration of the personnel in the delivery room—and even the magnetic field from the overhead light—Dean argues, have more influence than any celestial governor. And even if some "radiation" from a planet does impinge on the newborn, how could such an effect persist throughout that person's life?

Moment of Birth

Astrology portrays the newborn as some sort of sensitive photographic plate whose destiny is decided by the stellar influences that converge at the moment of his first cry. But some have asked: Why not draw the horoscope from the time of conception, rather than birth? It is, after all, the moment when all the hereditary factors come into play, when new life is sparked in the seed of the womb. Indeed, Pope Francis says an individual life begins at conception. How, also, is premature birth to be regarded? Or one caused by surgical intervention? If it was the doctor's decision that determined the hour of birth, would this change the astral influences, making the child's whole future life an artificial one?

The moment-of-birth issue sets off another question: What about twins? Do they have the same destiny? Biologists say that when twins share an idiosyncrasy, it is inherited. Sociologists say that when twins share an idiosyncrasy, it stems from the family environment. Astrologers say that when twins share an idiosyncrasy, it is their shared natal sign. Gay and straight male twins (studied by Sven Bocklandt at UCLA) pose a stumbling block for astrology.

The argument goes all the way back to the Stoics, who pointed out the entirely dissimilar fortunes and lives of those born at the same moment. In one story told by St. Augustine in *City of God,* two children are born at the same time in the same household. One is born of the master and the other of a slave. The position of the stars do not change their fate. "Nobody," admits Gauquelin, who is otherwise pro-astrology, "has succeeded in showing similarities in the lives of people born the same day of different parents. . . . Social conditions explain the pattern of a life better than the stars do" (St. Augustine 1967, 77). This is what Zarathustra meant when he spoke to King Asha of the "corporeal surroundings" rather than the "stars, moon, or sun."

Yet astrologers contend that there is a mysterious correlation between the personality of human beings and the configuration of their native constellations and planets. Engaging the doctrine of universal sympathy, they also claim the planets correspond to particular stones, metals, colors, numbers, and parts of the body. The Greeks even correlated the seven planets with the seven vowels and the seven tones of the musical scale. Such thinking is the very antithesis (the opposite fallacy) of the sci mat's indeterminism and randomness, for astrologers believe that nothing is random, that all things and events are part of one grand design: "as above, so below." Yet the macro-micro parallels they offer are more like a system of magic or fatalism than any sort of science. They claim the universe "is a gigantic human organism and man is a tiny image of it. . . . In this unified magical universe, mysterious forces are at work. . . . The impulses that move man—love, hate, lust, pity— are found on a much greater scale in the universe" (Cavendish 1967, 6). Poppycock! Alluring but phantasmagorical, overplaying the micro-macro card, astrologers believe that the part reflects the whole. But, as Michael Van Buskirk sees it in *Astrology: Revival in the Cosmic Garden* (Buskirk 1978), this makes man a mere "pawn in the cosmos with his life and actions pre-determined and unalterable."

*Since mind is nonphysical, it must exist in a different
dimension from the space-matter-time continuum of cosmic rays and intergalactic forces and atomic rhythms.
Consequently, mind and consciousness must be independent of planetary force fields.*

DAVE HUNT AND T. A. McMAHON,
THE NEW SPIRITUALITY

Planetary exploration in our space age finds no vibrations, no "ghost
fingers," no such determinism and no appreciable influence exerted on
the Earth by the heavenly bodies. This was predicted 130 years ago in
Oahspe, which cautioned,

There is no attractive force from one planet to another or from
its own satellites. Corpor [matter] itself has neither attraction of
cohesion nor attraction of gravitation, nor has it propulsion. It is
erroneous to say that the presence of this planet or that throws an
influence on mortals according to their birth under certain stars.
. . . Philosophers seek first to find the cause of things in the sun,
or, if failing therein, turn to the moon; or, if failing here, turn to the
stars and planets. Yet in all things it is the Unseen that rules over
the Seen.

OAHSPE, BOOK OF COSMOGONY AND PROPHECY 6:11

I think we have mistaken the bio-meteorological rhythms of the
Earth's own vortex (the Unseen) for stellar influence. "Man has the tendency to look towards the . . . planets and stars, forgetting the vortex.
. . . It's a vortexian current which causes the course of the planets. It's
the vortex which influences people" (Hendricks 1980). In fact, it has
been said that man living above the sixty-sixth latitude in Alaska,
Norway, Finland, and so forth must be uninfluenced by the cosmos,
for it is impossible to calculate what point of the zodiac is rising above
the Arctic Circle. No planet is in sight for weeks. (Astrology began in
much lower latitudes.)

Medieval physicians were adept at divination, prophecy, and astrology to address the ills of their patients.

Fig. 5.5. Medieval physicians. Three cartoons by Marvin E. Herring.

The fault, dear Brutus, is not in our stars, but in ourselves.
WILLIAM SHAKESPEARE, *JULIUS CAESAR*

Again, in *King Lear,* Shakespeare let fly his abhorrence of the

> excellent foppery of the world . . . that we make guilty of our disasters the sun, the moon, and the stars; as if we were villains by necessity, fools by heavenly compulsion; knaves, thieves . . . drunkards, liars and adulterers by . . . planetary influence. . . . An admirable evasion of whoremaster man, to lay his goatish disposition to the charge of a star!

One welcomes such fatalism with open arms, for it gives us something else to blame, not only for our own shortcomings, but for all ills that befall us.* St. Augustine, in *Confessions,* railed against divination by the stars, especially when the astrologer claims that this or that fault comes from the conjunction of Venus with Mars or Saturn or whatever, thus absolving ourselves of responsibility, shifting the blame to the "ruler of the heavens."

The Moon and Mercury have no atmosphere. Venus's surface is 85 percent lava. Mars is geriatric. Uranus has boiling oceans. Pluto is not even a planet. "Pluto is a chunk of ice which controls nothing" (Adler 2006, 50). Still, astrologers say that tiny Pluto, close to the edge of the solar system, is just as powerful as Jupiter. Even itty-bitty asteroids are used in some horoscopes. Unlike astronomy and classical physics, the concept of planetary influence on humans fails to take distance from Earth into consideration. How can such faraway worlds have an impact on us? Jupiter is more than 750 million kilometers from us, Saturn 1.2 billion kilometers, Neptune 4.4 billion kilometers! Do you mean to tell me these far-off orbs control our destiny? Nonsense!

---

*You will find me bewailing the same misplaced fatalism in the next two chapters, particularly in chapter 7 where I tear into reincarnation, that other occult favorite, on similar grounds; that is, placing blame in all the wrong places.

[H]ow vanishingly small are the forces exerted by celestial objects on things and people on earth; and how very small are the amounts of radiation associated with them received on earth.

LAWRENCE BOK, *OBJECTIONS TO ASTROLOGY*

Isaac Newton, the founder of theoretical physics (and a working astrologer himself), understood the motion of a planet in terms of forces that act on it from other planets at a distance. Newton also posited gravity originating in the center of the Earth, spreading out from there into space, as imaginary lines of force pulling objects toward the Earth. Maxwell and Einstein corrected this. The force arises in the field itself, acting not at a distance but impinging directly from the ionosphere to the surface of the Earth, a centripetal drive—a push, not a pull (see fig. 1.6, page 54).

Since all known forces (gravity, electromagnetism, the weak force, and the strong force) diminish with distance, how could far-off Jupiter and Saturn, huge masses of rock surrounded by gases, have a "subtle" effect on human beings?

The same [erroneous] philosophy holdeth in regard to the sun, and to Jupiter and Saturn and Mars, and all other planets . . . [Consider] the attraction (so-called) between two earth substances, as granite or sandstone or lead or gold or clay or water; it is far less than between two steel magnets. Wherein it will be observed that it is *utterly impossible for any attractive force to exert from one planet to another or even from a planet to its own satellite.* (my italics)

OAHSPE, BOOK OF COSMOGONY AND PROPHECY 1:56

In time, I believe it will be found that vortexya—not the Moon—rules the winds and tides. Kepler came up with the first supposedly correct explanation of the tides—the motion of water toward the regions where the Moon stands in the zenith. But to Galileo, the Moon pulling

Moon~Tides

"steel magnet" ha ha

Moon

9,000 miles

A man may prophesy of the Moon by calculations of the tides. But to attribute to the tides the cause of the Moon's position would be no more erroneous than to attribute the cause of the tides to the Moon.

It is utterly impossible for any attractive force to be exerted from one planet to another, or even from a planet to its own satellite.

Under the most extravagant supposition of power, her [the Moon's] magnetic attraction is more than 200,000 miles short of the mark.

Were the Moon a steel magnet, it would not exert power more than 9,000 miles.

222,000 miles to the Moon

By the tides, he proves the cause of the Moon, or by the Moon the cause of the tides. In other words, anything under the Sun that is corporeal, rather than search in the subtle and potent unseen worlds.

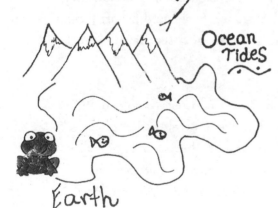

Ocean Tides

Finding a coincidence in the tides with certain phases of the Moon, they have erroneously attributed the cause of tides to the Moon, which is taught to this day as sound philosophy!

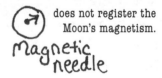

does not register the Moon's magnetism.

Magnetic needle

Earth

Fig. 5.6. Does the Moon in Aries cause discord among humans? Is the twentieth day of the Moon good for curing warts? I doubt it. The Moon is a passive body, merely a vestige of the condensation that took place when the Earth was formed. It has no power over other orbs.

the tides (from a distance of 240,000 miles!) did not exist at all, except as superstition. The same skepticism is found in Oahspe's Book of Cosmogony.

The cause of the tides is erroneously attributed to the power of attraction in the moon . . . which is taught to this day as sound science. . . . Were the moon a globe of magnetic iron ore, it can be shown approximately how far would extend its power of magnetic attraction. . . .

Under the most extravagant supposition of power, the Moon's magnetic attraction is more than 200,000 miles short of reaching the Earth. And even if the Moon did exert a pull,

Its magnetic attraction would not be on water [tides] or clay, but on iron and its kindred ores. [Surely] it would manifest more on the magnetic needle, or other iron substances, than on the water of the ocean. . . . Suspend a ball of magnetic iron alongside a suspended cup of water . . . [and] there is no magnetic attraction between them.
BOOK OF COSMOGONY AND PROPHECY 1:48–51

It is, moreover, unlikely that some subtle force of the Moon could pass through regions of immense turbulence, such as those that exist fifty thousand miles above the Earth. Finally, if all the Earth's weather comes from the lowest layers of atmosphere (the troposphere, no more than ten miles above ground level), why should the ocean tides not be subject to the lower vortex as well?

The sooner we scrap these implausibilities, the better will we grasp the workings of the Earth's own dynamo. The variations of ocean tides follow disturbances in the ionosphere,* the power station of the vortex. Researcher James McGill, in this connection, has reminded me of the Egyptians' thirty-three-year weather cycle, which is connected to the tides; and that the thirty-three-year cycle, called a "spell," originates in the vortex, not in the Moon. Scientists know it as the meteor cycle or even the drought cycle. The PDO is roughly a thirty-year cycle (see chapter 4 and appendices A and B).

---

*I believe earthquakes also follow ionospheric changes, but no editor (or geoscientist) will hear me out.

Where they have prophesied ebb and flood tide to be caused by certain positions of the moon, they have erred in allowing themselves to ignorantly believe the cause lay with the moon. A man may predict what time a traveling wagon will reach town by its speed of progress; but the correctness of his prediction does not prove that the wagon pushed the horse to town. . . . It is equally erroneous to say that the presence of this or that planet throws an influence on mortals, according to their birth under certain stars. It is this same astrological ignorance that attributes to the sun the throwing off . . . of rings to make planets of.

OAHSPE, BOOK OF COSMOGONY 3:17 AND 6:11

The power of a magnet decreases in proportion to the square of the distance from it.

Myth # 1

Even if the Moon was made out of carbon steel, it could not exert its magnetism more than nine thousand miles.

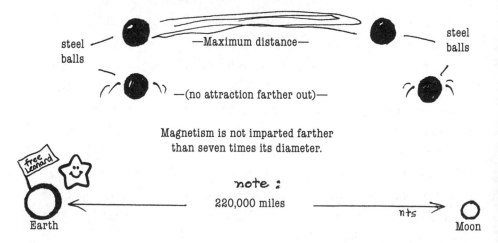

steel balls — —Maximum distance— — steel balls

—(no attraction farther out)—

Magnetism is not imparted farther than seven times its diameter.

note :
220,000 miles

Earth — nts → Moon

Fig. 5.7. Moon tides. Could the fair Moon, so far aloft, raise the mighty tides of Earth and bend her crust?

We hear, for example, of Earth "torn away from the young Sun by some passing star" (Gamow 1948, 191) or breaking off from Jupiter (Velikovsky 1965), or that our Moon was formed from a giant bulge of land broken off the area that is now the Pacific Ocean (site of the submersion of Pan!).

Despite the Moon's impotence, lunar effects are often cited as proof of astrology's claims. Is a woman's menstrual cycle (averaging 28 days) governed by the Moon's cycle (whose phase is 29.5 days)? Estrous cycles in the animal kingdom are actually quite variable: 11 days for cows, 24 days for monkeys, 37 days for chimpanzees, 5 days for rats, and 28 days for opossums. Well, the 28-day period for the opossum is thought to be merely a coincidence, "and if it is a coincidence for opossums, why not for humans?" (Frazier 1986, 233).

Rather than the Moon, shouldn't vortexian science be factored in? Behavioral extremes, in other words, may swing with the force field and its fluctuations. A greater influx of vortexya at certain specified times may be the unseen cause of myriad reactions. The so-called lunar effects may have more to do with Earth's atmosphere/vortex, which circles it once a month. The question of "lunacy" is one of the most popular superstitions of America's folk science, claiming the Moon's influence on the human psyche. Moonstruck: one often hears mention of the lunar cycle triggering madness and other human disasters—suicides, assaults, accidents, mania, crime, depression, and so forth. All these calamities supposedly increase with the Moon's enhanced phase. But can we really blame the breakdown of normality on our rocky satellite? In a sober piece of research, no evidence was found linking the phases of the Moon to homicidal attacks (in Ohio, during the 1970s). Indeed, "nationwide data for both homicide and suicide show no relationship to lunar phase . . . [while] a survey of the literature found that no conclusive statistical evidence existed . . . [for] any kind of lunar effect on human behavior" (Sanduleak 1986, 238–40).

Concerning the incidence of crime, James McGill, in his "new zodiac" based on a'ji (cold and dark times), has taken a look at homicide rates in the United States from 1900 to 1998, comparing them to worldwide temperature. His conclusion is that a'ji is one unseen trigger of aggressive and violent behavior. In his survey, years of homicide peaks coincide with down-spike years (in temperature): 1907, 1911, 1917, 1933, 1945, 1974, 1981, and 1992. This shows some agreement with Wheeler's scheme, which sees a'jian cold linked to revolutions and other instances of societal unrest.

Full Moon madness, I venture to say, may be largely encouraged by its own "mystique": built-up tensions find their outlet at the socially expected time. That depends, of course, on the culture. Americans are a bit Moon-haunted; it is a cultural thing. If in New York City there are more births just after the full Moon, in Germany, researchers have found no relationship between phases of the Moon and number of births.

In many lands, the full Moon marks a time not of lunacy, but of celebration. In Cambodia, full Moon is taken as an omen of good luck. In Thailand, Buddhist monks celebrate Makha Bucha on the full of the Moon to commemorate the life and teachings of Buddha. For some Amerind tribes, like the Yuchis, the time of the full Moon is devoted to group participation in festivities and rituals—not a time of lonely craziness, but of singing, dancing, and rejoicing. The Siamese Semang people dance every month at the time of the full Moon. In the Congo, pygmies put on a great feast at the time of the full Moon.

The phases of the Moon (Mas) were honored by the ancients, each phase marking a change of the angelic guard, a change of the watch. The tablet of Se'moin in Oahspe's Book of Saphah, which succinctly records prehistorical terms, names, and rites, reads:

> M'git, the third quarter (Panic). A holy day of rest. Let My chosen keep the four holy days of rest during each moon, for on these days My guardian angels change the watch.... Sub'da'don, a holy day (Panic). The day of the moon's change. Sub'da (Fonece). Sabbath (Ebra). A

moon's birthday. Because each moon has four quarters, so I give to you four Sab'da (holy days of rest), which shall be days of worship (Abram).

OAHSPE, BOOK OF SAPHAH

A cursory comparison of different traditions also reveals that the symbolism of the stars and constellations varies from culture to culture, meaning there is no inherent connotation of the signs. Capricorn, for example, was a goat for the Chaldeans, a bear for the people of India, and a unicorn in China. The Great Bear constellation was seen by the Egyptians as a bull. The constellation Taurus, the bull, was a boar in ancient Turkey.

There are also different zodiacs. The Etruscans, for example, divided the heavens into sixteen parts (see figure 5.8). Even today, some systems of astrology are diametrically opposed to others. A European astrologer, for instance, interprets a horoscope differently from a Chinese astrologer. Neither do astrology columns agree on what is supposed to occur, nor do horoscopes agree about "the proper method of drawing the houses in the chart. Some use the system invented in the eleventh century, others the system invented . . . in the fifteenth century. . . . The inevitable result is that different astrologers give different interpretations of the same planetary positions" (Cavendish 1967, 201).

How's your love life? The vaunted compatibility of certain signs seems to have backfired. When 3,000 happily married couples in Michigan were studied, along with another 470 couples who had divorced, "those born under 'compatible' signs married and divorced just as often as those born under 'incompatible' signs" (Kurtz and Fraknoi 1986, 220).

Career predictions fare no better. Statistical analysis, using the chi-square test, failed to confirm astrology's claim that certain sun signs enhance leadership abilities. Although Mars, the supposedly "bellicose planet," did score well for leadership (in a George Washington University study of enlisted Marines), that planet's reputation for

Fig. 5.8. The Etruscan dome was divided into sixteen parts.

courage and aggression did not hold up. No such correlation was found. Mars, incidentally, was not originally associated with war; among the ancients it was thought of as the wise guardian of the solar system.

Nor have politicians and scientists matched up with the expected signs of their alleged planetary benefactors. In a Case Western Reserve University study encompassing sixteen thousand–plus scientists and six thousand–plus politicians, "the distribution of these signs were as random as for the public at large" (Kurtz and Fraknoi 1986, 220). Also falling into a random pattern were the results of a Harvard study conducted by astronomer J. Allen Hynek, based on birth dates drawn from *American Men of Science*.

One survey, after consulting volumes of biographies, put the "Libra = artistic" claim to the test. With more than two thousand famous painters and musicians under consideration, it was found that Libra did not rule over these people more than any other sign. The hoped-for correlation did not exist. Libra actually had a smaller complement of artists than other signs. Michel Gauquelin, who quotes these results, himself surveyed the zodiacal signs of journalists and other professional groups, finding "no evidence of seasonal patterns" (Gauquelin 1967, 80–81). His response to another researcher (Edmund van Deusen), who had matched up journalists to the sign of Scorpio, was that van Deusen had not corrected his data for the effects of socioeconomic levels and country of birth, factors that must be taken into account before a genuine cosmic effect can be stated with certainty.

Nevertheless, averred van Deusen, in defense of his craft, "Arians *are* self-centered" (quoted in Dean 1980, 110). Hey, everyone is self-centered to some extent.* As in most horoscopes, that insight is pretty generic: "Today is a good day to make friends"; "Make personal needs your own private business"; "Keep emotions and money where they belong"; "Don't be rushed into decisions"; "This is a good time to face limitations"; "Avoid self-deception"; "Play it safe"; "Protect home and family." Such is the laundry list of generically sound advice offered by astrology.

In one daring experiment, an identical horoscope was mailed to over one hundred people, regardless of their natal information. Many recipients nonetheless admired the pertinence and exactitude of the reading, cheerfully fitting the information to their own experience. It is like seeing the features of the parent in their child, only to find out that the child is adopted. This sort of perception has a name: analytic overlay (AOL). Also known as "The Barnum Effect," it is the tendency of test subjects to agree with an evaluation, even if it is a cliché of a general truth.

In a similar experiment, Gauquelin placed an ad in a magazine for a totally free personal horoscope. Every one of the five hundred

---

*Doesn't the narcissistic streak in astrology bother you? Einstein called it "the bondage of egocentric cravings." Ego-astrology, as I understand it, is not the road to self-knowledge or self-improvement. Honesty is. Conscientiousness is. Taking responsibility is. This means admitting our own weaknesses and doing something about them.

respondents received the same reading, which was in fact the chart for "DP," a notorious French serial killer whose horoscope had been generated by a computer program. (It spoke of his "instinctive warmth" and "right-minded" character.) Of the five hundred recipients, those who replied said they found the horoscope to be 94 percent accurate—AOL.

There are times, though, when an astrologer's accuracy does seem almost uncanny. Here, I think, the intuitive factor comes into play— a certain acumen and gift for interpretation. Many astrologers, I have observed, are acutely psychic, but give their clever "reading of the stars" all the credit for their insights. The "chart," as I see it, is only a prop. The reader's own perceptiveness allows her to tap into the psyche's hidden motives. And to this extent, the reading may be therapeutic. Counseling certainly has its place and its benefits, just as the "talking cure" of psychoanalysis does.

Nevertheless, there is something extremely outdated about astrology, which developed at a time when we had less insight into how heredity and environment shape a person's character. Rooted also in the obsolete school of predestination, astrology embraces a kind of fatalism that has long since been discarded in favor of the overwhelming interplay of genes and surroundings (nature and nurture).

In chapter 2 we saw how we are supposedly shaped by natural selection. In this chapter, we saw how we are supposedly shaped by the stars. In chapter 6 we will see how we are supposedly shaped by the unconscious mind (another school of fatalism). The famous psychoanalyst Carl G. Jung studied astrology for thirty years. It was more than a passing interest of his. In 1911 he wrote to Sigmund Freud about his deep involvement in horoscopic calculations. To Jung, the signs of the zodiac were "libido symbols."

After Jung's death, an art historian who had met Jung wrote of their meeting in the 1930s. Jung had told his visitor that he had learned a great deal about himself by studying his natal chart and that he often recommended it to his patients. Yet he admitted that it was merely a device, a means of projecting the imagination onto certain images, like using coffee grounds, tea leaves, a crystal, or even a Rorschach test. But, laughing, Jung said that if he admitted that to his patients, it would not work.

His visitor then replied that since he was not a patient of Jung's, the latter could drop "the mystifying language" in which he conveyed astrological insight to his patients. Jung disagreed: "What was good for his patients was good for everybody, and if I [the visitor] declined to calculate my own horoscope, this merely showed that I had a resistance to learning to know myself a little better" (Storr 1996, 102). But Jung himself put up a fierce "resistance to learning" about the *spirit world,* as our psychic safari of the next chapter reveals, moving further into the hinterland of mind.

CHAPTER 6

THE UNCONSCIOUS

The Wizards of Id

The unconscious mind, as one disciple of psychoanalysis grandiosely describes it, "opens before us an immense and unexplored new field of realizations, within which objective scientific investigation combines in a strange new way with personal ethical adventure" (von Franz 1964, 253), only confirming the belief that they don't call behavior science BS for nothing. Freudianism, an "ingenious fiction" and a "precious illusion" (to quote Richard Webster 1995, 180, 245), gives us the Almighty Unconscious but dares not give us a soul. Trading light for darkness, this doctrine has us governed not by reason or insight, but by a host of murky forces beyond our control. Isn't this just like the priestcraft we noted in the previous chapter whose wizards we must consult in order to divine the secret of our fate?

Why not just call those murky forces emotions or urges? Why the mystique? Why have natural emotions (fear, anger, frustration, hurt, guilt, distress) been branded so undesirable and abnormal? And why are they made to be hidden, a mass of totally unconscious drives like the id (the pleasure principle consisting of dirty secrets that survive from infantile impulses)? Blind raging instincts. A chaos of fixations. The purported insight gained from these unconscious drives "seem more examined than lived . . . spawning more recrimination than understanding" (Gross 1978, 11).

Disciples follow in awe of this unprovable "vast darkness of our unconscious mind," declaring, idiotically, oxymoronically, that "most of our knowledge is unconscious"—an obvious contradiction. *Unconscious*—last time I checked—refers explicitly to the absence of intellectual or cognitive functions.

Take dreams. To Freud, dreams were famously "the royal road to the Unconscious." Dreams, it is argued, prove just how deep the unconscious mind is. And not just dreams—almost everything in the family of the paranormal has now been chalked up to the unconscious by the experts. Thus have most behavioral "inexplicables" been laid at the door of the meaningless unconscious, while the soul force and the unexplored powers of the psyche are all but forgotten.

Carl Jung contended that "a medium in trance communicated the . . . personifications of unconscious personality elements" (Jaffe 1978, 30). By contrast, psi researchers have been able to show that the medium is, telepathically, conveying some other personality, that is, not self. Jung had to admit that the "question of ghosts and spirits" has not been settled, particularly with regard to the strange phenomenon of poltergeists and PK (psychokinesis). While conventional theorists were happy enough to label such fantastic episodes "exteriorized nerve force" (whatever that means), the question still remains: Why do these phenomena—bells ringing, lights going off and on, things breaking or moving—so often occur when someone has died? The "unconscious" cannot help us here, though Jung did try, arguing that when the emotional attachment to a deceased person "has been intense, the image [apparition] remains alive and forms a spirit" (quoted in Ebon 1978, 116).

But no, that's not how spirits "form." Strange lights in a room where someone has recently died is the handiwork of the deceased. New powers come into being after dropping the body and becoming a pure spirit. "When the dining room lights flicker, the family knows it is their dead father" (Stillman 2006, 116–17). Likewise, in England's well-documented Borley Hall incidents, "The only way for a spirit . . . to get into touch with a living person was by causing some violent physical reaction, such as the breaking of glass" (Price 1940, 50, 61). Experienced ghosthunters are not surprised to find more extreme ramifications in cases of murder.*

"The poltergeist's violent action can be their [the victim's] cry . . . a poltergeist . . . can be a deceased jealous relative or partner. Usually a medium has no difficulty finding the identity of the culprit . . . and placat[ing] the spirit . . ." (Davison 2006, 237–38). In this connection, we might note that the word *niniki,* among the Mimika Papuans, means both ghost and ancestor, an equivalence that is common among many of the world's people. The Yoruba shaman or *babalawo* (father of mysteries) uses divination to discover the identity of offending spirits, usually deceased elders, who have invaded the victim's mind. The *babalawo* does not explore the victim's childhood or unconscious.

In the well-known case of "Robbie" (the original subject of the popular film *The Exorcist*), the boy was, according to parapsychologist J. B. Rhine, a victim of classic poltergeistery: fruit flew in the air, tables tipped over, a heavy dresser slid across the room. Still, the official explanation was "a *psychosomatic* disorder and some kinesis action that we do not understand" (my italics) (Allen 1994, 10, 15, 222). *Psychosomatic,* like *exteriorized nerve force,* is just an easy label. Do these evasive terms shed any light on the problem?

THE AGE OF SPIRITUALISM

> *Is it the uncanny feeling that nonphysical intelligences might in fact exist, that causes the scientific community to strain so hard to find an [alternate] explanation?*
>
> DAVE HUNT AND T. A. McMAHON,
> *THE NEW SPIRITUALITY*

When a group of more advanced excarnates were given a chance to communicate, as documented in the little-known nineteenth-century booklet called *Spiritalis,* they declared: "We have pounded on walls, and caused huge tables to tip . . . and after exhausting ourselves, we have

---

*Chapter 7 goes into much greater detail concerning the spiritual repercussions of violent or sudden death.

flitted silently amongst you, and heard you ignore us or call us the devil . . . or mere will-power, whatever that is" (Newbrough 1874). These messages from beyond came through in 1874, at the height of the Great Age of Spiritualism (ca. 1848–1881).

Thomas Edison (born in 1847 to parents who were spiritualists) once said that he was inclined to believe that our personality hereafter will be able to affect matter. Perhaps Edison proved in death what eluded him in life. At the time of his passing, on Sunday, October 18, 1931, at 3:24 a.m., three of Edison's associates, in their own homes, noted that their clocks had stopped—at precisely 3:24. Edison had been fond of the popular song "Grandfather's Clock," with this line: "But it stopped, short, never to go again, when the Old Man died." The grandfather clock in his own laboratory stopped that night at 3:27. At work, his people had always called him the Old Man (Ebon 1971, 136). There are numerous stories of clocks stopping at the moment of a death: Frederick the Great (pendulum clock); Cheiro, the fortune-telling English seer, and so on.

In more recent times, a witness to the well-documented Miami warehouse poltergeistery attested: "I knew that something was happening in which humans [mortals] were not involved" (Susy Smith 1970b). In another case, this one from ancient times,

> The angel intruders let fly such knocks and poundings that they moved many a house on its foundation.
>
> OAHSPE, BOOK OF WARS 24:23

A related passage in Oahspe recounts the actions of warring angels (minions of the false gods), who organized against the faithists who would not bow down to these usurping gods. In their midnight raids, the food on the tables was stripped off—

> even as mortals sat down to feast . . . In other places, the angels of the false Osiris rushed in . . . gaining power sufficient to hurl clubs and stones and boards . . .
>
> OAHSPE, BOOK OF WARS 20:10

We will hear more about the false Osiris in the next chapter.

Such uproars continue unabated even in the modern world. Psychologist Edith Fiore treated a client besieged with house plants falling over and lights and TV turning off and on. In this case, Fiore was luckily able to identify the culprit—a woman, a prostitute, who had been stabbed to death in a motel room. She (that is, her spirit) was enraged. With mediumistic intervention, plus a little detective work, the victim's account was corroborated in a news story. The prostitute had been one of several murdered by a former policeman who hid their bodies in oil drums. "She wanted revenge! Ultimately, we talked her into leaving [the earthly plane] . . ." (Fiore 1987, 148).

Think not that by slaying a man thou art rid of him. . . . The soul never dieth.

OAHSPE, IHUA MAZDA, BOOK OF GOD'S WORD 18:25

The cases I have cited are but a few among countless examples of a Presence that is neither exteriorized nerve force, psychosomatic, or anything else from self. Often it is not the inner world at all that triggers paranormal phenomena; rather, the psyche is known to separate out and travel in spirit from one place to another on Earth and even in the heavens. "The etheric double may be hustled out of the body by shock . . . or an anesthetic . . ." (H. F. P. Battersby 1979, 51). Self separated from body is something entirely different from the workings of the alleged unconscious, whose stock-in-trade is the inner world.

It is not, as psychology teaches, the unconscious mind that projects itself to other venues or wanders through space-time, whether in dream, trance, hypnosis, or delirium. This wanderer is psyche per se, the subtle body in etheric flight. To call it the "unconscious" is to sail right past the remarkable faculties of the human mind. Occultists better understand the feats and foibles of that subtle body, many from personal experience, having experimented with projecting self out of the gross body. Some have returned with stories of unpleasantness encountered on the outer planes. For example, if the subtle body is seen and struck or shot at by a negative entity Out There, the physical body may return with actual marks, bruises from that astral skirmish.

The ability of psyche to bring back a wound from an out-of-body excursion (OBE) actually has a name—"repercussion." It is known to occur in the form of "stigmata" (in religious mania), as well as in conversion hysteria, possession, poltergeistery, multiple personality disorder (MPD), hauntings, satanic rites, autism, mediumship, and criminal psychopathology. Our skin, our envelope, is the first to bear the brunt of a "close encounter" with the powers of the Darkening Land. An injury

Fig. 6.1. Spirit separating from self in a dream. Courtesy of Virginia Howard.

inflicted on the etheric double, as A. E. Powell once put it, can appear as a lesion on the dense body.

When such marks are negotiated by psyche, orthodoxy reverts to the term *psychosomatic* to explain them. But does that explain anything? Or is such verbiage merely "the illusion of explanatory power"? (Richard Webster 1995, 299). In situations where psyche ventures out (often unknowingly) and returns with "war scars," there is more involved than autosuggestion or tricks of the mind. Though unseen, a force beyond self is involved. Consider the paranormal scratches of one mind-traveler who described her return from an astral tussle, her back sore and stinging. Examining herself in a mirror, she saw scratches, as if she had been clawed by a gigantic cat.

Quite a lot of confusion underlies these matters. The term *double,* for example, has another meaning (besides "etheric double") in the esoteric literature, for it may also refer to a kind of spirit companion. In Iceland and Norway, a child born with a caul (a membranous sheath over the head) is considered very lucky; he will be attended by a blessed spirit, acting as his double and warning him of danger. Only those with the gift of second sight can actually see their double.

But here's where the confusion arises: while occultists and parapsychologists were coining concepts like the "fluidic double" to account for certain mediumistic powers, the spirits themselves, if given the chance to communicate, would roundly challenge such notions. The acclaimed nineteenth-century spirit "Katie King" declared that she could go in and out of the medium readily, but people must understand that she is not the medium, not her double! "They talk a deal of rubbish about doubles; I am myself all the time," the spirit declared (Fodor 1974). Apart from such sophisticated controls as Katie King,* there are in the spirit world

millions of angels who know not how to get away from the earth . . . making themselves as *twin spirits* to mortals. These spirits oft show

---

*In a controlled setting, trained mediums allow themselves to be (temporarily) possessed for the sake of science and experiment.

themselves to mortals, but are believed to be doubles [i.e., part of self]. (my italics)

OAHSPE, BOOK OF FRAGAPATTI 35:7

A twentieth-century study of spirit possession in Egypt shows how true this passage may be. There is in Egypt today the widespread belief that a spiritual double of every person is with him from birth. And if "these spirits oft show themselves," it is a fact that the Egyptian double sometimes "becomes visible," as in the *zar* proceedings, still practiced in Egypt, Sudan, and Ethiopia. These rites involve trance behavior, counseling, and, most notably, the propitiating of troublesome spirits. Participants are usually women, for throughout the Arabian Peninsula it is mostly women who are attacked by *zar*. But these spirits only become visible to one who is clairvoyant, or in dreams (Crapanzano and Garrison 1977, 177).

Thanks to her talent for "leaving self," the clairvoyant is more prone to such "attachments": "The excursion of the soul leaves an empty house which may attract a parasitic tenant," as Arthur Conan Doyle described it in his wonderful book *The Edge of the Unknown* (1930). Otherwise put: although the materialist teaches that there is but one body in man, "There is also a spirit body which *separates* out of the gross material" (my italics) (Newbrough 1874, 25). And when the spirit of the medium or clairvoyant detaches itself, the body house becomes vacant, suctioning in that parasitic tenant.

The term *twin spirit* came into usage along with the *double* (or doppelgänger) to accommodate what seemed to be some sort of ethereal companion to the self. Most often these turn out to be dark angels, seeking to

engraft themselves on mortals, becoming as a twin spirit to the one corporeal body. . . . By fastening on, they dwell in that corporeal body, oft driving hence the natural spirit.*

OAHSPE, BOOK OF BON 14:9

---

*This occurs in the condition known as multiple personality disorder.

One well-known American novelist, having tendered a very candid memoir of his own mental breakdown, discussed a feeling that a number of people like himself have had while in deep depression—a trance-like state, in which one's own consciousness goes into abeyance, that is, the "empty house." To our novelist, this state of being away is marked by the uncanny perception of a second self, a wraithlike observer who, not sharing the dementia of his double, is able to watch with dispassionate curiosity as his companion struggles.

Among Arabic peoples, these wraithlike observers, called *djinn,* are the operators behind trance states, which are not too different from the well of "deep depression," as our novelist describes his trancelike state of supreme discomfort—a helpless stupor. The *djinn* are discarnate beings who are able to speak and act through mortals. Cousin to the Arabic *djinn* is the African *ejeni or ijeni,* generally translated as a "possessing spirit." This, in turn, resonates with the Moroccan *jnun,* a class of somewhat capricious spirits who may render themselves visible in a variety of forms. Symptoms of *jnun* victims, as seen in the ethnographic record, include depression, sudden blindness, and paralysis of limbs and face; that is, good old-fashioned hysteria. But autosuggestion? Psychosomatic? I don't think so.

Nevertheless, in the early days of psychoanalysis, Dr. Pierre Janet and most of his colleagues in Europe reduced such cases of hysterical paralysis to the subconscious, just as J. M. Charcot and Sigmund Freud labeled it "unconscious symptom formation." This, according to Richard Webster, was "one of the most significant misunderstandings in the entire history of medicine" (Richard Webster 1995, 72). Freud, as Webster sees it, invented "a mental entity whose specific function was to harbor our hidden thoughts, memories, and impulses. He called this hypothetical entity 'the Unconscious' and treated it as an autonomous region of the mind . . . [but it] was based on a chain of mistaken medical reasoning about hysteria." And so I ask, along with Webster and Martin Gross, is there really an unconscious? "Our Unconscious becomes real," says Gross, "only after revelation by a therapist-priest who can explain [its] mysteries . . . as reasonably as the clergy once explained the secrets of the Unseen God" (Gross 1978, 44). This is secular priestcraft.

Despite its moot foundation, the unconscious became the quick fix of modern psychiatry, made to account for every unknown factor in the great puzzle of mind, the catchall, the cleanup bin in which we toss everything we can't figure out about behavior, and, into the bargain, sweep away all evidence of the soul's existence and life after death. However, in therapy, the well-known multiple "Sybil" rebelled against this pat formula: "You call them [her alter personalities] the unconscious and say they are part of me, but you also say they can take me where they please. Oh, Doctor, these others drive me, possess me, destroy me."

"It's not possession, Sybil," the doctor smoothly assured her. "It comes from *within* and it can be explained not by the supernatural but in very natural terms." (my italics)

"It doesn't seem very natural to me," Sybil was quick to answer. Interesting that "in dreams, Sybil was more nearly one [i.e., herself] than at any other time" (Schreiber 1973, 321, 345). What does this interesting observation do to the theory of alters coming from the unconscious, the seat of dreams?

> *There are thousands of well-documented tragedies, which demonstrate that psychological explanations involving the "unconscious" for the phenomenon long known as "possession" are pitifully inadequate.*
>
> DAVE HUNT AND T. A. McMAHON,
> *THE NEW SPIRITUALITY*

Misdiagnosis

Consider the 1970s case of Anneliese Michel, whose name was changed to "Emily Rose" in the American film that told her tragic story. Anneliese, a bright German college student, was given Dilantin for the seizures that are so typical of diabolical possession—kicking, lashing out, striking blows at her loved ones, and so forth. But she only got worse. As Felicitas Goodman, her biographer, explains, anti-convulsive medication (perilously) relaxes the brain, thus opening the door to the "dwellers on the threshold," the European equivalent of the Arabic *djinn*. And indeed, "The drug won out," as anthropologist Goodman concludes sadly, "and the demons came back" (Goodman 1981).

Related to these pitfalls of Anneliese's case is recent talk that certain modern antidepressants can trigger suicide. Since World War II, psychoanalysis and her handmaiden, Big Pharma, have taken over. One result: in the United States there may be some serious overprescribing for depression, ADHD, and anxiety. Pharmaceuticals came into vogue to suppress symptoms of unknown origin, that is, psychic disorders not yet canonized by the high priests of the *Diagnostic and Statistical Manual of Mental Disorders* (DSM), psychiatry's bible.

A recent issue of *Discover* magazine reported on the latest firestorm over the new edition of DSM, which specifies the parameters by which people can be diagnosed with particular mental disorders. The revisions in DSM-5 triggered bitter controversy, centered on "worries that psychiatric disorders will be misdiagnosed . . . mislabeling normal people as psychiatrically sick" (McGowan 2013). Critics argue that too many, too powerful antipsychotics are being given to people with mild symptoms. Dr. Allen Frances, head of the task force that had written the fourth edition, warned doctors to use DSM-5 "cautiously, if at all" (quoted in Park 2013, 16); while the lead editor for the earlier DSM-3, along with other top names in psychiatry, formed "a chorus who worried that the DSM-5 authors were in bed with the pharmaceutical industry. Big Pharma: the pill peddlers" (McGowan 2013, 49).

Possession can be seen as one extreme on the paranormal spectrum; it does not call for the unconscious, but rather for spiritual entities *of which* we are unconscious (that is, unaware)! Oftentimes, "The person in the body is unconscious of the influx," said spiritualist A. J. Davis (Brandon 1984, 13).

The scenario may also entail our own perambulating soul, of which we are also unconscious!

Freudian analysts say the unconscious wish is a major driver of behavior. But most of us, I believe, are actually aware of these supposedly repressed desires, often censoring them consciously. Richard Webster protests this "hypothetical region of the self . . . this sunken continent" (1995, 251) of the mind. Although these forbidden emotions may simply be inhibitions, Jung made them a "shadow" life, which can be "transformed into demons when they are repressed" (Webster 1995, 251). But Jung, as we will soon see, was of split mind: when he wasn't calling demons our own repressed anxieties, he talked about how "his house became crowded with spirits, which he concluded were the spirits of the dead" (Storr 1996, 91).

Neither is our "ruthless egotism" (Freud's phrase) any part of the buried unconscious, but simply a factor of personality, a predisposition at the lowest end of the character spectrum. Freud tended to equate the "evil self" with the unconscious or the id, a seething collection of cruel and unclean impulses. This is a decidedly negative view of humanity. Freud stripped man naked and found him mean and ugly to the core. "The negative character of psychodynamic theory, with its emphasis on abnormalities . . . is a magnificent legacy of Freud's own neurosis" (Gross 1978, 241).

As Richard Webster sees it, some parts of the imagination are more shrouded in darkness than others. However, the unconscious as a secret inner realm dominated by repressed sexuality or rage simply does not apply to uninhibited people. Psychoanalysis, we know, came out of the stuffy Victorian era, with its many taboos. Is it not now obsolete?

Superannuated, the prevailing theory of mind nonetheless remains the psychoanalytic one, the Almighty Unconscious serving as the convenient escape clause and even smokescreen for the scientifically expurgated soul. Cast aside is the perambulating psyche, so prone to visit Otherwhere—"dislocation of the soul," as Conan Doyle once phrased it. This subject actually intrigued Freud. But then, as now, in this age of scientific materialism, the serious study of survival and immortality was off-limits. However, "If I had my life to live over again, I should devote myself to psychical research," Freud is famously quoted as saying. Writing in the summer of 1921 to Hereward Carrington (a psychic investigator), Freud said: "I do not belong with those who reject in advance the study of . . . occult phenomena as being unscientific, or unworthy. . . . If I were at the beginning of my scientific career, instead of at the end of it . . . I might perhaps choose no other field of study: in spite of all its difficulties."

Freud, Jung, and others, according to the Brazilian researcher Luis Rodriguez, did not probe the human psyche deeply enough. To Rodriguez, an occult influx came from unknown personalities outside of oneself. The victim of psychosis, for example, was actually an overwhelmed medium (a "sensitive"), which neither she nor the therapist recognized as such. Rodriguez's view has been backed up by studies in Brazil, Nigeria, and elsewhere, where medicine men/healers routinely treat the insane through channels of mediumship. They see the more bizarre symptoms as outside forces flowing through the victim and not from the unconscious. (In this connection, I highly recommend *Case Studies in Spirit Possession,* edited by Vincent Crapanzano and Vivian Garrison [1977].)

Almost a hundred years ago, the astute Dr. Carl Wickland (a dear friend of Conan Doyle's) asserted the impossibility of subconscious mind playing any role whatsoever in true mediumship. The unknown intelligence that speaks through the medium—or through the multiple, or even through the psychotic—is independent of her. It is not her unconscious.* Yet we will now see what happens to the mind when it is rendered unconscious—by shock, injury, grave illness, high fever, surgery under anesthesia, pure terror, or any other cause, even excessive daydreaming. In such cases, the built-in barrier against Otherwhere is lowered. Normally, we are pretty well insulated against receiving outside mindstuff. A bad blow, though, can lower that threshold.

---

*Granted, the ambitious or unscrupulous or ignorant "psychic" may indeed draw on his and her own imagination. We see quite a bit of this in the media today.

Fig. 6.2. Dr. and Mrs.
Carl Wickland (and dog).

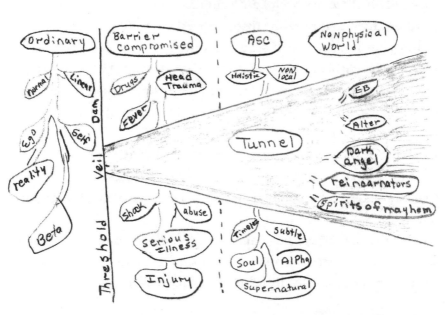

Fig. 6.3. Diagram of the Outer Darkness. Drawn by Janice Henslee.

Such prodigies have been hastily ascribed either to past lives (see chapter 7) or to the Almighty Unconscious. Instead, we are dealing with the threshold of psi, a place where normal and paranormal converge. We also find the *ab*normal here, where the "real world" leaves off and mindstuff flows uncontrollably earthward from the unseen darkness of the atmospherean world.

Some have called that threshold the "dam," others the "shield," without which a person is subject to spillover—that is, the influx of any astral entity desiring to seize his vehicle (that's not carjacking, it's mindjacking). Almost any trauma can compromise the shield—abuse, assault, fright, fury, fever, shock, accident, total exhaustion, even drugs and serious fantasizing.

When the human aura or dam becomes weak or otherwise leaky, nonmortal (excarnate) stuff, usually not good, can penetrate from the Outer Darkness. The two worlds, the two realities, are normally separate, but their breach—a chink in the dividing wall—can incur paranormal as well as abnormal manifestations. Behind most psychic wonders is just such a rupture in the dam, a rift in the veil, most especially if there has been head trauma. I agree with Freud's teacher, J. M. Charcot, that puzzling disturbances of mind have their origin in trauma, but not necessarily sexual trauma.* And certainly not because the trauma "enabled a particular idea . . . to become lodged in a [separate] part of the brain . . . removed from every influence, strengthened, and finally powerful enough to realize itself objectively through a paralysis" (quoted by Webster 1995, 67). Webster rightfully calls Charcot's reasoning "completely speculative." Nor can we agree with Freud's notion that hysterics suffer from reminiscences of traumas that they have intentionally repressed. The memory loss is automatic, not volitional.

Yet the barrier can be compromised without trauma of any kind. This occurs in a state of mind once labeled as "sloth of will," wherein the person is chronically disinclined to make decisions, assert self, think

---

*Sexual trauma, however, *is* overwhelmingly at the root of MPD.

for self, and would rather have someone else do it for him. This is a welcome mat for attaching spirits.

Nevertheless, psychologist Robert Mayer is one among many who have assured us that "any demons that may be lurking around are our own *unconscious* invention" (my italics) (1988, 165). Though it is the best explanation that the sci mat can come up with, it is still a serious blunder to declare that some unconscious, insensible part of ourselves creates or invents things, invents people, invents demons! Thus has our talented psyche and our very soul been replaced by a figment of imagination, the experts discounting foreign intelligences on the pretext that supernormal knowledge is (supposedly) unproven and that spirit origin is (supposedly) undemonstrated. Neither of which is true.

Instead of a soul, psychology gives us brain mechanisms and the soulless subconscious. The Almighty Brain will then explain everything, if we throw enough money at it. The anatomy of the brain, in my opinion, has very little to do with the idiosyncracies of human behavior. Brain is a biological entity. Mind is something else entirely.

> *There is something, a something, which does not require a brain . . . which knows many things that the five senses cannot supply.*
> T. C. LETHBRIDGE, *ESP: BEYOND TIME AND DISTANCE*

In his remarkable book *Psychic Warrior,* former Army officer David Morehouse attributed his disturbing visions (after suffering a head trauma) to "something from another dimension" (1998). These "others" were, as he termed it, "the dark ones," destructive forces that were, much to his chagrin, capable of taking over. His psychiatrist, though, would brook no such entities; she diagnosed the problem as something bubbling up from "deep in your limbic system." A brain thing. No, psychiatry will not abide the disembodied state. Yet, David's handlers, in remote-viewing experiments for the military, did peg his dissociated states as unfolding "in the ether." Out there. Otherwise. Not in here. Not a part of self and certainly not his unconscious demons. Something that should have stayed closed got opened up when that bullet struck David's helmet.

Even esotericists evoke Almighty Brain. Stan Gooch, for example, in *The Origins of Psychic Phenomena,* linked poltergeists, subpersonalities, and paranormal stuff to the unconscious mind, exploring "the functioning of the dream-producing part of the *brain*" (my italics). It was simply assumed that all these strange impressions and different personalities were coming from self, from the unconscious, from within. But consider this: if the deeply embedded unconscious is behind our "demons," how can it spread to others, like the exorcists themselves, who have been attacked by these kinds of entities on so many occasions?* Attacked by the unconscious? I seriously doubt that.

If consciousness is real and unconsciousness is unknown and shadowy, why did Jung contend that the contents of dreams are probably "greater than the conscious mind" (Jung 1974, 95–96)? Why do theorists say great talents come from our subconscious mind? Does it make any sense to call inspiration a product of the "unconscious," ascribing to the unconscious intelligence, purpose, reason, problem solving, and superior insight? Have we forgotten that *inspiration* means, literally, "taking in spirit"?

> Wherein thou art quickened in spirit, behold, it is the heavens upon thee that stirreth thee up.
>
> OAHSPE, FIRST BOOK OF GOD 28:13

Why, asks one critic, have we made "our conscious mind a second-class being . . . a mere puppet of the unknown true self" (Gross 1978, 43–44)? In the case of neurosurgeon Eben Alexander, who created a media sensation with his 2012 book *Proof of Heaven,* everything he experienced as a free spirit (in his near-death journey out of the body) suggested an expanded functioning of the mind. "The more I read the

---

*One such case is that of Father Peter, in Malachi Martin's *Hostage to the Devil* (1992, 72, 80).

scientific explanation of what NDEs are, the more I was shocked by their transparent flimsiness" (2012). One by one, Alexander dismissed the pseudoexplanations of what he underwent, all of it jacked up in jargonese: REM intrusion, endogenous glutamate blockade, DMT dump, reboot phenomenon, and feigned-death strategy. In fact, they are just words and guesses—not explanations at all.

Neither is any light shed by the cosseted Uncertainty Principle: "We can never determine the reality of any particular dream," says a leading dreamologist (Bulkeley 1995, 142). Does this sweeping and, as I think, incorrect generalization apply to the following dream?

In Swaffham, England, at the marketplace, there is a sign that announces: "The tinker of Swaffham . . . did by a dream find a treasure." At the Swaffham church, there are memorials, ever since John Chapman, back in 1486, installed an elaborate prayer desk for the townspeople, as a thank-offering for his good fortune.

Chapman had been a poor man who happened to be told in three separate dreams to go to London Bridge, where he would be led to a fortune. Everyone thought he was crazy, but he set out on the arduous journey and waited at the bridge a long time. Finally, a man approached him and asked what he was doing. Chapman, point blank, explained that he was there because of his dreams, whereon the stranger told him, "I had a dream myself, a very vivid one," and he proceeded to tell Chapman his puzzling dream in which he was told to hurry off to a town called Swaffham, find the garden of a tinker named Chapman, dig under the tree, and discover a crock of gold!

Well, the stranger thought the whole business a different kind of crock. Chapman, though, swiftly returned home, dug in his garden, and found two pots of gold pieces, which made him a wealthy man. (The story was retold by Justine Glass in They Foresaw the Future *[1969, 197–98]). Does the unconscious or inner world have anything at all to do with such remarkable dreams? Don't these matching Swaffham dreams, instead, suggest the interplay of informing spirits?*

Among the world's pygmies and little people, informing spirits were and are their otherworldly helpers, their teachers. The Malaysian Batek get remedies from the spirit world, their unseen mentors the very angels who helped the early humans and taught them how to live, for they showed which fruits were safe to eat and which herbs were good to use for their ailments. Today, the Batek healer, called the *bomoh,* still obtains his songs from celestial beings while he is in trance or asleep. It is not his "unconscious" that reveals these things. Similarly, among the Bay of Bengal Andamanese, a person becomes a shaman by contact with spirits—either in dreams or by "dying" and coming back.

Otherworldly inspiration is a carryover from the sacred little people, who had been the first on Earth to practice spiritual communion.

And God's angels taught the chosen these things.
OAHSPE, FIRST BOOK OF THE FIRST LORDS 3:7

The little people went on to teach others how to receive instruction from the invisible beings. Today, medicine songs are given to the little African Bushmen by the Great God who sends his messengers (spirits of the departed) to a sleeping person. And this is how our forefathers became wise, for

the angels . . . talk to him [man] in his sleep, and show him what is for his own good.
OAHSPE, BOOK OF SETHANTES 10:5

In America, among the Iroquois, these dream helpers are portrayed as a healer tribe of little people. The bearded and blue-eyed little Nunnehi would come in dreams to the Cherokee, showing them how to protect their corn from predator birds using a certain spell. The Cherokee medicine man consulted the bearded little people (Yunwi, Nunnehi) for remedies and plants, herbs, and roots needed for the ailing.

The souls of men are partly inspired and agitated from without. . . . There is a divine energy in human souls . . . and how great is the energy of the mind when abstracted from the bodily senses . . . as in ecstasy and sleep. . . . When the soul of man is disengaged from corporeal impediments, and set at freedom . . . being relaxed in sleep . . . it beholds those wonders which, when entangled beneath the veil of the flesh, it is unable to see. (Cicero, *On Divination*, 1997)

Informing spirits reach us in our sleep more readily than in our waking hours: then the barrier is down. Sleep is an altered state of consciousness (ASC). Within these understudied ASCs lies a universe of surprises. A thousand questions unanswered by psychiatrists, priests, educators, law enforcement, and behavioral scientists may be clarified by lifting the veil. ASCs place one at the threshold of the nonphysical realm. And dwellers there be, at that threshold.

Many NDErs (near-death experiencers) have described that threshold as a tunnel. For the multiple personality, the tunnel is the precise spot where the interchange of personas takes place: "When one of them wanted to come out" (speaking of the alters attached to a multiple patient), "he or she went down the *corridor* to that spot and then came out" (my italics) (Mayer 1988, 172). One multiple lost time while screaming angrily at her mother. Suddenly, "I felt as if I were in some dark tunnel. . . . Then I blacked out" (Schoenewolf 1991, 26). "Deep in the Tunnel," recounted Truddi Chase, a supermultiple (with ninety-two alters!), when "the Gatekeeper gave the signal," an alter emerged (Chase 1987, 104). Driven to the tunnel by supreme terror, Kit Castle (yet another multiple, with seven alters) "had gone somewhere . . . that was utterly still and dark and completely outside of time" (Castle and Bechtel 1999, 235). And though it may be used as a refuge by the traumatized, the tunnel is not—as many suppose—a make-believe place or a contrivance of the unconscious. It is real. It is the bardo in which mind separated from body finds itself.

*Fig. 6.4.
Hieronymus Bosch's*
The Ascent of the
Blessed *shows
spirits in a long
tunnel that leads to
the light.*

THE UNCONSCIOUS MIND

To man I gave life and spirit also. And the spirit . . . was separate from the corporeal life.

OAHSPE, BOOK OF JEHOVIH 6:10

Beginning with J. M. Charcot and Pierre Janet, hypnosis was used to dredge up repressed trauma, the supposition being that all hypnotic disclosures belonged to the personal (unconscious) history of the subject. Yet, information brought up in hypnosis, which is an altered state, sometimes comes from elsewhere. Not from self. How does it feel to be elsewhere? After psychologist Charles Tart's hypnotic session with Anne, she reported having a sensation of relinquishing control, a most unusual physical sensation of disintegrating. Soon this feeling passed and "my body was *gone,* and I felt like a *soul* or a big ball of *mind.*" This is the precise sensation of the spirit body, not of the unconscious. Another Tart contributor remarked that to put it all down to unconsciousness is very convenient. . . . I, for one, do not believe in unconsciousness any more than in Santa Claus.

In all Nature there is no such thing as unconsciousness.
WALTER RUSSELL, *THE SECRET OF LIGHT*

The unconscious was invented and brought to market in order to establish the new field of psychoanalysis, replacing the scientifically unacceptable idea of soul or spirit: "The Unconscious [is] man's anticlerical equivalent of the *Soul*" (Gross 1978, 10). But let us realize now, a century later, that the unconscious itself has no scientific reality. "There is not a shred of scientific evidence to support the mythical powers attributed to the unconscious" (Hunt and McMahon 1988, 119). The unconscious stands as nothing more than the agnostic solution to our ephemeral powers. Like every other doctrine tackled in this book, the unconscious serves as a kind of intellectual cover-up, blocking truths of which we are ignorant or to which we are opposed.

A cover-up usually created more problems than it solved.

VINCE FLYNN

The theory of the unconscious solves no problems and only raises more questions. "At best it is a cop-out and at worst a piece of sophistry" designed to cover up our ignorance, our spiritual illiteracy, with the hollow, pedantic language of academia. "No one could calculate the billions of hours and the prodigious fortunes spent in consulting psychologists and psychiatrists by the untold millions of faithful believers in this highly promoted myth . . . the deep silk hat from which the magicians of the mind pull out all their cures and rationalizations!" (Hunt and McMahon 1988).

Whether real or not, the Almighty Unconscious has served nicely as the proverbial rug under which we may sweep all the unsolved mysteries of human behavior. It is the morgue of unexplained facts, the graveyard of enigmas. Most of all it is the last redoubt against the forbidden supernatural, as if to touch on the unseen powers were to throw us back to the Dark Ages of medieval sorceries.

The overworked unconscious has held out against the spirit world for a long time. Almost a hundred years ago Dr. Carl Wickland, a psychiatrist and a paranormal researcher, noticed an article in the Chicago papers written by a Dr. Lydston, describing a patient who, under anesthesia (that is, in an ASC), sang well the "Marseillaise," the French national anthem, despite having no knowledge whatsoever of the French language. Dr. Lydston explained the phenomenon as one of unconscious memory. Rising to the occasion, Dr. Wickland sent in a letter pointing out that such prodigies were "met frequently in psychic research and . . . despite materialist scientists, these cases clearly proved the posthumous existence of spirits and their ability to communicate through mortals." The dauntless Wickland then asserted that the French-singing man must be a psychic-sensitive, controlled at the time by some outside intelligence (a French speaker). Postscript: having read Dr. Wickland's rebuttal, the patient called on him and declared: "I don't know anything about French, but I do know I am bothered to death by spirits" (Wickland 1924, 42–43). These freeloading spirits crave embodiment, expression, experience (of which craving we will hear much more in the next chapter).

In East Africa, instances have been recorded of people possessed by spirits speaking in English or Swahili, although they neither understand nor speak those languages. In one African case of possession, a woman spoke fluently a language of which she previously knew not a single word. Both the natives and the anthropologist, in such cases, regard these as instances of entity possession. In these cultures, this is the "normal explanation."

But we need not "go bush" or travel to exotic lands to explore possession and its strange earmark—xenoglossy (speaking unknown languages). It exists in our midst, only we call it by other names, such as alternating personalities (MPD). In a twentieth-century case, a young American woman, M, began speaking with a Spanish accent, also writing (automatically) in Spanish and Italian. In keeping with the usual roots of overshadowing, M had had the psychic door opened by an early trauma. And in keeping with the diagnostic clues for possession, M was particularly vulnerable to psychic invasion, being "timid, frail, and easily exhausted." Hallucinations and automatisms were in the mix, the sound of strange singing issuing from M's own throat, "as if something had possession of me." Now the new personality was "out," and, in the phenomenon called "co-consciousness," was actually able to converse with M, confiding that she was "the departed soul of a Spanish woman." Although the analyst, a professor from Washington University, thought the whole business was a fabrication of M's subconscious, the alter herself insisted that she was in no way part of M, "but the spirit of a long-dead Spaniard" (Crabtree 1985, 41–42).

The obvious question is this: Can foreign (unlearned) languages be dredged up from someone's subconscious imagination? Of course not. When the automatist performs any such feats, can it be shown that unlearned skills—say, rapid typing—can be coaxed out of the subliminal self? No, it cannot, and, no, MPD is not a "split" personality but a real departure—self in abeyance, self vacating the premises, self driven out, usually by a trauma too painful to endure, too dreadful to "stay" there. And that vacancy makes possible the intrusion of others. (In Irish tradition this is just what discarnates are called: the "Others.")

It should come as no surprise, then, that multiples are prone to obsession by the undead who just may happen to speak some other language. The most striking example I know of is Billy Milligan's Yugoslav alter, Ragen, who spoke in Serbo-Croatian or in halting English. When Ragen "took the spot," Billy's voice became sibilant, like "someone who had been raised in Eastern Europe." Ragen was among the twenty-four alters housed in Billy's body (Billy being the first multiple recognized as such in a U.S. court of law; a few of his alters were criminals, called "the Undesirables"). Ragen's Serbo-Croation, of course, was flawless, and his English heavily accented. There is no way Billy, an all-American kid from Ohio, could have accidentally stashed the knowledge of Serbo-Croatian in his subconscious!

> We are all different people.
>
> RAGEN, CITED IN DANIEL KEYES'S
> *THE MINDS OF BILLY MILLIGAN*

"That such personalities are independent entities could easily be proved," argued Dr. Wickland, "by transference to a psychic intermediary, as experiments have so abundantly demonstrated. Any attempt to explain [these] experiences on the theory of the Subconscious Mind or auto-suggestion . . . would be untenable, since it is manifestly impossible that Mrs. Wickland [his wife, who was a medium] should have a thousand personalities." It is also of particular relevance "that the majority of these intelligences are oblivious of their transition [death]. . . . It does not enter their minds that they are spirits, and they are loath to recognize the fact" (Wickland 1924, 43–44).

Xenoglossy, like Billy's or M's or the man singing the "Marseillaise," is an abnormal automatism that occurs independent of one's volition and also quite apart from one's alleged unconscious. However, in the context of controlled mediumship, speaking for a discarnate is viewed by some as a psychic "gift," right along with ESP, precognition, PK, and so on. When trance mediums hold forth in unexpected languages, the ability comes from their spirit control. Describing her trance state, England's psychic medium Winifred Tennant recalled, "It seemed as if somebody else was me, as if a stranger was occupying my body, as if another's mind was in me." That mind was the discarnate communicator (the spirit control) taking the spotlight, while the medium's own self was suspended in time. "So strange to be somebody else. To feel . . . somebody else's mind inside your mind" (Cummins 1965, 105–7). And how similar this is to the (mediumistic) sensations of multiples, but with this difference: the multiple is told by the therapist, time and again, that it is all coming from her unconscious!

A multiple called Toby complained that nightmares kept her up all night; people were chasing her, hurting her. When Dr. Robert Mayer set out to track down Toby's "inner personality," he came upon "not one but a trinity. They claimed they were not mere alter personalities but angels from the spiritual world." Too bad Dr. Mayer, like the majority of his colleagues in the mental health field, was admittedly "no expert on the spiritual world . . . the afterlife. . . . Quite frankly, the subject confuses me." Instead, his goal was to find "some scientific explanation for such phenomena." Yet, once the trinity surfaced, he asked, "Who are you?"

"We come from beyond . . . we are not part of her."

The dialogue continued, as the trinity asked the doctor what he thought happened to the soul after the body dies. "They treated me," Mayer admits, "with patronizing disdain": *"Your spiritual awareness is that of a flea"* (Mayer 1988, 7, 81, 155–57).

Intrigued but not moved, Mayer ran straightaway for shelter under the canopy of consensus—that demons are a product of the unconscious. Though this view has strength in numbers, it is threaded with the worn-out strands of the materialist fallacy. Are spiritual realities unscientific? True, they are invisible. But we know them by their effects, just as parts of scientific knowledge are built up by studying not things, but effects, without benefit of direct observation.

The multiple, in a word, absorbs other lives. Three possibilities exist to explain this strange condition. Either (1) the "personalities" come from the unconscious or (2) they are older incarnations of oneself (chapter 7) or (3) they are ex-mortals (excarnates) lost in the Outer Darkness and drawn to the "host" for a variety of reasons.

In the case of Christine Beauchamp's four alters, Sally, the dominant one, had a will of her own and could hypnotize the other alters. Sally had the appearance of an invading entity, an obsessing one; she was "not a cleavage of the medium's self" (Fodor 1974, 280). Sally insisted that she was not the same as Christine, that her own consciousness was distinct from that of Miss Beauchamp.

Replace the term *unconscious* with *psyche* and the hidden forces come into view. Replace the word *memory* with *overshadowing* and we are that much closer to the truth. In a deep analysis by Malachi Martin (1992, 253–54, 312), the possessing spirit was found to be someone named Uncle Ponto. He was asked:

"Is Uncle Ponto you? Are you Uncle Ponto?"

"We are spirits. . . . Powers. Dominations. Centers. Minds. Wills. Forces. Desires."

Similarly, in Dr. M. Scott Peck's work (2005), which combined methods used by the exorcist with the psychoanalyst's arts, the incognito discarnates were probed and, with some success, Peck dislodged these unwanted guests. Significantly, his technique worked well with patients diagnosed with MPD.

> *While some clinicians deeply suspect [that] my cases of purported possession are actually cases of MPD, I have come to suspect . . . that their purported cases of MPD might actually be cases of possession.*
> DR. M. SCOTT PECK, *GLIMPSES OF THE DEVIL*

THE WATCHERS FROM THE GATE

Let's consider the human brain for a moment, to ground these comments in observed phenomena and to get a grip on altered states of consciousness (ASC). The brain's beta waves, as measured by EEG (electroencephalogram), are low voltage and more or less the normal frequency of the mind when it is involved in everyday activity and logic, outwardly focused, and wide awake. Slower than beta, but higher voltage, are alpha waves, which, by contrast, are associated with meditative states, deep calmness, dozing, and the imaginative faculties. In a way, beta is sensory while alpha is parasensory. Beta is ego looking out on the world. Alpha is psyche looking in; it refers to more subtle processes of thought—dreamlike, sensitive, hypnotic. If beta betokens the five senses, alpha (as well as theta and delta) betokens the sixth sense and beyond.

Exploiting the alpha state, the new age has discovered that it enhances creativity. (It also may help alleviate a number of somatic difficulties, from bladder trouble to epilepsy.) One catch, though, is a tendency toward hallucination: some alpha traines have had negative experiences and "unbidden images," for the "dam," the "shield," has been lowered. In alpha experiments there have been a few alarming reports of entity possession, including the frightening *fratzen*—demonic faces seen in the mind's eye. The conscientious alpha trainer well knows that trifling with one's brain-wave frequency carries certain risks. With this in mind, such training is not recommended for unstable personalities, drug addicts, depressives, schizophrenics, or anyone else who is already too borderline (too close to the threshold) for further excursions into the unknown. Indeed, it could make them worse; they could be psychically catapulted into realms beyond their control.

These realms are the equivalent of Robert Monroe's Second State, the altered state of consciousness that permits psyche to venture out of its cage of flesh. When Monroe's OBEs were tracked by Dr. Tart, unusual and unique EEG tracings were obtained. In alpha, one makes "brilliant discoveries" (Stearn 1973, ix), solves problems, and is enhanced in "perceptiveness." The linear, verbal, analytic mode imposed on us by beta and the real world is no longer at play; instead, the holistic, the gestalt, is catalyzed in such a way as to allow us to see complete solutions to creative problems. Nonverbally, mystically, often in images, the ineffable

Unity is grasped. Once intellect per se is withdrawn, once the mundane restraints are loosened, the "watchers from the gate," as Sigmund Freud once put it, step aside, and ideas rush in. Even with simple relaxation (good music, a walk, a vacation, whatever), "flashes of creative insight may occur" (Krippner 1975, 14).

The point is that alpha is an altered state; it is not the unconscious. Most artists are aware of it. But it has its pitfalls. If Freud's "watchers from the gate" (i.e., ego, beta) become distracted, certain ordinary functions may be thrown off. Take dyslexia, for example. Experts in ADHD have collected the names of great talents who were dyslexic: Thomas Edison, Hans Christian Andersen, Woodrow Wilson, Tom Cruise, Émile Zola, Galileo, Leonardo da Vinci, Michelangelo. Each paid a price for his (Second State) genius with an (ordinary state) dysfunction.

> *Multiples were the best subjects for hypnosis and were generally psychic.*
>
> Dr. Gerald Schoenewolf,
> *Jennifer and Her Selves*

The pitfalls may be considerably more severe. With multiple personalities, whose trauma has catapulted them far from the "gate," awesome talents may indeed accrue, but at the expense of the integrated, inviolate self. The gate is wide open to mindstuff, aptitudes, and talents from Elsewhere. Our multiples are psychics, channels for disembodied minds who can rush in pell-mell. These are their alters and they are, emphatically, not part of their unconscious.

"Multiples," thought Marshall Fielding, "are creatively brilliant." One MPD specialist sees them as "gifted child[ren], at least that's what we've found in most cases" (Chase 1987, 155). Billy M. had been a child prodigy; his EEGs, incidentally, looked like those of "two different people."

Sigmund Freud has had more than his fifteen minutes of fame, and it is time for the unconscious to give way to altered states of consciousness. What occultists once called the third eye also needs to be understood in less mystical terms; we can no longer afford to romanticize or mystify the extrasensory apparatus in humans. Neither can we afford to obfuscate or censor or despise the otherworldly faculty inherent in humankind. The universality of its imagery—tunnel, dam, timelessness—is no coincidence. The sooner we explore this borderland, the sooner can we clear up a myriad of enigmas, held too long in the suspended indecision of reluctant science and spooked society. Let us brave the next step—the threshold of psi. Let us connect the dots.

Xenoglossy. Blackouts. Lost time. Alternate personalities. The struggle at the threshold. And still psychiatry would have us believe that MPD is a "highly creative survival technique" (Barry Cohen et al. 1991, xx); that any demons are the product of the so-called unconscious; that the "voices" heard are hallucinations; that PK is psychosomatic. Thus endorsed by the materialist fallacy, the obfuscation of MPD is a fait accompli. Because invisible, the unseen participants are consigned to imagination, to fraud, or to fragmentation of self. "I tell you, they were people!" Truddi Chase persisted, when her therapist explained that abuse victims often suffer splitting or fragmentation into separate selves. But bewildered himself by the distinctiveness of her alters ("They sounded like people"), the therapist had to admit, if only to himself, that "Fragmentation [was] a 'trash can' term and that he was only biding his time by using it" (Chase 1987, 92).

Thoroughly whitewashed by the materialist bias, today's science of mind is utterly unprepared for diseases of the soul. Analysts fear that the acknowledgment of psychic activity would undermine the respectability of the "science" of psychology, as if *spiritual disease* were one step down from reality, from objectivity. But are "the arcane mysteries of the Viennese seer" really a science? Don't these Freudian mysteries "touch more on the magical and the mystical than on the framework of science"? Without the stamp of science, they "might be considered occult speculation not unlike the mysticism of Kabala. . . . Freudian theory may be proven no more scientific than astrology or phrenology" (Gross 1978, 16, 45, 141).

CARL JUNG

*It is amazing how timorous most people are of being asso-
ciated with any sort of psychic experience. . . . The fort of
ignorance is defended by an organized body of men finan-
cially and otherwise concerned to buttress the ideas on
which their own reputations have been founded.*

H. F. P. PREVORST BATTERSBY,
MAN OUTSIDE HIMSELF

And this is precisely where the professionalization of knowledge has
gotten us: facts fitted around the SM, the permissible theory, to the
exclusion of all that lies in the Unseen. The weak-kneed, obscuran-
tist policy, forever invoking the mysterious or inexplicable to make up
for its own failings, has spurned the most compelling evidence, that
of ASCs and the wonders produced by the psychically awakened—and
disturbed—self.

Carl G. Jung, one of the founders of psychoanalysis, once had a
dream-sent ADC (after-death communication) from his recently
deceased father. The persistence of the "visit," for it was repeated, moved
Jung to grasp the reality of it. In the dream, his father appeared, claiming he was recovered and coming back from his vacation. This made
Jung apprehensive that his father would be put off that he, Carl, had
moved into his father's room (thinking he was dead). When the dream
recurred two days later, it troubled Jung: It seemed "so real. It was an
unforgettable experience, and it forced me for the first time to think
about life after death" (Jung 1974).

Actually, Jung's rocky life brimmed with the occult. He himself was
mediumistic. For example, one night he woke up with the sensation of
a blow at the back of his head. The next day he heard that one of his
patients had shot himself in the head the night before (see the discussion on repercussion, page 247). Jung also had dream visions. Early in
1914, he saw all of Europe engulfed in a great flood, with water rising
up to the mountains. Thousands drowned. Then the water turned to

blood. A few months later the Great War began (August 1914). In later years, Jung would write of an experience he had in 1920 at a haunted cottage where, while he was trying to sleep, something brushed against the walls. There was a repulsive smell, something dripping, rustlings, creakings, bangings. As good a haunting as it was, Jung still tried to explain it as "exteriorization of the psychological content of his unconscious" (Jaffe 1978, 94). Split mind followed Jung to the grave.

In Zurich he investigated eight mediums. It is no secret that Jung had been brought up in a mediumistic family. His maternal grandmother once remained in a trance for three days, during which time she accurately described people unknown to her. Other relatives were involved in occultism: Jung's mother and her father were both ghost seers. His mother kept a diary in which she recorded her presentiments and psychic experiences, like strange events occurring in the house. Jung's grandfather held intimate conversations with his deceased wife. The atmosphere in which Jung grew up, though strongly Christian, was also full of séances, psychokinesis (PK), and poltergeist activity. As a young man, Jung himself had the power of PK, moving objects around by "mental force."

One famous story has the Swiss psychiatrist and his mentor, Sigmund Freud, meeting in Vienna, the latter still skeptical about matters spiritualistic. At this meeting, Jung became increasingly annoyed with the Jewish sage's cynicism. At the same time, Jung began to feel a red-hot sensation in his stomach. Suddenly, a loud report issued from the bookcase. Ah, exclaimed Jung, right here is an example of the phenomenon called "catalytic exteriorization" (better known as psychokinesis). Freud dismissed it as "sheer bosh." Jung protested, adding that he hereby predicted a sequel to the event. Whereupon there was another loud report from the bookcase! Ten years earlier Jung's own solid oak table had, of its own accord, split right across; and two weeks later he found that a strong steel knife of his had broken into pieces. (Poltergeists have been known to smash things to smithereens.)

Jung had his first psychic experience at age three. Later, during university days in Basel, he avidly attended séances and lectured to a student organization about the immortality of the soul. The twenty-two-year-old student affirmed the reality of spirits on the evidence of telekinesis, messages of dying people, hypnotism, clairvoyance, and prophetic dreams. Jung's doctoral thesis was titled "The Psychology and Pathology of So-Called Occult Phenomena." Jung, in fact, had his own spirit guide—Philemon—whom he acquired while on the brink of suicide after breaking ties with Freud (1913). Indeed, it was Philemon who (mis)led him to his "discovery" of the collective unconscious and cosmic archetypes.

Split mind: Jung kept a lid on his psychic involvement most of his adult life. "Many theories I have kept private all these years," he wrote in a deathbed letter, left at his bedside for a lifelong friend. To save his career, Jung came up with the collective unconscious, a poor and unprovable substitute for the spiritual realities of which he was aware but was afraid to espouse.

As Anthony Storr (1996) relates in *Feet of Clay,*

> The whole of [Jung's] later work can be read as an attempt to discover a substitute for the religious faith which he had lost. . . . His variety of analysis . . . promises a secular form of salvation. . . . Jung's father was a pastor in the Swiss Reformed Church. Two of his paternal uncles were also ministers. In addition, his mother was the daughter of a theologian, and five other members of her family were in the church. Jung found the atmosphere at home oppressive. . . . Like Nietzsche and others who have been brought up in a strongly religious atmosphere, Jung found it difficult to live without a faith.

Two important factors kept Jung from adopting the faith of spiritualism. First: he was a passive (undeveloped) medium himself with no real awareness of his dark and invisible handlers (his "controls"). This made him vulnerable. His exposure to the spirit world was often negative. Deep in the psyche, he thought, was the "stuff of psychosis," the makings of insanity. Only toward the end of his life did he actually admit that haunting spirits nearly drove him insane. Hypersensitive, he was a "candidate for schizophrenia" (Storr 1996, 89). Any experiments or inquiries into this unknown realm were for him "a risky business," a dangerous path that had the power to lead to the "quintessence of horror." There were years, as Jung himself recounted, when he teetered on the brink of psychotic breakdown (1913–1918), thanks to unbidden ASCs. These he incorrectly called the unconscious contents of his mind—"they could have driven me out of my wits." Jung's paranormal experiences deeply frightened him as he whirled about in the winds of spirit, like Nietzsche, whom he admired but who had died insane.

More to the point, Jung's occultism threatened to ruin his budding career. Viewed as academically embarrassing, spiritualism was a ticket to infamy in the rational world, the world of professional success. So Jung began to mask his psychic history at every turn, using euphemistic words and phrases to soften the facts and being quick to contradict himself if the strength of his involvement slipped out. He would postulate that strong emotions—instead of spirits of the dead—caused apparitions. (This is the "thoughtform" school, which argues that strong images attached to deceased persons are activated to form spirits simply by the intensity of one's own emotions.) The theories proposed to defeat the presence of spirits are sometimes more far-fetched than the idea of eternal life itself. But any other theory, no matter how flimsy, would be more acceptable in the eyes of materialist science. Thus did Jung conclude, publically, that spirits had their origin in the unconscious. He labeled them, rather meaninglessly, as the exteriorized effects of "unconscious complexes"—repressed desires (id) and fears.

To these he added archetypal imprints. Indeed, Jung's famous collective unconscious—and its archetypes—were a stand-in, a substitute, a whitewash for something that he knew came from the realm of spirits. To eradicate that realm, he labeled paranormal phenomena "inexplicable" and went on to declare, "I see no proof whatever of the

Fig. 6.5. Carl G. Jung

existence of real spirits." Yet according to his private secretary, Aniela Jaffe, in old age Jung "was no longer able to maintain with certainty his original thesis that spirits are exteriorizations [from self]" (Jaffe 1971, 6, 9).

Instead of spirit-familiars, Jung staked his reputation on impersonal archetypes—lifeless symbols—a wholly fictitious kind of "racial memory."* All the mystery of dreams, say his followers, reverts to the "ageless, *cellular wisdom* held within the hidden dimensions of your own mind." Jung invented such fanciful constructs as "anamnesis," signifying the process that recovers these buried collective memories. Oblivious of his youthful involvement as well as his old-age turnaround, some of his disciples have gone on to provide ever-fancier names for Jung's archetypes, such as "universal mind" and "cosmic reservoir."

---

*There is no known way by which this memory can get into that person's germ-cells and thence into the brains of his descendants argues de Camp (1975, 100) against this whimsical idea.

One Jungian (Jean Houston) argued that spiritual entities did not really exist but were "projections and creations of the immensity that is the personal and collective unconscious . . . 'goddings' of the depths of the psyche." One pair of critics retorts, "Why God is outdated superstition and 'goddings' (whatever that may be) is modern science, is not explained." The same critics believe that the entities Jung encountered "were not archetypes at all. Rather than representing powers innate in the collective unconscious, these entities were in fact hostile beings independent of human consciousness" (Hunt and McMahon 1988, 68, 122).

But the ambitious psychoanalyst, especially after breaking ties with Freud* to make it on his own, could ill afford to associate himself with despised spiritualism. All that crazy stuff had to be swept away, reinvented, renamed, disguised.

Nevertheless, there were (and are) clearly mediumistic aspects to the behavior of multiples, schizophrenics, phobics, hysterics. It is in all cases a product of the dissociated mind. Psyche unleashed. Loose cannon. Why were hysterics performing the same feats as clairvoyants and mediums? "Jung stated that spirits and psychic phenomena should . . . be regarded as unconscious autonomic complexes . . . [or] exteriorized forms of unconscious complexes" (Jaffe 1978, 91). But it was just a label, and in no way succeeded in unriddling the hidden wellsprings of the mind, the true psyche. Although Jung, in his own words, acknowledged that he had "examined cases where the test-person replied to specific stimulus words . . . as if *a strange being* had spoken through him," he nonetheless concluded, "such words belong to the unconscious complex. . . . *Complexes behave just like independent beings.* . . . However, in all this I cannot see any proof for the actual existence of spirits" (my italics) (quoted in Ebon 1978, 112).

No, it was not safe for Freud or Jung to follow up on what they suspected or even knew: the continuing existence of those gone before,

---

*Jung had been Freud's favorite protégé, his "crown prince." Part of Jung's disagreement with Freud centered on Freud's overemphasis on sexual themes.

the lingering and perambulating soul. The time was not yet, the world not yet ready. Their baby, psychoanalysis, would never have been born, no less survived and thrived, had its founding fathers submitted to the revolutionary but scorned findings of psi research, which had indeed discovered the everlastingness of the human spirit.

One of Jung's recurrent dreams finds him in his house with a new, mysterious wing added on. When his dream self makes it through to this unknown part of the house, he comes upon a zoological laboratory with many specimens. Right there is a curtain and something "haunted" behind it. He presses on and comes upon his dead mother's bedroom. Then another door opens—this time into a vast, ostentatious hall such as in a well-appointed hotel, with pillars, sumptuous wall hangings, fine tables, and a brass band playing music. All the ambience of fun-festing and worldliness.

The interpretation? All this showiness, say the disciples of Jung, represents "the conscious mind, the rational world of daylight" (in contrast to the mysteries of the house's dark inner wing). But I say, baloney! The haunted place, his mother, the curtain (the "veil")—all represent his occult past as well as the livingness of the "dead." And whereas the lab specimens stand for the gross body (the animal body, which becomes lifeless at the time of death), the haunted room of his mother betokens the soul in transformation, that is, into a spirit. The sumptuous hotel with its rich appointments obviously stands for respectable society, success, and acceptance, all of which Jung would have thrown away had he opted for spiritual explanations of phenomena.

Fear of the unknown, both professionally and personally, eclipsed the genius of Jung, whose contribution is now best placed in the history of ideas (or the graveyard thereof), right along with the "cultist views of the Freudian revolution" (Gross 1978, 141). Yet, Jung was a learned man and a deep thinker; it irked him that our "rational age" had closed the door on psyche's "voyage of discovery." But as the tortured path of Jung's thinking unfolds, we see how desperately he tried to escape his own convictions, grasping for almost any other theory no matter how suspect. Hence, the meaningless and unprovable collective unconscious, which Philemon had taught him.

England's Arthur Guirdham, a psychiatrist, picked up on Jung's doublethink: "Though Jung shirks the question of discarnate entities, what does he mean when he speaks of his mentor Philemon from whom he derived so much? Was Philemon merely a particularly vociferous archetype? How can an image give tongue?" (Guirdham 1982, 62) Toward the end of Jung's life, he himself suspected that Philemon and other entities his family had encountered were not archetypes at all, but independent, perhaps even negative, beings.

Jung discussed these matters with New York's James Hyslop, a professor of logic and an expert on spirit possession. Hyslop had culminated an illustrious career by pioneering the survival hypothesis in relation to mental health. He found "unmistakable invasions by discarnate agencies" in a number of his so-called MPD cases. After speaking with Hyslop, Jung felt that spirit presence—not the unconscious—was the best explanation of "all these metapsychic phenomena." It was a clear case of Occam's razor: "I am bound to concede he is right," Jung wrote, in reference to Hyslop's mediumistic theory. "In the long run I have to admit that the spirit hypothesis yields better results in practice than any other" (Jung 1973, 43).

But the damage was done and has not been undone. With Jungian philosophy as a guide, theorists up to the present moment continue to float ideas such as the akashic record, akin to the Jungian collective unconscious, as a reservoir of mental images and forces accruing throughout all history and supposedly instilled in all people.* As an example, while Edgar Cayce's powers, as it was thought, came from his subconscious, there was a problem. Some said the information he uncovered lay in the akashic record. Subconscious, of course, is inner. Akashic record, though, is outer—somewhere in the great beyond. Well, which

---

*Although I object to the akashic record as a universally available pool of information, it is nonetheless my belief that impressions of long ago can be read, but only by the pure prophet developed in ethe, who is then capable of overtaking the vibrations of things long past.

was it? I believe Cayce was an open channel, a live wire, a magnet for spirits. He spoke flawless Italian while entranced, though he was not familiar with that language.

Recently I read an e-mail from a member of an online forum, claiming that

> when one attunes and can travel in spirit by going to the high heavens and stating the password, that person can read from those libraries and learn actual events in their most true events [sic]; while the Akashic record is done [sic] by meditation to focus on that book to receive a singular answer according to the question presented.

Pure (ungrammatical) moonshine.

Jung's "fantasies of chosenness" were not unlike Freud's "extravagant ego" (Gross 1978, 243) and "fantasies of greatness" (Webster 1995, 392, 334). Psychiatrist Anthony Storr thought that Jung's belief that "he had privileged access to a realm beyond consciousness . . . is close to a grandiose delusion." Storr reminds us of Jung's "coterie of close disciples . . . [who] waited hopefully each morning to hear if the great man had had another significant message from the unconscious." All this seems to suggest Jung's "insane self-absorption." Jung took his own dreams and visions as prophetic, with "special insight not granted to others. . . . It is narcissistic . . . Jung seems to have believed that he was the vessel of a higher power" (Storr 1996, 97, 104, 91), just as Freud, messianically (and narcissistically), identified himself with Moses and had "a love affair with fame" (Webster 1995, 33). Freud contended that his theory was the third and last great revolution in science, the first being the Copernican revolution and the second being Darwinism. Why, then, when his theories were finally accepted in America, did Freud blurt out, "It was like the realization of a fantastic daydream"? (Gross 1978, 10, 222).

"We are still far from understanding the unconscious or the archetypes," say Jung's followers, "[which] spring from innate mental tendencies in man . . . [and] have an enormous impact on the individual,
forming his emotions and his ethical and mental outlook . . . affecting
his whole destiny" (von Franz 1964, 376–83). What a bunch of bull.
Jung himself, tapping into Darwinian evolution, calls these "archaic
remnants . . . aboriginal, innate and inherited shapes of the human
mind. . . . This immensely old psyche forms the basis of our mind. . . .
In ages long past, that original mind was the whole of man's personality [but] as he developed consciousness, his conscious mind lost contact
with some of that primitive psychic energy . . ." (Jung 1964, 57, 88).
Pure speculation, and racist, too. He lumped together primitive tribes
and mental incompetents as persons under the same illusion: "To primitive man . . . the spirit world has a positive existence. This is . . . a naive
perception . . . projected from his unconscious. . . . Mental diseases have
also great influence in causing belief in spirits. . . . It appears, of course,
to the naive mentality that such voices come from spirits . . . [but] it is
only . . . a psychical fragment. . . . Spirits viewed from this standpoint
are unconscious autonomous complexes which appear as projections"
(quoted in Ebon 1978, 103–8).

This unproven line of thinking would have us believe that the
human psyche retains important traces left over from previous, "naive"
stages of its evolutionary development. The unconscious has supposedly
preserved primitive characteristics, archaic things from which the mind
eventually evolved—illusions, fantasies, and fundamental instincts.
Even the greatest intellects (provided they are atheists) have parroted
these questionable ideas: "We are descended," stated the brilliant atheist
L. Sprague de Camp,

> from . . . hairy little ground apes . . . and the professional criminal is
> the nearest thing we have to 'natural man,' since his life is the clos-
> est to the nomadic, perilous, predatory existence [of] our ancestors.
> . . . Men were programmed to endure risk and hardship by millions
> of years of this life. . . . There is no mystery in the fact that some

people would rather make one crooked dollar than two honest ones. (de Camp 1996, 301)

But this is a matter of lower self, not phylogenesis or any other evolutionary carryover. The low points of human character cannot be blamed on Neanderthal man or any other subhuman type.

But the idea of evolutionary holdovers appeals to sci mats. James Stein, for one, claims "There is a great deal of the brain that is a holdover from previous stages of evolution" (Stein 2012, 127). This unproven idea seems to be on a par with Jung's assertion that "the contents of the collective unconscious are not acquired during an individual's life, but are congenital instincts and primordial forms . . . the archetypes" (quoted in Ebon 1978, 110). In the same vein, von Franz asserts that the unconscious is "our instinct world": "The unconscious makes its deliberations instinctively" (Jung 1964, 67). Freud, a follower of Darwin, also believed in the animal origin of human instincts, humanity's ineradicable animal nature. To Freud, sexual and sadistic impulses were merely a residue of our animal past, now residing in the unconscious. Darwin wrote in his autobiography that we developed from a mind as low as that possessed by the lowest animal. Humans were domesticated animals, our instincts deeply rooted in animal behavior, and "savages in the natural state" are more like apes.

Holdover, instinct, baggage: are we really hardwired with primitive brains and emotions? I don't think so. We are, methinks, born neither good nor evil, only shaped by our own constitution and surroundings. Animals have instinct; human behavior is more determined by surroundings. When we are talking generally about our "instincts," often we mean good judgment or a gut feeling or intuition or perhaps merely reflex.

Our excessive passions, our tendency to violence, lust, or depravity—these are not instinct—they are only expressions of lower self! We can't blame our bad behavior on evolution!

To make the infant into a born sadist is one of the most hideous notions I have run across, as false as Freud's notion of a "death instinct."

(One Freudian argues that private plane accidents are deliberate on the unconscious level.) Humankind, opined the learned anthropologist Ashley Montagu during a 1946 interview with Albert Einstein, was not innately aggressive: "There was no such thing as a drive toward destruction . . . as Freud had postulated. . . . Indeed, human beings *had no instincts at all*" (my italics). Though I completely agree with Montagu on this point, Einstein begged to differ, recalling Freud's personal reply to him, to the effect that man is incapable of suppressing his aggressive tendencies, but with this codicil: "In some happy corners of the earth, they say . . . there flourish races whose lives go gently by, unknowing of aggression. . . . This I can hardly credit; I would like further details about these happy folk." As far as Montagu was concerned, Freud had not done his homework: "The details were available in Freud's day for the Australian aborigines, the Veddas of Ceylon, the Hopi and Zuni, the Pygmies of the Congo. Many such gentle, unwarlike peoples do exist" (Montagu 2000).

As we move now into chapter 7, we will pick up this thread, finding that human beings, rather than strapped with ancient baggage, are born a blank slate—blameless and innocent as the lamb. Corruption is in our minds, not our souls.

REINCARNATION

Body Snatching for Karma

Carl Jung seems to have invented his archetypes and memory pool just in time for the modern revival of the ancient doctrine of reincarnation. Followers today, having observed so-called recollections of past lives, have taken to calling it genetic memory. This "protoplasmic memory stream" is supposedly "in our genes, which we inherit." What!? "Your cells remember, even if you don't" (Stearn 1973, 174, 15, 67).

> *Speculation that the body "remembers" the past is blatant anti-science.*
>
> Martin Gross, *The Psychological Society*

Reincarnation seems to have been dreamed up by the good idea fairy. It makes sense, proponents exhort. "Pity that so many have fallen in this trap," wrote England's James Webster, author of *The Case Against Reincarnation*. "The virus is caught by so many, all and sundry. The attraction seems to be rather like the flies that are drawn to the allure and perfume of the Fly Catcher plant, little realising the danger" (personal communication).

Not far from me, here in Georgia, an Atlanta past-life therapist claims to eliminate depression and other problems by clearing out her patients' cellular memory. Memories from previous incarnations, she asserts, are stored in the file cabinets of our subconscious mind. The body, as the argument goes, remembers everything: "The soul imprint is on every single cell in the body" (MacLaine 1986, 324–25). Such are the entirely hypothetical pronouncements of certain new age "healers," some of whom then offer (costly) procedures, such as acupuncture, to release those horrid imprints. The claim entirely ignores the findings of modern science, whereby the myelin sheathing of the fetal brain and even of the infant is not sufficiently developed to carry any memories whatsoever.

> *Children come into this world with no baggage except for their innocence.*
>
> WILLIAM PETER BLATTY, *LEGION*

Reincarnation is the philosophy that has the infant arrive in this world loaded with baggage from past existences. This metaphysic stands in contrast to doctrines such as Kabbalah (Jewish mysticism), which speaks of the soul given in a pure state. It was also enlightenment philosopher John Locke's belief that the newborn arrives clear and beautiful, fresh from the hand of the Creator. Yes, a tabula rasa who will become a person in his own right with the unfolding of the years.

> The first of man, the newborn babe, I created a blank. . . . Yea, from the hour of conception it is a new star in the world, and it magnifieth itself forever.
>
> OAHSPE, BOOK OF KNOWLEDGE 2:24

The little one, rather than saddled with an ancient personality, wakes up in this world for the first time, knowing nothing, understanding nothing, not yet a person, like the bud that is not quite a rose. All he needs is time to become himself.

If astrology (chapter 5) says the constellations determine our temperament, reincarnation says it is a former self that shapes us, even limits us. If astrology holds that certain planets imprint traits on us at the moment of birth, reincarnation contends your past self enters you at the moment of birth. Now here, finally, is a theory—reincarnation—that at least grants us a soul (as opposed to the stark materialism of Freud). But without knowledge of the soul's journey and of the mechanics and administration of the spirit world, it is impossible to determine the truth or fallacy of this doctrine.

Its subject is, after all, what happens to us after death. Where does the soul go? Who are the experts here? Who do you call? Who are the final arbiters of the afterlife? Theologians? Psychiatrists? New age gurus? Metaphysicians? Our best bets are seasoned psychic researchers, particularly the seasoned spiritualist. As an academic researcher, the late Ian Stevenson ranks high. But even he, who was perhaps the most qualified reincarnation investigator of the twentieth century, had to admit that "We know almost nothing about reincarnation" (Stevenson 1987, 260).

The learned spiritualist is not as diffident or clueless as Dr. Stevenson. The well-trained adept is quite familiar with the broad sweep of overshadowing by discarnates, and this includes the illusion of past-life recollections. The keynote to the debate is this: the doctrine of reembodiment is based on the premise that only past lives can explain our personal knowledge of other existences, like Billy Milligan's or Sybil's, as we saw in the last chapter. Time and again the argument

Fig. 7.1. The afterlife.
Cartoon by Marvin E. Herring.

is put before us that there is no alternative to the rebirth interpretation. Therefore it must be true.

But there *is* an alternative: the overshadowing, the power of otherworldly inspiration and intrusion. The wide-ranging forays of the human psyche—beyond the physical—have been very well documented in the phenomenon known as NDE—near-death experience. Many forms of such ASCs (altered states of consciousness), as well as dormant mediumship, have been mistaken for visions of one's own former self. Granted, in most of us the psychic faculty is but a seed, an untapped potential, lying fallow until something happens to expand it. This latent force—this overlooked, slighted, even suppressed and despised aspect of self—will open a door that has been closed for too long.

With it we will put the psi back in *psychology.*

We will learn how to handle the strange mindstuff that triggers "disease of the soul" (Philo). We will untangle the tangled web of irrational disorders, remove the stigma of the uncanny, and put to rest— finally put to rest—the freakish and phony black art of reincarnation.

"The individual is preserved," opined Protestant theologian Paul Tillich, "but only in the realm of essences." Spiritualism proper, which deals in those "essences" and has never endorsed the theory of plural

Fig. 7.2. Rebirth. Cartoon by Marvin E. Herring.

lives (though too often it is confused with it!), is, simply stated, the belief in immortality of the soul, not the body. Big difference.

Quite plainly, it is only the spirit part that lives on (in the "realm of essences"), improving by and by, becoming stronger and wiser and more useful. When reincarnationists claim that the essential part of the person becomes reembodied in new earthly lives, the argument is precisely the opposite of the spiritualists', which sees the essential part, the spirit part, ascend and make its halting progress through the stratified heavens.

> One great light I have bestowed upon all men, which is that they can progress forever. Though the waters of the ocean rise up and make clouds; and the clouds fall down as rain and run to the rivers, and from there back into the ocean, and this repeats a thousand times, ten thousands of times, yet that water has not progressed. Neither have I given progress to a stone, nor to a tree, nor to an animal; but I have given progress to man only.
>
> OAHPSE, BOOK OF FRAGAPATTI 21:18

This, my friend, is a one-way trip. There is no turning back. The survival hypothesis posits an afterlife, a real and active postmortem existence that unfolds on (nonphysical) planes of livingness just beyond our grasp.* Spiritualism's central dictum is the existence of a hereafter, which is not a staging ground for the coming and going of souls but a continual unfolding, the eternal progression, onward and upward always, worlds without end.

SPIRITUALISM AND REINCARNATION

We climb from world to world in the continuing Road of Life.

HOPI MYSTERY PLAY

---

*See my blog, "Tourist Guide to the Afterlife."

Confusing spiritualism with reincarnation (they are most definitely not the same) has become a real problem. Consider the Bridey Murphy case. "The eventual debunking of the [Bridey] story," wrote one scientist, referring to the well-known reincarnation hoax in the 1950s, "helped to shape my view of supernatural and paranormal phenomena. . . . I'm not going to believe in reincarnation, or ghosts . . . or out-of-body experiences . . ." (Stein 2012, 26). Throwing out the baby (spiritualism) with the bathwater (reincarnation), Dr. Stein was talking about the sensation started by Morey Bernstein's book *The Search for Bridey Murphy*, in which a Colorado housewife named Virginia Tighe was taken (mistaken, actually) for the reincarnation of a long-dead Irishwoman named Bridey Murphy. It turned out that someone named Bridie Murphy Corkell had been a neighbor of Virginia's when she was a little girl.

Three months after the publication of Bernstein's book, America's leading expert in parapsychology, J. B. Rhine, was asked what he thought of Bernstein's claims: "Leading a person back through hypnotic regression—as was done in this case—is the wrong road to take. Science should first attempt to discover whether there is a *spirit personality*. . . . Remarkable persons . . . while in trance . . . could receive messages from spirits trying to reach living people" (my italics) (Rhine 1956, 300).

The fact that Virginia Tighe had the ability to enter immediately into deep trance (less than 12 percent of the population can do this) suggests that she herself was a psychic-sensitive, capable—even unbeknownst to herself—of making contact with departed souls, the spirit personalities of which J. B. Rhine spoke. In fact, Bridey herself comes off as a typical earthbound soul whose best shot at some "action" is through the sensitive, the clairvoyant, the mortal receiving station. Catholic, Bridey said she was surprised that she had not gone to purgatory upon her death in 1864. Instead of beginning the upward journey, she stuck around. "I stayed right in that house until John [her husband] died." After which she proceeded to overshadow her brother Duncan:

Did you ever let Duncan know that you were there?

He wouldn't answer me.

How did you try to speak to him?

I would stay there by the bed and talk, and he would never see me (Bernstein 1965, 132).

I must say, it was with considerable cunning that the "splinter" sect or "combination" of reincarnationists established itself in the tailwind of true spiritualism, shanghaiing the curious and gullible and inverting the meaning of *immortality*. All this seems to have been orchestrated by certain spirits, the same earthbound factions that, centuries ago, abrogated the higher heavens and set themselves up as God of All in the lowest heaven.

Thousands of millions of the spirits of the dead ... know not the plan of the resurrection to higher heavens ... not even the organizations of the kingdoms in the lowest of heavens ... and [these] do inspire mortals with the same darkness.

OAHSPE, BOOK OF ESKRA 1:4–7

The immortal soul flies out in empty space
To seek her fortune in some other place.

OVID, *METAMORPHOSIS*

But when a believer in reincarnation dies, his spirit falls inexorably into the otherworldly precinct occupied by those angels who teach reincarnation. And he becomes their servant, thus helping to make their heavenly kingdom large and powerful.

Thus do ye prepare mortals to become slaves in your heavenly kingdoms ... building unto your own personal glory. For ye take advantage of the [credulous] mind, to bend it away from eternal progress, that ye may inherit it as your dutiful subject.

OAHSPE, GOD'S BOOK OF ESKRA, LII, 22–23

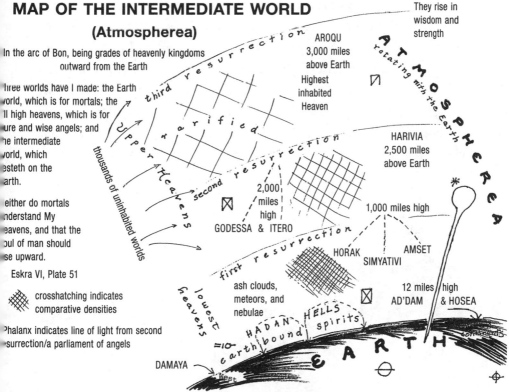

*Fig. 7.3. Top left and right: Resurrection cartoons by Marvin E. Herring.
Bottom: The resurrections, after Oahspe Plate 51.*

Pythagoras (582–507 BCE) was a reincarnationist who fancied he had died at the siege of Troy. The Pythagorean brand of reincarnation, teaching the transmigration of souls into animal bodies, was (more than two millennia later) elegantly routed by William Shakespeare:

> What is the opinion of Pythagoras concerning wild-fowl?
> That the soul of our grandam might haply inhabit a bird.
> What thinkest thou of his opinion?
> I think nobly of the soul, and no way approve of his opinion.
>
> (FROM *TWELFTH NIGHT* WITH SIMILAR VIEWS
> EXPRESSED IN *AS YOU LIKE IT*, III: II
> AND *THE MERCHANT OF VENICE*, IV: I)

Reincarnation today sees rebirth as a golden opportunity, the choice of when and where to "manifest" presumably made according to the soul's purpose. In a rather well-known case, a reader of past lives once told the fantasy novelist Taylor Caldwell that she had been a writer before, in a previous life, but not terribly successful, partly due to her gender. "This is why it was decided that in this lifetime success would be yours. . . . You chose to come back into an incarnation at a time when the work of women has been accepted" (Stearn 1973, 255).

Fig 7.4. Reincarnation.
Cartoon by Marvin E. Herring.

According to these beliefs, then, we are choosers. We choose our parents, we choose our century, we choose our circumstances. We choose to come back and learn and evolve (to paraphrase James Van Praagh). But if this process is optional—a choice—what happens to those who do *not* choose to come back? What about the billions of people on earth who do not believe in reincarnation and will take their chances in the "upper country," rather than return for another bout in the body? Significantly, Professor Stevenson openly admitted that reincarnation is hardly reported in societies that do not believe in reincarnation. Conversely, most of his cases come from societies that do believe in reincarnation. I'll say this: those who believe in it are most likely to try to come back.

One member of Stevenson's team, psychiatrist Jim Tucker, took up the problem by proposing that some people have past lives "while others would be here for the first time" (Tucker 2005, 199). Doesn't that imply that the first-timers have their own soul? Of course it does. But that leaves me wondering about the mind-bending logistics of it all, and the probability that we each have our own soul. Given the scenario of optional reembodiment, some people are presumably reincarnated and others not. Right? So, exactly what soul inhabits the "nots"? It would be their own natural spirit. And it follows that that natural spirit must be pushed out or replaced when a reincarnater comes along for a brand-new life in flesh.

I find it interesting that the business of entering new bodies—and living again—was not as universally credited in classical India as thought. What do you think of this amusing, but instructive, Sanskrit fable from the *Panchatantra*?

One day a king was one day passing through the marketplace when he observed a hunchbacked merry-andrew whose contortions and jokes kept the bystanders in a roar of laughter. Amused with the fellow, the king brought him to his palace. Shortly thereafter, within earshot of the clown, a necromancer taught the monarch the art of sending his soul into a body not his own.

A little while after this, the monarch, anxious to put in practice his newly acquired knowledge, rode into the forest accompanied by his fool.

Presently, they came upon the corpse of a Brahmin lying in the depth of the jungle, where he had died of thirst. The king, leaving his horse, performed the requisite ceremony and instantly his soul migrated into the body of the Brahmin, while his own lay lifeless on the ground.

At the same moment the hunchback deserted his body and took possession of the king's!

Shouting farewell to the dismayed monarch, the merry-andrew rode back to the palace, where he was received with royal honors.

But it was not long before the queen discovered that something was amiss. When the quondam king (in the Brahmin's body) finally arrived and told his wife of his plight, a plot was devised for the recovery of his own true body.

Just so, the queen asked her false husband, the merry-andrew, whether it were possible to make her parrot talk. In a moment of uxorious weakness, he promised to make the bird speak. Laying his body aside, he sent his soul into the parrot. But immediately the true king jumped out of his Brahmin body and back into his own. Then, with the help of his queen, he proceeded to wring the neck of the hapless parrot.

In other words, those who parrot the underhanded philosophy of the necromancer may risk being throttled. In any event, grabbing other people's bodies, in the end, will not save our own neck!

The new age has been with us for a while, apparently offering various alternatives to religion. The failure of the old religions has left a vacuum. The recovering Catholic, as the phrase goes, or the renegade Christian, or even the born atheist still craves a taste of the sublime, of that which is permanent and full of glory. But with the widespread rejection of formal religion has come another god, another kind of blind faith—the church of materialism itself, the tabernacle of the flesh— and, with it, the rite of reembodiment. "The secret source" of reincarnation's popularity, surmised William Butler, a nineteenth-century thinker, is materialistic prejudices. What he meant is the glorification of man by resurrecting him in the flesh. And in doing so, we confine all our beingness to the level of Earth, the physical. Karma can only be worked out in a body. Thus does the spirit world lose all importance except as a tawdry waiting room, a holding tank, for souls hoping to return to materiality in brand-new bodies. And all the great truths of the Ongoing are cast aside.

The theory of reincarnation makes karma the kingpin: all suffering can be explained by past-life mischief, without bothering with present-day inequities and human misery. Karma will explain all. Why do we suffer? So that we may atone for our peccadillos committed in a former existence! For instance, the family that produced a Down syndrome child was told that the little boy's past-life sins created the "need to incarnate in a deformed state" (Langley 1967, 38). The scars of our passions, it is explained, "reach down the centuries" (Langley 1967, 38), and the resulting torment is the only path to redemption for our blighted selves.

Ignoring any monstrous causes from this very life, the reincarnationist tells us to trace the root of our problems to past-life wrongdoing. And it is a law, "fixed and unalterable" (Langley 1967, 38). But this fictitious logic breaks down when we consider the further claim that we have no knowledge of our previous existences. How then can we learn from putative sins we cannot even remember?

Karma was the connecting thread in the life readings given by Edgar Cayce and subsequent imitators. Many times, states a Cayce biographer, "He had to exorcise the dark tenets of Predestination and Original Sin from the hopelessness and confusion in the minds of the people who came to him for help" (Langley 1967). Cayce, in other words, had something better to offer guilt-ridden Christians. But let us ask: is there any difference between the reincarnationist's "fixed and unalterable" karma and the very gospel of original sin and predestination, which the Cayceites so deplore?* After all, the newborn babe, in their philosophy, is shorn of innocence and purity by the daunting law

*To Arnold Toynbee, the two doctrines are almost identical: "One variant is the Christian conception of Original Sin; the other is the Indic conception of *Karma* . . . and [they] agree in the essential point of making the spiritual chain of cause and effect run on continuously from one earthly life to another" (Toynbee 1961, 447).

of karma. He has sinned. His birthright is, in this sinister philosophy, sullied by the taint of alleged crimes and misdemeanors in a past life!

Instead of fledgling hope and fresh life, the new arrival is branded a repository of decrepit and vile karma that needs to be purged! Who needs original sin if we have karma to ruin our destiny! First, they decry the Christian concept of sin, then bring it right back, disguised as our own fault, our own karma. One reincarnationist has tried to explain the difference: "All men are certainly born with an inheritance of sin, but it is original with themselves and not with the allegorical characters of Adam and Eve." Enough to turn one's stomach, this writer goes on to offer as an example the plight of slaves brought from Africa, who might well "think on their own [past-life] sins" (Cerminara 1967, 208). Does this, then, justify slavery? Neither is this approach any different than the medieval belief that persecution and hardship are divine punishment for our sins.

One reincarnationist goes on to enlighten us that "the cosmic justice pattern of the Law of Karma might bring a sense of justice to . . . a world . . . indifferent to the way it distributed wealth and misery" (Weisman 1977, 42). I note that this brand of justice is peddled by white upper- or middle-class reincarnationists who, washing their hands of human misery, are too attached to the status quo to do anything about it in the here and now. Let the future (another lifetime) take care of all those social inequalities that they bemoan. Let some future embodiment work out the balance sheet, thus keeping racism and every other social shame alive and well.

Thus does the theory betray its own spineless complaisance, whitewashing all the troubles of this sorry world. Not to worry, future lives will iron out all the disparities. Alfred North Whitehead was once asked his opinion of Buddhism, a religion known to embrace the rebirth doctrine: "It is a religion of escapism," he replied. "You retire into yourself and let externals go as they will. There is no determined resistance to evil. Buddhism is not associated with any advancing civilization" (Whitehead 1956, 155). Apart from its escapist value, I really don't understand the siren call of a concept that flings us from life to life, in relentless unfulfillment of justice, leaving us a half-complete being—a slave of rebirth, bondsman of the past, beggar of the future.

FROM INN TO INN

Reincarnation is an alluring form of escapism which belongs in the wilderness of human thinking, the jungle of false concepts.

HAROLD SHERMAN, *YOU LIVE AFTER DEATH*

So we can just sit back, relax, light up a Lucky, and let karmic justice take its slow course. No need to lift a finger for your fellow man, because "superior spirits" have whispered "the answers" in your ear, telling you reincarnation is the great equalizer. In our many lifetimes, says professional reincarnationist Genevieve Paulson, we will experience all ways of being, as well as all races. Really? Not according to Miss Paulson's fellow reincarnationist, James Perkins, who in *Experiencing Reincarnation* (1977) discusses the Japanese people and their repeated incarnations in the same group: "This tendency to reincarnate in the same race is even more evident in the Jewish people," says Perkins.

"A man has a soul, and it passes from life to life, as a traveler from inn to inn," was H. Fielding Hall's (1859–1917) pleasant dream. But wait a second. The traveler is not really himself. I am not really me. No, I'm somebody else freeloading in "my" body, the regurgitation of some old soul. We are recycled souls using this life to work out some ancient karma! The deformed, bizarre, and illogical dialectic of rebirth is truly stupefying.

And if rebirth is true, when did it begin? In the "last" lifetime? The one before? How far back must we go to find the original soul? Neanderthal times? Adam and Eve? The big bang? One celebrity reincarnationist solves the riddle: "We have all experienced many, many incarnations in many forms from the beginning of time" (MacLaine 1986, 313).

So, unless we go back to the "beginning of time," even our past self was not himself; he was only someone else living in his body. And if I was a different gender in a past life, what am I now? A man in a woman's body?

One Western writer informs us that a member of India's untouchable caste was "born into so low a stratum of society . . . [to] expiate the arrogance and evil of a former incarnation" (Cerminara 1967, 219). So that makes everything copacetic. It is significant that the rebirth doctrine as "carrot" attained a certain foothold in the land of India. What could be more welcome in this land of fixed caste than a doctrine that lets the turn of the wheel give "a chance for the development of the neglected or abused"? (Head and Cranston 1961, 24).

With karma we are handed the biggest voodoo of them all, the greatest folderol of all time, the slyest sophistry known to humanity: that the immortal soul can (and should) enter a mother's womb for a new round in the school of life. For return we must, on the pitiful wheel of rebirth known to the Hindu as *samsara*.

One common criticism of reincarnation focuses on the unlikely

Fig. 7.5. Reincarnation. Two cartoons by Tony Webster (courtesy of James Webster, The Case Against Reincarnation: A Rational Approach, *2009)*

glamour and fame of one's past life (as Cleopatra, Joan of Arc, Napoleon, and so on). Defenders of the faith have attempted to counter the charge by citing cases of a humble, poor, drab, uninteresting prior self. Plain old Bridey Murphy would be a good example. Yet a greater, more serious reproof exists—the frequent occurrence of sudden death suffered by those who have now "come back." This is a most telltale statistic—a life cut short. The case histories of so-called past lives are top-heavy with sudden death. Why? Often quite oblivious of their abrupt transition, such spirits do not rise up and may—even inadvertently—lodge themselves in the aura of the living.

SUDDEN DEATH

If you sample the reincarnation literature today, you find that perhaps seven or eight out of ten such "previous personalities" met a violent or untimely demise. Now this is something to ponder. Contrast this 75 percent with the mere 7 percent of the general population who die in a sudden or violent manner. How can reincarnation be natural, as proponents claim, if this lopsided statistic puts the lie to its normality? Clearly there is something else at work. That something else brings us back to the invisibles, the unquiet dead whose traumatic or abrupt demise becomes the source of a great array of ghostly disturbances (like the murdered prostitute whose "return" we examined in chapter 6).

Speaking of murder, let me cite a book on children's past lives written by Carol Bowman, a psychotherapist and reincarnationist. One can scarcely find among her various case histories a commonplace sort of death for the "previous personalities" mentioned: "'killed in a quarrel" . . . "killed by an angry customer" . . . "gunshot" . . . "killed by the train" . . . "fatal wound" . . . "shot in the back of the head" . . . "killed in a gang shooting" . . . "robbed and killed" . . . "hit by a car" . . . "one of the Marines killed in the barracks explosion in Beirut in 1983" (Bowman 2001, xvii, 16, 211). Truth to tell, this is standard fare in the reincarnation literature! It hardly varies: past scenes of horrific crime, accident, slaughter. Where is the (statistically) expected 93 percent of natural deaths?

It's not there. As I say, something else is going on.

Bowman further informs us that a child will matter-of-factly tell them [her parents] she remembers being shot, or dying in a car accident or in a war. She even quotes her own children's stories: "My daughter . . . remembers dying long ago in a house fire, and my son gave a realistic description of dying amid the horror and chaos of a Civil War battle." It is much the same in Helen Wambach's survey, where allegedly reincarnated subjects are simply giving voice to the sudden dead. One of Wambach's subjects told her, "This current lifetime . . . seemed the direct result of my childhood traumatic death in . . . past life" (Case A-440, death at age four in a bombing attack in Rumania, 1942); another said, "I felt I had just died in a war" (Case A-148) (Wambach 1979, 47, 58). Et cetera ad nauseum.

Nevertheless, sudden death, especially in war, is precisely the factor behind the "return" of discarnates to the earthly plane.

> There are such as are slain in war, who . . . dying in the heat of passion and fear and anger, become wild and [earth-]bound on battlefields, or, mayhap, stroll away into deserted houses . . . and are lost. . . . Of these there are hundreds of millions; and they are in all countries and amongst all peoples in the world.
>
> OAHSPE, BOOK OF JUDGMENT 32:9–10

In a French case, an eighteen-year-old girl named Marie was put into a hypnotic sleep. She was regressed to a time before her birth in which she "was" (i.e., channeled) a fisherman's wife named Lina. It seems that the death of Lina's little child, followed by the loss of her husband in a shipwreck, drove her to despair: she jumped into the sea from a high cliff and drowned herself. But Marie's mesmeric vision was not past-life recall at all, only hypnotically induced clairvoyance, bringing her to the threshold of the unquiet dead, most likely a case of temporary spirit obsession. "Reincarnation is spirit obsession and the squandering of one's true spiritual destiny," wrote Faithist Bob Bayer (personal communication).

A past-life showman type, in another case, said he wanted to help people by unlocking "all their lives." In one of his sessions, a woman began to experience the pain of an ill-fated, prolonged delivery aboard ship on the high seas; she died in childbirth. Vivid as the impressions were, the woman in this "guided recall" was not sure what she was seeing and feeling. Her instructor, gently initiating her, "knew" it was a past life (Weisman 1977, 126). Not in my view. She had simply encountered someone else's tragic end, in the deep. It is often the nature of spirit influence to be felt as if it comes from self.

In Ian Stevenson's extensive studies, too, a grossly disproportionate number of the previous personalities had died violently. "Many have ended in murder," a fact Stevenson considers potentially most important, since the statistic far exceeds the incidence of violent death in the general population. Yes, of course. But then he attempts to explain it, or rather explain it away, by reasoning that violent deaths are more memorable than natural ones. Among Stevenson's cases of children claiming reincarnation, the following are the ways the incoming personality had died: shot by brother; killed in World War II, drowned in a flood at age eleven; heart attack; murdered by two men at a fair; stabbed and robbed; hit and run; died at eighteen of severe diabetes; "died young"; hit by a car at age six; car ran off road at age eighteen; promised his niece that he'd come back as her son.

In one striking case, a little Belgian boy fancied himself his own uncle who had died in World War I in a hail of machine gun fire. Hearing the click-click of an old-style camera, the three-and-a-half year old screamed, "Don't! They killed me that way last time" (Stevenson 1987, 55, 118, 160). To the spirit-release worker, though, it is a case of entity obsession and should be cleared!

Another "life reader" blames the tension in postwar generations on "a carryover from the frightening conditions under which many of them recently met death in World War II" (Quinn 1975, 48). In the same vein, "past-life regressionist" Bettye Binder claimed to have had countless dealings with casualties of the Second World War. Reborn shortly after the end of that conflict, these souls came back quickly because they died young and felt cheated or had left unfinished business. "Murder victims incarnate almost immediately," agrees Sybil Leek in *Reincarnation: the Second Chance.*

Did they really "come back"? Or did they ever leave the earthly plane, lingering in the Outer Darkness, "the gray condition, . . . clinging desperately to physical existence" despite the loss of a body (Perkins 1977, 101). Interesting that among one northwest Amerindian group it is held that those mutilated and slain find it difficult to reincarnate. Instead, they simply lead the ghostly life, never quite able to either make it up to the spirit world or back to the earthly one. They are the same kind of earthbound entities we get in stubborn poltergeist cases and hauntings (as discussed in chapter 6). In a word, sudden death and earthbound go hand in glove throughout the enormous literature of ghosts and hauntings. And this situation is possible thanks to the suddenness itself: many do not even realize they are dead or know where they are. Why has this truth been so patently ignored by the merchants of rebirth? Because they, too, are sci mats, unable to conceive of life outside of a flesh body.

Dr. Stevenson thought past-life stories of children were particularly convincing. In Thailand, for example, he studied a very young boy who had told his parents that his past self had been robbed and killed in a town some distance away. It seems that this victim, after sitting (as a spirit) in a tree near the murder site for several years, finally spotted a man he thought he should follow home. He did. The man's wife was pregnant and the murder victim was "reborn" through her. As the investigation deepened, the father of the boy confirmed being in that town on the day referred to, while his young son provided many veridical details of the man who had been robbed and killed.

INVISIBLE PEOPLE

Surely the circumstances could just as well indicate that this child was haplessly overshadowed and obsessed by that earthbound spirit who had been left bewildered by his brutal demise. But that the spirit became the very boy is another story—a theory waiting for "evidence." Stevenson's many anecdotes, rather than proving reincarnation, actually document the typical stunts of entities lingering on the earthly plane. Indeed, Stevenson's cases "suggestive of reincarnation" fall apart completely when *rebirth* is understood as "body snatching."

A Hindu man, for example, is poisoned to death; instead of waiting the theoretical ten or twenty or fifty years, he reincarnates immediately. Not only does the time lapse fail to fit the theory, but the fresh body is not that of a newborn. The spirit of the poisoned man has overtaken a three-and-a-half-year-old boy who has apparently died of smallpox, leaving his body vacant. And sure enough, upon reviving, the little boy identifies entirely with the life and habits of the poison victim. He is no longer quite "himself" and instantly recognizes all "his" former family members when taken to them (Langley 1967, 231). The scenario, though, is seriously akin to ghostly intrusion, the mortal victim (or "host") serving merely as a target of opportunity for the pathetic, hovering, homeless spirit.

None of this can be understood or appreciated without a grasp of overshadowing, the subtle ways in which the unquiet dead can impinge on the living. Over the years I have amassed a bulging file on the (negative) overshadowing of criminal personalities. I call the subject psi crime.* Richard Chase, the notorious "vampire of Sacramento" (can anyone doubt that this mutilating cannibal was demon-driven?) had exhibited the famous "homicidal triad" marking the criminally obsessed youngster: bedwetting, animal torture, and arson. Neighbors frequently saw a zombielike Richard rambling "aimlessly around the grounds; he would wander unsteadily from place to place." To the voices coming from "invisible people," Richard would answer, "Stop bothering me" or "I'm not going to do that" (Markman, *Alone with the Devil*).

---

*See, for example, my article on Seung-Hui Cho, the Virginia Tech slayer: "School Shootings and the Paranormal. Case Study: Cho, the Question Mark Kid," *Journal of Spirituality and Paranormal Studies* 32, no. 4 (October 2009): 211–19.

Among Richard's reading material, detectives found news articles about violent deaths and Old West outlaws. It happens that in 1976, one year before Richard's reign of terror began (six homicides), he became convinced that he was the reincarnation of one of the Younger brothers, the bank robbers who rode with Jesse James. Psychiatrist Ronald Markman devoted a chapter of his book, *Alone with the Devil,* to the completely deranged Richard Chase, concluding that he was motivated by subhuman forces. But I am afraid those awful forces are very much within the human family, though at the lowest end of the spectrum. (See the information about *drujas,* in chapter 8, especially the caption for fig. 8.1 on page 323.) No, we cannot escape or explain away the horror by calling it "subhuman"—a label that explains nothing. "Nonperson" (emptied of self, by some prior—probably traumatic—misfortune) may be closer to the mark, and is indeed the Russian term for sociopath. This nonentity is flypaper for the vagabond spirits that flit about in the darkest corridors of hell.

THE WORST TORTURE

Past-life regression therapists sometimes have clients who come to them with fears of lightning or of being trapped and suffocated or of falling from a height or jumping. The reincarnationist then concludes that "hundreds of documented cases of . . . phobias . . . stem from the way they died in a past life" (Bowman 2001, 17) or that "the emotional trauma of a violent death can carry over from one life to the next" (Tucker 2005, 123). But I have to ask: Why should an abnormal state, such as a phobia, become the paragon of reincarnation, if rebirth is normal and universal? Cases of reincarnation should (statistically) not be crowded into the confines of abnormal psychology, should they? They should be natural and normal. They should be everywhere. But they are not.

If the "evidence" takes us anywhere, it is manifestly to the casualties of sudden death and the victims' persistent haunting in the gray area of psyche's outer reach (see fig. 6.3, page 255). I offer this interpretation as an alternative to the reincarnationist's touted business of "family return,"

Fig. 7.6. Reincarnation cartoon by Tony Webster.

whereby the departed relative makes a giant U-turn in the sky and goes right back to his earthly home to reincarnate as a newborn member of his own family! You can be your own grandson! You can practice "mother switching," wherein the deceased mom is reborn to her own daughter!

Such possibilities exist, says the reincarnationist, because some souls choose the circumstances of their next incarnation. I wonder why some souls take advantage of this cosmic law and others do not. Theorists who have cottoned to this choosing idea tend to overrate the concept of free will. Malachi Martin, for example, asserts that victims of possession choose their possession. Roger Depue, for another (and most of his colleagues in the FBI Behavioral Science Unit), believe that sociopaths also choose the evil path. It is very much a Western concept, this limitless choosing. But note that in the *wuqabi* cult of Ethiopia, the possessed person "has no recourse but to accommodate the spirit. Possession is instigated by the spirits; people cannot choose spirits and cannot choose to be possessed" (Crapanzano and Garrison 1977, 195).

The overwhelming evidence identifies would-be, wannabe reincarna-
tors as none other than those who died and went into spirit believ-
ing they could come back. During Dr. Carl Wickland's "thirty years
among the dead" (the title of his book), the spirit of William Stanley
spoke through the mediumistic Mrs. Wickland:

> I was a Theosophist and I wanted to reincarnate . . . I got into a
> child's body and crippled it; and also crippled my mind and that
> of the child. . . . *I did not know how to get out.* . . . The state I
> was in was the worst torture anybody could have. . . . I want to
> warn others never to try to reincarnate through a little child. Leave
> reincarnation alone. It is only a mistake. . . . There is no such thing
> as reincarnation. One gets all mixed up trying to enter another's
> body (my italics).

PANGS OF THE DEAD

The reincarnationist, decking himself out in the garb of spiritualism,
has loftily pronounced on matters of soul, karma, heaven, and cosmic
laws. These are all crotchets. In reality, the doctrine of rebirth offers
precious little illumination when it comes to the truths of life in spirit.
How presumptuous, how misleading, is the sweeping claim that the ills
of today stem from some bygone existence. Past-life readers, profession-
alizing the sham, slickly apply this fake formula: Do you suffer from an
allergy to dust? But of course! You died, in your former life, in a mine
explosion's cloud of dust. Nonsense!

In his fascinating book, *The Psychic Dimensions of Mental Health,*
England's Dr. Arthur Guirdham, though a reincarnationist himself,
showed that the sensitive empath (the host) is all too prone to the tra-
vails of others, both living and dead. He should know—he's one (an
empath):

"Sometimes I acquired the symptoms . . . of patients who began to
recover at the same time that I assumed their symptoms." The good

doctor had also been assailed by the pangs of the dead. During his stay at an inn near Otmoor he was overtaken with violent shivering, cold, and jaundice for two days running. Afterward he found out that Otmoor was one of the last places in England blighted by malaria. Only then did he realize that he "had experienced all the symptoms of malaria. . . . It seems to have been a case of picking up the psychic vibrations" (1982).

To Guirdham we can also turn for an understanding of why we pick up certain "psychic vibrations": "We share," he says, "a common psychic life and exchange with those on the same wavelength, not only thoughts and feelings, but the syndromes of disease." None of this, of course, is reincarnation. It is the unseen influence, the overshadowing, a form of mediumship sometimes involving an ongoing relationship with the departed.

> *We would associate, after death, with persons of attainments similar to our own.*
>
> IAN STEVENSON,
> *CHILDREN WHO REMEMBER PREVIOUS LIVES*

There is, after all, no firmer principle of spiritual science than "like unto like," what Guirdham calls the "same wavelength." Herein we make our destiny, pull in our fate, willy-nilly creating our own spiritual gravity. America's great medium, Arthur Ford, when asked to explain spirit-familiars, simply pointed out that they are "drawn to you because you are doing something that they were doing" (Editors, *Psychic* 1972, 28). Particularly in the arts and creative work we hear the occasional story of overshadowing by a talented forebear. A St. Louis author named Dorothy Wofford, for example, wrote poetry that turned out to be a facsimile of verses written by a seventeenth-century poet named Anne Broadbent. The matter was investigated and it became clear that Wofford shared some unique characteristics and tastes with Broadbent. Despite the suggestion of reincarnation, researchers concluded that the St. Louis woman's writing was a case of unconscious mediumship or slight possession by the long-dead poetess.

"Suppose an inventor passes on before completion of his invention," hypothesized one of Dr. Wickland's discarnate "guests," in an effort to illustrate this kind of influence.

> He will not give it up. He studies it on the spirit side of life . . . he finds some [mortal] sensitive and impresses the invention on his mind. Then that one . . . perfects it, and gives it to the world. If I impress a sensitive with an idea, in one sense I reincarnate—not in his body—but by impressing him with what I want done. . . . Everything that is invented on earth has first been invented in the spirit world. (1924)

RAPPORT

It is relevant, I think, that Joseph Dunninger, the famous mentalist, was convinced that inventors (intrigued with his telepathy) were flocking to him because they themselves received impressions and ideas "that seem to come from some outside source" (Ebon 1971, 136). This is inspiration in its most literal sense. It isn't past life and it isn't "memory" and it isn't the subconscious. It may not even be insight. But what it does imply is an ASC. Albert Einstein, for one, cheerfully admitted that the theory of relativity came to him "mystically." Isaac Newton, for another, made a point of literally "sleeping on" his most troublesome mathematical problems. (The dream state is but another ASC.) Likewise, Thomas Edison was a "flylight," deliberately taking catnaps at his desk to escape his (beta) brain, oftentimes waking up with the solution to a stubborn problem. And like Socrates of ancient Greece, the American statesman Josiah Quincy believed he was inspired in important crises by his own "daimon"; he would sometimes declare that the happy choice of his wife had indeed been made in heaven.

But when the artist or composer or author or inventor claims to be the incarnation of some great genius of the past, let us smile, remembering that the word *genius* itself (variants: *genie, jinn*) suggests a subtle form of spiritual rapport, mind to mind, less rare than we think and also less bizarre than the riddle of reincarnation. Joel Whitton, Ph.D., a reincarnationist, observed that the patient "instinctively accepted these characters as himself" (Whitton and Fisher 1986, 85). Well, yes, of course: most people are not aware of the invisible guest (= *geist* = *ghost*); and if a spirit comes to you, you might "absorb his forces, and then think it is an echo of your own mind" (Newbrough 1874, 55).

We need a better understanding of these guests and visitors. Susy Smith, who was a well-known twentieth-century psychic, author, and ghosthunter, was once afflicted with a malevolent spirit in her sleep. Waking up in a fright, she realized, "I was myself somehow confused or identified" with the intruding entity (Susy Smith 1970b, 153). In a similar haunting, an old ghost made his appearance in a householder's sleep: "In the nightmare, Jim saw himself as an old Indian, wrapped in a blanket, choking to death" (Scott and Norman 1985, 434). But it was, of course, not himself, simply some spirit *en rapport* with Jim.

IMPRESSIONS

Shortly after the tragic death of his son, America's Bishop James Pike was interviewed and questioned on these phenomena. In cases of "reincarnation," Pike said, when a person thinks he's recalling certain things from past lives, he is perhaps being "possessed by the deceased person" (Editors, *Psychic* 1972, 148). Medical personnel have commented on a similar form of possession (not reincarnation) that may come into play after surgical procedures involving organ transplants. England's Sue Allen, for instance, familiarizes us with a few such cases where the "recipient of the donor organ starts to have memories [read: *impressions*] that are not theirs or cravings . . . that they did not have prior to the operation" (Allen 2007, 123). There may be striking behavioral changes as well. Allen, a de-possession therapist, is then given the task of releasing those donor spirits, wandering spirits, lost spirits. Not past lives.

When the reincarnationist mind-set comes upon a "totally different personality"—whether through hypnosis or "reverie recall"—it is then assumed that the subject has regressed beyond the latest birth to a past lifetime. But shouldn't we first try to find out if the information is coming from some lost soul wandering about in the Outer Darkness?

Though reincarnationists often claim to have eliminated every other possibility to explain such "memories," they have skated right past the most important clue: the telepathic powers of the psyche. When you go on a psychic fishing expedition—like the Ouija board or "encouraged" regression—there is no telling what big fish you might net. The Outer Darkness has a population problem equal to that of us earthlings.

And how would reincarnation address the formidable puzzle of MPD—people beset by sometimes a dozen or more distinct personalities? Imagine eighty-six different personas reincarnating in one body! In such instances, invading spirits often want the host to remain unaware, weak, anorexic, depressed, and confused, so they can continue to play their part through them, supplying perhaps strength, cleverness, vitality to their host, or even absorbing their pain. The shadow person then convinces everyone involved—sometimes even the therapist—that he is a "coping mechanism." Moreover, the usual strategy is for him to be integrated or fused with the core personality. This is SOP in most psychoanalysis.

On the flip side of that coin, the therapist himself may be gulled into a past-life interpretation. As an example, entities attached to "Catherine" called themselves "master spirits" when Catherine started seeing Dr. Brian Weiss for her chronic anxieties. And when she began "recalling" scenes from those entities' lives, the psychiatrist was so enthralled that he missed all the outstanding signs of obsession and possession, and dashed full tilt into the world of reincarnation.

> Such angels as engraft themselves on mortals, becoming as a twin spirit to the one corporeal body, shall be known as re-incarnated spirits. . . . By fastening on, they dwell in that corporeal body, oft driving hence the natural spirit.
>
> OAHSPE, BOOK OF BON 14:9

It is precisely the illusion of reincarnation that keeps unwanted spirits anchored in the host's body, with no one to address them or chase them out, no one to challenge their presence where it does not belong. These clinging entities—which need to move upward into the spirit realm—are known in India by their Sanskrit names: *nidhanas* and *tanhas,* fixated on the earthly plane, desiring rebirth.

SPIRIT OBSESSION

As a drowning man clingeth to a log, so cling the drujas [earthbound spirits] to mortals.

OAHSPE, BOOK OF LIKA 22:11

Professor Stevenson thought that one of his subjects, "M," had been reborn as the son of a liquor-store owner because of his past-life love of drink. Even before young M could speak, he showed the "previous personality's" fondness for alcohol. As such, the convenience of his mother's abundant supply seemed "a sufficient explanation for the previous personality's apparent rebirth as her son" (Stevenson 1987, 242). But compare, if you will, this apparent rebirth with similar cases of "frank possession," in which the victim, a nondrinker, suddenly becomes an alcoholic, or begins to crave substances he never cared for previously (Fiore 1987, 70).

Having learned that Professor Stevenson rejected spirit overshadowing in favor of previous personalities, I contacted his successors at the University of Virginia and asked where I might find the professor's argument against possession. Cordially, they referred me to one of his books, *Children Who Remember Past Lives.* Eagerly I turned to the chapter indicated but found only two short paragraphs devoted to the matter. Stevenson, I was told by a mutual colleague, was a lovely guy, and he had considerable knowledge of the paranormal. Reincarnation, he told his readers, should be adopted only after we have eliminated all the alternative explanations. But how did he eliminate possession? Hastily (in a mere eight sentences) and ineffectively. First of all, he argued that, since children usually forget about their past lives by age seven or eight, this militates against the presence of a possessor. But how so? Perhaps it could equally militate against reincarnation! The argument is never developed or explained. As we have seen, children are most impressionable in their earliest years, and soon grow out of the natural clairvoyance of infancy (which condition enabled them to get wind of other lives).

Secondly, Professor Stevenson contended that possession cannot explain "stimulation of the subject's memories by visits to the family of the previous personality." Why not? Is this a valid basis for ruling out the presence of an entity? Not at all. It might even argue for it.

Finally, Stevenson (paradoxically?) cites the inability of his subjects to know about changes that have taken place since the death of their previous personality. Again, the reasoning is obscure, and the assumption is made that a possessing spirit would have such knowledge. Why assume that? One prominent theosophist, in fact, has pointed out that "people in the heaven world are not in direct touch with . . . affairs on earth. They could not know [what] might be taking place in this world." What's more, the same writer quotes Stevenson's case of Shanti Dev (in Delhi, India), who, when taken to her (former-life) village, "began pointing to changes that had taken place in the past few years" (Perkins 1977, 170, 115). Stevenson's contradictory argument therefore leaves me flummoxed.

Today, spirit obsession is still with us, but under a different name. What used to be called "demons" is now called "neurosis." In our collective ignorance of spiritual matters—in our drive to banish the superstitions of the past and in our own fear of unseen powers—we hang on to the idea of man as a finite, boxed-off being, strictly made up of flesh and blood and nothing more, denying, even contemptuously, his subtle nature, his spiritual self, his very soul. The stigma of demons or hellish personas infesting self is repugnant to our rational sensibility. We reject it but then rechristen it, erroneously calling all these spirit-freeloaders "alters" (MPD) or, alternately, past lives. And by ignoring the true source, effectively banning it, we empower it, for we thus enable the reckless spirit of reembodiment, leaving it to do its mischief unnoticed. For example, when an ambitious psychic named Carl V., as

Malachi Martin tells the story, set out to discover his own past lives, he was caught instead in the clutches of diabolical possession!

In one case of MPD, the multiple Henry Hawksworth disliked his alter Johnny. Johnny is not me, he protested, yet he wears my body and ruins my world. Sometimes the patient himself (not the doctor) correctly diagnoses the problem. In *Through Divided Minds,* Dr. Mayer mentions a multiple patient who heard voices berating and cursing him: "the patient was convinced he was possessed by a dybbuk, a Jewish demon that would yell at him" (Mayer 1988). (*Dybbuk* traditionally refers to the spirit of a dead person.) Dr. Adam Crabtree, in a similar case, describes his patient feeling controlled by "an entity other than himself. In other words, he feels possessed" (Crabtree 1985).

When the author Taylor Caldwell was regressed to find out her supposed past lives, she came out of trance, commenting, "I feel as if I'm two people." The various lives of the eccentric Miss Caldwell give us pause. A quirky woman, the famous novelist, I believe, was keenly clairvoyant—not reincarnated. "Tremendously psychic," breathed her biographer Jess Stearn in *The Search for a Soul,* though he proceeded to bill her as a reincarnator of the highest order. But even Taylor did not go for the hype: "I have read many books on reincarnation, and the very thought of it horrifies me." But Stearn liked it, felt "it was an exciting thought." He believed that he himself had come from Saturn (to help Taylor), also having lived in the nineteenth century as the poet Robert Browning! Taylor's recalls, of course, were prompted: "The hypnotist had suggested she remember her past life." Predictably, the settings were history's most alluring milieus: Atlantis, Lemuria, Egypt [sic], Jerusalem, China, Mexico, Peru, France, England, and Florence, Italy. Yet it was not through "recall" that flashes of the past came to her, but through her own (undeveloped but potent) mediumistic powers.

An undeveloped sensitive can very easily be taken unawares.

ENA TWIGG, *THE WOMAN WHO STUNNED THE WORLD*

The key word here is *undeveloped*. At the dark end of the ASC spectrum are the victims of negative overshadowing. Playing host unwittingly to earthbound souls, they are usually psychic, but they don't know it or have suppressed it. Or they have been told that their visions and voices are hallucinations. Neither have they learned any techniques for controlling their familiars, thanks again to their guardians in the mental health profession who are clueless about (or outright opposed to) the psi factor that underlies dissociative disorders.

Contrast this Western attitude with the Chinese view: the Chinese word for victim of spirit is *hiangto,* which also means "psychic medium." It is the same in India, where *shiva-shakti* refers to the possessed person as well as one with psychic tendencies. Knowledge is power, and, conversely, denial of psychic talent is a fine formula for escalating trouble and unsolved mysteries. Past-life recall, little more than a dressed-up form of spirit possession, represents the low end of mediumship and—albeit inadvertently—the glorification of earthbound entities.

These "attachments," as Dr. Wickland once explained, summarizing his *Thirty Years Among the Dead,* are due to the "millions [who] remain . . . in the earth sphere, still held by their habits or interests." How do they break into the human aura?

> Many discarnate intelligences are attracted to the magnetic light, which emanates from mortals, [especially clairvoyants] and, consciously or unconsciously, attach themselves . . . finding an avenue of expression through influencing, obsessing, or possessing human beings. . . . These earthbound spirits are the supposed devils of all ages . . . by-products of human selfishness, false teachings, and ignorance. (Wickland 1924)

Today, lagging behind tremendous advances in other fields of knowledge, we are beset with almost universal ignorance of the psi factor and the next world. The situation invites wild speculation. I remember, a few days after September 11, 2001, a new age woman told me that all the victims of the attack were "okay," having been carried off to a place of peace and safety in the next world. *Yeah, right.** It reminded me of what my mother said when I told her I was writing about reincarnation. She could never forget the comment a young friend had made to her about victims of the Jewish Holocaust. The remark from that friend, a reincarnationist, was that they, the victims, were "all walking on Kings Highway" (our part of Brooklyn). The blithe and almost mindless remark had remained in my mother's memory for decades.

I would like to quote portions of an e-mail attachment I received from a colleague who is the director of a twenty-five-year-old spiritist (reincarnationist) society in the United States. The passage is representative of the logic (illogic?) typical of the reincarnationist spin on "soul growth" (grammatical errors left as is):

> Heaven and Hell go against reason. . . . How can the justice of a God be reconciled with a doctrine that would condemn wrongdoers eternally to Hell? Would not the Creator Forgive us Infinitely and Give us the means to improve and redeem ourselves no matter how long it took [meaning rebirths, of course]? Reincarnation is one of Hope. . . . Reincarnation is a universal law. Other religious beliefs . . . do not answer the question: Why does a good person suffer? Only reincarnation can answer this, for we do not know what faults or crimes that good person may have had or committed in a past lifetime for which now they may be atoning for . . . in order to spiritually and morally progress.

Whew! *Redemption. Atonement. Forgiveness. Hope. Universal law. Creator. Spiritual progress.* Never doubt that this new reincarnation is a religion, a fisher of souls (though followers deny it, pointing out that they have no priesthood or ceremonies). The lure, the hook, the promise is "soul growth." The beauty of this process, they say, is that each incarnation offers us a fresh opportunity to learn from our mistakes.

---

*After the Newtown, Connecticut, massacre of little children (on December 14, 2012), I was told exactly the same thing by another wannabe new age priestess.

"We have infinite opportunities to truly understand whatever it is we need to learn about the human condition. . . ." (Bowman 2001, 138). But that is the part that bothers me. No hurry. You don't have to work it out in "this" life, in the here and now, because "what remains undone in one life can always be completed in the next incarnation" (Whitton and Fisher 1986, 99). Exit responsibility. At least, we can put things off for a lifetime or two. It's like school semesters: if you flunk out this term, just sign up for the next semester (i.e., get another body) and try again.

> *We keep doing it, until we do it right.*
> SHIRLEY MACLAINE, *DANCING IN THE LIGHT*

With former lives in the therapeutic spotlight, a curtain is thrown over this one. The real seeds of distress, buried in the flotsam of this very life, are recklessly abandoned for the sake of uncovering karmic determinants, the meat and potatoes of past-life therapy. "If the energies in a particular situation do not make sense . . . then a past life karma is probably involved. . . . Regression therapy . . . [is] on the increase especially when people cannot seem to find reasons in this life for their . . . problems" (Whitton and Fisher 1986, 54, 94). This is nonsense, a wild goose chase. With our sights set on the phantom life, we sacrifice the uniqueness, even the identity, of this one.

Phobias. Homosexuality. Distrust. Inferiority complexes. Deformities. Guilt. Secrets. Loneliness. High blood pressure, marital infidelity, nail-biting—you name it—past lives will account for it all. Never mind that a (likely) trauma or privation in this very life remains undiscovered and untreated! No, with the focus on past-life awareness, on deep hypnotic regression, on far memory, the client is now the sport of ancient hang-ups and fabled identities. Even Professor Stevenson recoiled at the "foolish implausibilities that disfigure most hypnotically induced previous personalities" (Stevenson 1987, 45).

Sybil Leek argued that a more advanced animal can reincarnate as a "low-level human being, and to become fully enlightened beings we

*Fig. 7.7. Reincarnation.
Spoofing previous life as a bug.
Cartoon by Marvin E. Herring.*

must undergo evolution of consciousness [by being] born into bodies time after time" (Leek 1974, 93, 21).

Thus the aim of reincarnation is the progressive improvement of spirit, which needs the trial of material life to move ahead in its evolutionary progress.

> *Each incarnation becomes more highly specialized as more permanent growth takes place . . . by means of which Man accomplishes his spiritual evolution into Superman . . . ever unfolding toward perfection.*
>
> JAMES PERKINS, THEOSOPHIST,
> *EXPERIENCING REINCARNATION*

Yet according to this version of evolution, humanity (which is no youngster) should already be in its moral apotheosis, having suffered countless incarnations in the ever-advancing development of soul. One would think that the world should be in great shape by now.

I also wonder if it is fair to a child to impose such pet theories on him or her. Regressionists think so. One of them tells us of a "gentle, beautiful little girl with very morbid streaks and unreasonable fears. Would it not help the parents to know she had had several violent, traumatic lives . . . ? It would be much easier to understand and guide her" (Wambach 1979, 127). So, do we keep this terrific insight a secret from the little girl, or do we tell her "You had several violent past lives"? This is appalling. I think we know enough about fears and phobias to say that their antecedents lie not in previous embodiments, but in the vulnerabilities and unrecognized (often denied) disasters that took place in early life. This life and none other.

The syndrome of the "nonperson" (see page 304) gives us a good idea of where, say, "inferiority complex"—the ravishment of normal ego (self-image), usually by our own caregivers—comes from. But to the reincarnationist, the girl who "felt a deep sense of inferiority" (in the Cayce files) was once "an American Indian girl who felt hopelessly inferior." These agents of sham, with their pat answers, seem to have no compunction about ignoring the very real trauma in this life and blaming instead some far-off drama of the past, cloaked in mystery and make-believe antecedents.

Pangs of guilt? Not to worry. Past-life readings explain away the

Fig. 7.8. Past lives.
Cartoon by Tony Webster.

roots of our flaws and failings. No wonder reincarnation finds adherents, for it manifestly exploits the great reluctance to recall our own traumas and faults. Psychic John Edward has commented on the occasional sitter who comes to him and says, "I don't want to hear anything negative." In "The Aetiology of Hysteria," Sigmund Freud also affirmed patients' "reluctance to think about or mention certain noxae and traumas" (1896). As long as the trauma (or disaster or wrongdoing) is buried or disregarded, fairy tales like reincarnation can step in. Patients in past-life therapy seem to speak more freely about their problems, almost as if speaking about someone else. (Aha!) Accumulated psychological pressures are thus lifted, with pleasant assurances that one is now "coping" with the devilry by bringing it to the surface. Good old abreaction at work. A chance to vent. But suppose that this surfacing stuff comes not from a previous embodiment at all but from some entrenched spirit that has (undetectedly) taken hold.

I have written this chapter to convey a simple message: this sublunary life we've been given is a one-shot deal. There is no second chance, not on Mother Earth, and little choice, either. The only choosing I am aware of is to make—or not—the most of it because it's the only life we have.

> Remember, O man, these are the lessons of your Creator, which He gave unto you. . . . This is the beginning, and you shall be your own judge and master. Wait not for a savior to save you; nor depend on words or prayers. . . . But begin to save yourself by purifying your flesh, by purifying your thoughts, and by the practice of good works done unto others, with all your wisdom, love, and strength. For through these only is there any resurrection for you either in this world or the next.
>
> OAHSPE, BOOK OF JUDGMENT 13:34

I have come to understand that the spiritualization of our race is upon us, but will not blossom until the psychic faculty, the inner All, is fathomed, appreciated, and mastered! Hasten the day. A high level of performance in this life is all that is needed, now and forever. A million dreamed-up past and future lives cannot compare to one nobly lived.

In ancient Egypt, many women had familiar spirits.

And they prostituted themselves in counseling with the multitude on earthly things, thus inviting spirits of the lower heavens who would not raise up from the earth; and when young babes were born, they were obsessed, and these evil spirits, in justification of their sins, taught reincarnation.

MOSES, IN OAHSPE, BOOK OF SAPHAH: OSIRIS, 116

In the next chapter we will take a closer look at these familiar spirits and how they infested and finally brought to ruin the first great civilization of ancient Egypt.

PART THREE

·············

SOCIETY

CHAPTER 8

BEYOND
THE FERTILE CRESCENT

Antiquity Denied

> *On every continent there are . . . artifacts that do not fit the current paradigm because they are too old, too advanced. . . . Everywhere we look, the scientific, artistic, and engineering accomplishments of the ancients seem to have reached their peak early on and then suffered a decline.*
>
> RICHARD HEINBERG,
> *MEMORIES AND VISIONS OF PARADISE*

The mathematicians foretold the great cities and nations that would rise up; how this one and that one would move to battle; how their great cities would fall in ruins . . . so that even the memory of them would be lost.

OAHSPE, BOOK OF SETHANTES 9:13

I spelled out in chapter 7 the potentially devastating effects of reincarnation on people, on individuals. But what about whole nations? In search of the causes of societal decline, I came across one stunning account describing gangs of reincarnating spirits marauding through ancient Egypt, bringing it to ruin.

Woe came to them, for the land was flooded with hundreds of millions of drujas [earthbound spirits] . . . [who] watched for the times when children were born, and obsessed them, driving away the natural spirit, and growing up in the body of the newborn, calling themselves "reincarnated." And they taught, as did their master, Osiris the False, that there was no higher heaven than here on the earth, and that man must be reincarnated over and over.

OAHSPE, BOOK OF WARS 51:10–12

Soviet scholars, noting 35,000 years of Egyptian history from calendrical counts, have supposed that their development was halted around 10 kya due to "the fall of an asteroid" (Kolosimo 1973, 92). Here again is that hypothetical and out-of-the-blue terminator, invoked unnecessarily but maybe for lack of any other explanation. Yet tens of thousands of years before the fall of Egypt, the sages had foreseen the cycles of ruin that would overtake society, triggered by both spiritual depravity and the rush to war. Such cycles are partly cosmological in rhythm (see "a'ji," chapters 3 and 5), giving us to understand the almost inevitable waxing and waning of civilization, its cadence of rise and fall.

Fallen were those who built cities in the Andes covering more than ten square miles, something the modern world did not see until the nineteenth century. Lost also were thousands of cities in the Old World.

Arabia and Persia . . . [were] full of colleges that taught the sciences, and the arts of painting, engraving, and sculpture, and astronomy, mathematics, chemistry, minerals, assaying, and the rules for inventing chemical combinations.

OAHSPE, BOOK OF WARS 21:10

Archaeology says that all people, until 9 or 10 kya, were simple hunter-gatherers. The truth is that even in deepest antiquity (well before

Fig. 8.1. Ancient sign of drujas, wandering spirits of darkness, the lowest grade of excarnates, whose behavior and effects we touched on in the last two chapters. I think the words drug *and* drudgery *come from* druja, *as does the Persian* druj, *meaning "spirit of falsehood."*

10 kya), people of advanced learning inhabited the same Earth as tribes of the most rudimentary culture. They coexisted. And concerning their dates: some of the most ancient cities we have found are merely the top layer of multiple strata of ruins, like Eridu and Ugarit, which sat on several layers of earlier sites. Troy was also found to be layers of city on city. "So too, the Maya often buried city upon city, piling up level after level" (Frank Joseph 2013, 112). We get a similar sequence of several occupation layers in the Indus civilization.

We need to dig deeper.

ALBERT C. GOODYEAR III
(UPON FINDING 38,000-YEAR-OLD TOOLS
AT AMERICAN ARCHAEOLOGICAL SITES)

No one who is hooked on deep antiquity and lost civilizations has failed to notice two recurrent themes: a vanished golden age and a great flood. Regarding the former, there is good reason to believe that war and strife were unknown in some earlier utopia: "There was no sin," states the Mayan *Chilam Balam*. "In their holy faith their lives were passed." And there was wisdom in that golden age. As the *Popol Vuh* goes on to recount, the first humans knew all in the sky and in the Earth. Who were these clever people? They were the sacred tribes of yore—in all places. Our Mayan priests and pundits were matched in the Fertile Crescent by the legendary "seven sages" who ruled over Babylonia's great cities and were so erudite as to be considered mirac- ulous, even immortal, by the unlearned. Often enough, these learned people became the "gods" (read: wise men, equivalent also to India's Seven Rishis, who prevailed during an earlier enlightened age).

Such heaven-born people are common to the legends of America, China, Greece, and Egypt. "Exceedingly wise" was the Mesopotamian Noah. Indeed, it was the very sons of Noah who spread that golden age, that paradise, throughout the world. Survivors of the Great Flood (the sinking of the continent of Pan in the Pacific), these sacred tribes are one and the same as the sons of Noah, being named Shem, Ham,

Jaffeth, Yista, and Guatama, and their offshoots. A priestly race, they were the first people on Earth to forge a covenant with their Creator, and they taught the same to the wild tribes in every part of the Earth that they settled.

Ignatius Donnelly (1882, 54) found these sons of Noah in the three oldest races of higher humanity, the seers of the prehistoric world: (1) the Indo-European (from Jaffeth), (2) the Semitic (from Shem), and (3) the Hamitic/Cushite (from Ham). But by 13,000 BP, the only peoples in the world who retained the sacred teachings of the original sages were the Hebrews and the Vedics (from Shem), and the American Algonquins (from Guatama). Yista refers to the civilizers of Japan whom we will meet below as the Jomon people.

In that lost golden age—remembered as a time of grace and harmony—legend tells us that happiness and abundance were the rule and crime was unknown. There were no thieves; the door of everyone's house was left open. Are these various stories simply utopian fantasies, as orthodox history tends to brand them? If so, why do such stories come from every region on Earth: Australia, Mesopotamia, Egypt, India, Japan, Africa, the Americas, China, and Greece? Tales of a lost epoch come from Greece* (in Hesiod's *Theogony*), Japan (in the *Kojiki*), and India (in the *Ramayana* and the *Mahabharata*). And in ancient Egypt, *Zep Tepi* meant "first time," signifying the beginning of civilized life. It is remembered as a golden age. But after Apollo, the "reign of the gods" ended; this had consisted of several peaceful millennia after the Flood (ca. 24,000 BP to 18,000 BP).

Nor was there any war in any land under the sun.

OAHSPE, THE LORDS' FIRST BOOK 2:18

---

*Athens, according to the Egyptians, was built seventeen thousand years ago, which would antedate the Fertile Crescent by ten thousand years!

In Europe, this was the golden age of the Solutrean, marked by the first great Cro-Magnon painting, engraving, and sculpture.

The Chinese, for their part, recall an idyllic age in a land far across the sea, ruled by the gods for eighteen thousand years (as noted in the *Record of Tchi* and the *Huai-nan-tzu*). Though living at the same time as stone age tribes (cave dwellers), these sons of Jaffeth in China were a breed apart, standing as the head of learning, teaching the applied arts and industries. The doughty protohistorian James Churchward identified them as "Uighur," sons of Mu (the submerged land in the Pacific, a.k.a. Pan or Lemuria). Churchward gave them the arts of astronomy, mining, textiles, architecture, mathematics, metallurgy, agriculture, medicine, and writing, fully ten thousand years before the so-called dawn of civilization in the Fertile Crescent.

The Uighurs, in Churchward's reconstruction, developed a vast empire, eventually reaching Europe through the Caspian and Asia Minor. Thus do we find its offspring in Turkey's city of Troy. Foebe was lordess and tutelary spirit of Troy, which city, legend holds, was built by Apollo eighteen thousand years ago. Hence it became known as the domain of "Phoebus Apollo." This decidedly Paleolithic date for the settlement of Troy by one of Noah's great sons matches Churchward's time frame, establishing the occupation of Asia Minor and the Caucasian Plains before 14,000 BP. (More on Asia Minor, which is today's Turkey, below, as we touch on the recently discovered and paradigm-shaking Gobekli Tepe, proof positive of the above claims and dates.)

Sharing this unconventional time frame are North America's (Guatama's) basketmakers and moundbuilders, as well as Central America's high culture, whose earliest kingdoms Churchward inferred from the Troano manuscript. Even their names retain a hint of Mu, the lost motherland, also called Pan.

Sites in Early Central America Named After Pan
- Sta. Izabel Ixta*pan*: one of the earliest known cultures in Mexico, dated to 18,000 years ago
- San Felipe Hueyotli*pan* and Zaha*pan* River: in Puebla's Valsequillo

region, which holds artifacts that are at least 20,000 years old
- Caula*pan*: in Puebla with a 22,000-year-old stone scraper
- Kich*pan*ha: a Mayan site with the first signs of writing
- Tlalla*pan*: the land that the legendary Quetzalcoatl came from

Some Atlantists have argued that atomic warfare caused the destruction of most of humankind in prehistoric days. (The Atlanteans, they say, got into a high-tech war and self-destructed.) Others favor the theory of a celestial body ravaging the Earth, ending the wondrous civilization and plunging the survivors into barbarism, their knowledge forgotten by all except a few initiates. However, all these disaster theories ignore the cyclical nature of civilization itself, particularly the three-thousand-year danha cycle. Remember c'vorkum in chapter 1? In its travel through the roadway of the stars, the Earth passes alternately and regularly through regions of darkness and light. Humankind's learning, as the Earth sails through a stretch of darkness and density, tends to be lost or corrupted, only to see a new dawn of discovery in the coming light of danha, which recurs approximately every three thousand years.

The prophets of old divided time into cycles of three thousand years. . . . And they found that at such periods of time [at the dawn of a new cycle], some certain impulse came upon the people, causing them to try to be better and wiser.

OAHSPE, BOOK OF COSMOGONY AND PROPHECY 7:6

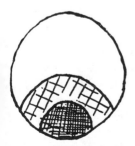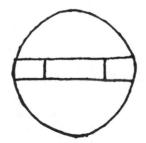

Fig. 8.2. Plate 62, from Oahspe. Three signs of light. The center figure is a cosmic one; the horizontal strip running across the middle of the circle (which is the cosmos) represents c'vorkum, wherein the solar system enters regions of light (symbolized by the two vertical lines) approximately every three thousand years.

The [Egyptian] priests themselves had a legend that their
knowledge had come from Thoth, the god of wisdom,
during his three-thousand-year reign on earth.

LANDSBURG, 97

To give one example of danha: in the beginning of the three-thousand-year cycle of Thor a.k.a. Thoth (fifteen thousand years ago, on the heels of the three-thousand-year cycle of Apollo),* humanity prospered for a long period. Then darkness came upon the races of men, millions of them returning to a state of savagery. So that, at the end of Thor's three thousand years, all the lands of the Earth were at a low point, accounting, as we shall see, for the disappearance of Cro-Magnon art and culture as well as for the enormous, inexplicable cultural gap unbridged at the close of the late Paleolithic.

Some scholars think the precession of the equinox can be used to track the rise and fall of civilization; others credit the rise of the Neolithic Revolution with the "warm gap" at the end of the last ice age (chapter 3). But these ideas have not been borne out. Neither did the Great Flood (the submersion of Pan) wipe out a golden age. Contrary to current speculation on Atlantis, the Flood destroyed not a high culture, but a low one—the outsized barbarians, the Philistines who were busy exterminating the sacred people.† It was indeed the sins of these giants—as many traditions recount—that caused the Flood in the first place! The Skokomish Indians of Mt. Rainier in Washington, for example, relate that the Great Spirit was displeased with the evil on Earth, and safely secluded the good people (the Ihin Noachics) before causing a great inundation. Likewise, the Arikara Indian legend holds that the Creator caused a deluge to get rid of the unruly giants, but saved the

---

*Apollo, Thor, Osiris, and so forth are the names of the gods who ruled the Earth for the duration of those cycles.

†The Flood itself came at the very end of a three-thousand-year cycle; the final years of a danha cycle are typically wretched, bloody, and degenerate, as was the period just preceding the fall of Pan 24 kya.

little people by "storing them in a cave." In the same vein, the Pima creation myth involves the salvation of chosen beings from the Flood.

The thread of oral history agrees on this much: the survivors of the Great Flood were a humble, virtuous people—and that is why they were chosen to survive. "Destroyed [is] the Land of our Fathers. . . . The first great waters came . . . [but] the good were chosen and saved," recount the Mayan priests, who also say that their ancestors had learned in advance of the coming deluge: "Go and choose the good and the wise and bring them . . . to a new land." *Mayab,* in fact, means "Land of the Chosen Ones."

They were chosen—and preserved—for at the time of the Deluge, 24,000 years ago, the Earth was overrun by tribes of darkness; the people of light were either exterminated or back-bred. Just prior to the flood, Neph, who was then the god of Earth, said:

Mortals are descending in breed and blood. . . . The earth and her heavens have gone down in darkness. The Ihin [little people, in the lineage of Noah] had been destroyed . . . *except on Whaga* [a.k.a. Mu, Pan]. (my italics)

OAHSPE, SYNOPSIS OF SIXTEEN CYCLES 2:1–7

"Descending in breed and blood" is a reference to the retrobreeding that took place both before and after the Flood. In the time of Thor, as noted, man "dwelt with beasts, falling lower than all the rest" (The Lords' Fourth Book 4:10).

We got an idea of this widespread retro-breeding (or hybridization) in chapter 2. People wonder why Cro-Magnon art disappeared around 12,000 BP, which, as I have described, came at the end of Thor's cycle. This is when humanity returned to a state of savagery. These Cro-Magnons of Magdalenian culture (the Lascaux cave artists) simply drop out of sight, not because of some imagined cataclysm that annihilated them or any other bolt from the blue, and not because of "the violent close of the last glacial epoch" (Frank Joseph 2013, 242), but simply because they back-bred so extensively with Neanderthals.*

Retro-breeding itself is an effect of darkness (hundreds of years of a'jian occulsion toward the end of Thor's cycle). Around that time Egypt also took a dive, under Osiris the False, as we saw at the beginning of this chapter. Until then, for hundreds and hundreds of years, the Egyptians had been the most learned people in the entire world, especially in knowledge of the stars, the Sun, and the Moon, as well as in adeptism and miracles. But when they became obsessed with (earth-bound) spirits, they lost sight of higher learning.

But, you may ask, where did those great adepts of Egypt get their knowledge? How far back can science and learning be traced? As I see it, the first true civilization arose more than 40 kya. "A tradition of communicable knowledge of the heavens," said Cambridge anthropologist Richard Rudgley "has existed for over 40,000 years" (Rudgley 1999, 100). At that time, humanity knew all about not only astronomy but also sailing, mining, and agriculture.

> Apollo told us when to plant, and when to reap.
>
> OAHSPE, BOOK OF OSIRIS 9:15

At the vanguard of this civilization were the Ihins, and it was by the angels that they were taught and guided in all things.

> He had books, both written and printed. In schools, the young were taught about the sun, moon, stars, and all things that are upon the earth and in its waters. This was, therefore, called the first period of civilization on the earth.
>
> OAHSPE, SYNOPSIS OF SIXTEEN CYCLES, 1:21
>
> (REFERRING TO A TIME 40 KYA)

It is also from the Book of Enoch that we hear of the constellations first discovered, divided, and named by the antediluvian patriarch

---

*All these gene exchanges are explained in much greater detail in my last book, *The Mysterious Origins of Hybrid Man*.

named Seth. Who exactly was Seth? The teachers of the golden age—
those survivors of Pan—were called the family of Seth, denoting the
Sethite priesthood (Noah's lineage) who are remembered as the first
astronomer-sages. Legend has them settled on mounds (see page 359),
"in the vicinity of Paradise," living peacefully and apart from others
on (and in) their high mounds. One such was Adapa a.k.a. Oannes
(Babylonia), the antediluvian priest-sage who civilized the people of
Mesopotamia. It is commonly taught that civilization began here in
the Fertile Crescent (today's Iraq), some six thousand years ago, their
culture boasting the full complement of institutions, arts, and sciences.

There is only one problem with this Standard Model of human
history: there are no known predecessors. How could the arts of civilization have appeared so suddenly? Mesopotamian science, calendars, agriculture, writing, mathematics, and city life all seem to have
sprung up out of nowhere. The big bang of history! Also shot from the
Stone Age, Egyptian civilization seems not to have evolved from anything before. Rather, "architecture, engineering, medicine, science, and
big cities materialized within a century . . . almost as if they had been
imported from somewhere else" (Landsburg and Landsburg 1974, 90).

> *Searching through the historical record for the origins and
> the evolved civilizations, I was disturbed by the series of
> "suddenlies."*
>
> ALEXANDER MARSHACK,
> *THE ROOTS OF CIVILIZATION*

Let's change *suddenly* to *arrival*. Berossus, the Babylonian priest-
historian, recorded that some race of special intelligence had brought
knowledge to the Fertile Crescent, to the ancient peoples of Sumeria,
who, until then, had lived a humble if not abject existence. Indeed, one
Sumerian tradition reveals that they learned astronomy, medicine, and
mathematics from "strange outsiders" (like Adapa), while a Babylonian
legend has their forebears arriving in Mesopotamia after a long sea
journey from a destroyed homeland. This reminds us of traditions in

India recalling a great flood in the East, from which they escaped by ship, ending up in the Himalayas, and consequently repopulating the subcontinent. Here they established the Rama Empire with its seven Rishi cities.* In this connection, author Patrick Chouinard brings to our notice an ancient manuscript of the Rama Empire, found near the Silk Road, which includes a map of Mu (Chouinard 2012, 225).

> *The civilizations of India, Babylonia, Persia, Egypt, and Yucatán were but the dying embers of this great past civilization.*
>
> James Churchward, *The Lost Continent of Mu*

Fig. 8.3. Statue of Nebo, a Mesopotamian soothsayer, prophet, and the inventor of writing, depicted as a little man, that is, an Ihin sage.

---

*Rishi, like Sethite, denotes a priestly caste.

According to textbook wisdom, writing was invented in the Bronze Age along the Fertile Crescent. Hundreds of thousands of clay tablets were written in cuneiform (wedges that represent syllables) attributed to the Sumerians, perhaps 6000 BP.

And from this hub in the Mesopotamian Fertile Crescent, the art of writing supposedly spread to Egypt, Iran, and the Indus Valley, where cities like Harappa and Mohenjo-Daro started out fully literate. Their script (a system of about four hundred "graphemes") was already highly developed when it first appeared 5 kya. Yet India's Rishi cities of the Rama Empire might well be "tens of thousands of years old" (Chouinard 2012, 228–29). Discovered off the coast of northwest India in the Gulf of Cambay, an unnamed ancient culture—a submerged city with metal implements and pottery—preceded the Sumerian civilization, rather than borrowed from it.

And if scholars attribute Egyptian civilization to the Sumerians, Sumerian civilization itself could be traced farther east, to central Asia, as pottery and burial customs indicate. Artifacts of the Indus-Sarasvati culture—which extended more than a thousand miles north-south—were found in those Cambay ruins to be at least 9,500 years old, that is, pre-Harappan and certainly pre-Mesopotamian. This find, however, has enjoyed "virtually no international media coverage, even though it could rewrite not only the history of the Indian subcontinent but of the world" (Hancock 2002, 676–77).

Written inscriptions in Egypt, which Churchward traces to the Naacal Tablets of India, have also been seen to predate Sumer perhaps by a thousand years. In fact, Diodorus wrote that, according to the Egyptians, the *Chaldaioi* (of Mesopotamia) were colonists from their own land, Egypt, whose priests had taught them the science of *astronomia*. According to Richard Rudgley "in the case of Sumeria . . . the cuneiform writing system was built on an earlier token system which has so far been traced back 10,000 years. Hieroglyphs . . . of Dynastic Egypt are now known to have been used . . . a thousand years before history began" (Rudgley 1999, 261). Rudgley also mentions evidence of scripts predating Sumeria in Bulgaria and Yugoslavia.

The script of Indus society predates the Sumerians by a long shot, at least according to James Churchward, who thought India's cosmogonic writings were perfectly understood by our forefathers 15,000 years ago. Temple records, he says, trace India's Danava civilization on the Deccan as far back as 20,000 BP. And that ancient Indic script has itself been traced back to Rutas, a lost continent and home of a highly civilized race. (Rutas, according to Ceylonese chronicles, once had 100,000 large towns along the coast.) A form of writing surprisingly similar to the Indic script has been found in the Pacific (old Pan) at Easter Island and the Caroline Islands, while the origin of other Oceanic scripts remains to be explained, such as the hieroglyphics inscribed on sunken columns (east of Tahiti) and the undeciphered inscriptions on Fijian monoliths. Were these the alphabets of Pan?

FROM EAST TO WEST

They journeyed from the East.

GENESIS 11:2

Rather than dispersing from the Mediterranean (Mesopotamia) to Asia, there is much evidence to suggest the transmission of culture and alphabet from India to the West. Indeed, Hindu records speak of a migration to Babylonia at least 15,000 years ago. Isn't this why we find mathematics, writing, and astronomy appearing "suddenly" along the Fertile Crescent?

Six thousand years ago, the Sumerians, having migrated from the East, were speaking an agglutinative language, perhaps akin to proto-Chinese and also resembling the tongue of drowned Lemuria. Whereas the people of Mesopotamia have Persian/Iranian cultural roots, their language, being agglutinative, was unique in the ancient Middle East. In this

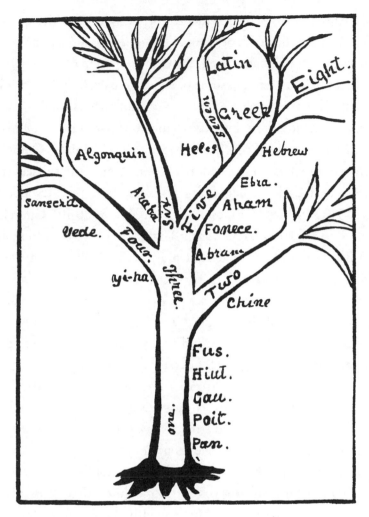

Fig. 8.4. Language tree showing Chinese as the earliest branch off the Panic languages.

respect, they were more like Native Americans (their Guataman cousins)*
as well as the Chinese (Jaffetic), whose syllabary, even today, uses signs
similar to the pictographs of the Sumerians (see appendix E, comparing
Amerind and Chinese vocabulary). The writing systems of China and

---

*As an example of agglutinative language, Lake Webster in Massachusetts was earlier known
by its agglutinative Indian name, Lake Chargogagogmanchaugagogchaubunagungamaug,
meaning, "You fish on your side, we fish on our side, nobody fishes in the middle."

Middle America arose independently of the vaunted Mesopotamian "cradle." And the rootstock of them all lay in a now-sunken land.

Older than the alphabet itself, the Chinese system of writing is made up of characters (not letters), some of which go back 8,600 years. Graves in central China with characters carved into tortoise shells antedate Mesopotamian cuneiform by more than two thousand years. This early date was confirmed by the International Cliff Painting Committee. The Damaidi rock carvings in northwestern China, first discovered in the 1980s, push back the beginning of writing from the Mesopotamian date (ca. 5500 BP) to a Chinese date of at least 8000 BP. Churchward had gone even farther back, dating Uighur writing to as long ago as 17,000 BP. The books of that golden age were discovered in the Gobi Desert at Khara Khota, fully fifty feet below the surface—one silver tablet engraved in this ancient script seemed to contain the names of kings and the dates of their reigns.* The Uighurs had not only writing but the full complement of civilized arts, all, in Churchward's opinion, before the history of Egypt or Sumeria commenced.

Along the mountainous frontier that divides China and Tibet lies the isolated Bayan-Kara-Ula range, a great enigma presented by remarkable artifacts excavated from its caves. For here were unearthed 715 granite discs, evidently twelve thousand years old and shaped something like a vinyl phonograph record with a hole at the center. The discs contain a form of grooved writing, while other disks bear pictograms of the constellations. Discovered in 1938, these cobalt-laden "records" curiously resemble the acclaimed Phaistos Disk of Crete, which is also imprinted in a spiral, as is Egypt's Disk of Dendera.

These similarities suggest a shared but lost heritage, perhaps a common homeland. Comparative linguistics has long entertained a single primordial language, spoken before the destruction of that homeland. That language, I suggest, was Panic. (Extensive word lists of the Panic

*Mural paintings in the Khara Khota, dated 18 kyr, were found in the tomb of a young king. His holy emblem was a circle divided into four quadrants, with the letter Mu at the center. The twice-divided "circle cross," as I have elsewhere written about and illustrated, is the eternal sign of the Great Spirit.

language, which is more than 24 kyr, can be found in Oahspe's Book of Saphah.) On rocks at Klamath Falls in Oregon, the area thought to have been settled by a colony of survivors from that homeland, are inscribed thousands of hieroglyphics, symbols of Mu, with an uncanny resemblance to Sanskrit and Greek. Latin and Greek words were even found there in the local dialect. The Modoc Indians call the area Valley of Knowledge. The name Pan itself turns up in many Greek words pertaining to the ancients, including a legendary and utopian archipelago called Panchaia, with its golden column on which were written the histories of the gods.

Writing, then, must be antediluvian. Along the Fertile Crescent, it was the epic hero Gilgamesh who engraved in stone the story of the Flood. Chaldean tradition, in the same vein, has the king ordering the history of all things to be committed to writing in advance of the Deluge. Mediterranean tradition also says that the chosen people (Sethites, who escaped the submersion of Pan) preserved the original inventions, histories, religion, and ethics, transmitting them to posterity on monuments of stone. In the Scottish Rite of Freemasonry, the Sethite Enoch (great-grandfather of Noah) is named as the builder of a granite column and a brass column, each engraved with knowledge of the arts and sciences intended to survive the coming flood.

And the rabbah made records in writing on stone, which they taught to their successors [ca. 24,000 BP].

OAHSPE, THE LORDS' FIRST BOOK 2:16

This motif—stone tablets of knowledge—recurs time and again in the great traditions of postdiluvial humanity. The Hopi Indians remember their own golden age, toward the end of which the True White Brothers prepared holy records into which the Great Spirit breathed all teachings and prophecies, after which the chief led the faithful into their new land. Today, the Hopi stand as one of the last remaining People of Peace from the prior golden age.

The Hopi story of teaching tablets resonates in the Old World with Josephus's account of the antediluvians who inscribed all knowledge on two pillars, which Hermes discovered after the Flood. Such pillars figure extensively in the legends of the patriarchs (the True White Brothers), legends that spoke of hallowed records inscribed in pre-Flood times on sacred pillars. Specifically, in the story of Lamech (son of Methuselah and father of Noah), this great patriarch impresses the knowledge of his forebears onto two mighty pillars.

Later tradition, a bit modified, has Moses's words of law written upon two tablets of stone. Related narratives explain that the holy descendants of Seth (who was Adam's third son) and of Enoch were charged with preserving the original religion, science, and arts of peace, and transmitting them to future generations on monuments of stone. In that way their inventions, wisdom, and histories might not be lost.

Even in Australia, huge limestone pillars near the Roper River are said to have been the work of a white race—a memorial of the Flood? At Panama as well, the majestic columns of Cocle are covered with inscriptions in an unknown language, left by intensely religious, peaceful, and very industrious people with unexcelled weaving and pottery. These artisans, carvers, and engravers in the Americas (Guatama) can be traced back to Noah—and the Hopi's ancestors—if the "Flagstaff engraving" in Arizona, similar to European Cro-Magnon work, is any indication. Technologically sophisticated, this Arizona engraving is said to be more than 20,000 years old, around the same age as the bowl found at Santa Barbara, California, with undeciphered letters, not hieroglyphics, inscribed upon it. Equally old is the DNA of California people, dated "on the order of 20,000 to 30,000 years old" (Frank Joseph 2013, 251).

The Druids, Hindus, Chinese, Egyptians, and Phoenicians all had traditions that placed the earliest writing to a time before the Deluge. The Hebrews, for their part, held genealogical records from the time of Adam to Noah. Noah's son Jaffeth gave rise to the northern Pelasgians, who were known as *Dioi* ("divine") because they alone, being the priestly caste, preserved the use of letters after the Deluge.

As for the ancient Mayas of Guatama, their books—made of long strips of native paper folded zigzag and bound between a pair of wooden covers—were all destroyed, after which knowledge of writing was apparently lost. The Mayas said that Itzamna, the "god" of medicine (really, the high priest) had invented writing. As elsewhere, this god or priest is a Noachic figure, that is, portrayed on icons as short, white, and with a prominent nose, rather like the antediluvian patriarchs of the Old World. If script was given to the Mayas (by god, angel, priest, or new settler), it would then explain why their writing appears full-blown, as far as archaeology can tell, just as their mathematics and astronomy have no known antecedents, except for the Olmecs, who are considered the earliest civilizers in Mesoamerica. But where did *they* learn it?

> *These same characters [alphabetic] found in the [Olmec]*
> *ruins of Monte Alban in Mexico and in Egypt . . . were all*
> *produced by the same race—the survivors of Pan.*
> EDGAR LUCIEN LARKIN,
> DIRECTOR OF MT. LOWE OBSERVATORY, 1900–1924

In 2006, news came of the "oldest writing in the New World," found by archaeologists digging at Cascajal on Mexico's Gulf Coast. The prized find was a twenty-six-pound stone slab with inscriptions (sixty-two glyphs) looking like a true written language. It has not been deciphered. How old are those glyphs? While Churchward dates Mexican writing to as long ago as 35 kya, some form of writing dated even earlier, 50 kya, has been found in Nicaragua, engraved in the roof of a cave (Corliss 1978, 11–12). These dates approximate the 40 kyr footprints of humans found in Puebla, Mexico (Frank Joseph 2013, 256).

Inscribed tablets of unknown origin have been unearthed in North America as well, at moundbuilder sites. For example, in the nineteenth century an inscribed slate wall was uncovered during coal-mining operations in Ohio. These undeciphered hieroglyphics had lain one hundred feet below the surface (Steiger 1974, 51). That's deep; that is, old. More engraved writing was found in a mound at Grave Creek, West Virginia—a sandstone tablet with written characters similar to Etruscan, Egyptian, Aegean, Scandinavian, Phoenician, Greek, and Tunisian.

Fig. 8.5. Grave Creek tablet (left); Bat Creek stone (below). "To the viewer who does not know it is alphabetic writing," wrote Americanist Joseph Mahan, "an inscription appears to be an accidental grouping of lines . . . [or] an ornamental design. The early form of this [Bat Creek] alphabet, known as Kufic, is found in rock carvings and paintings across the American Southwest" (Mahan 1983, 196).

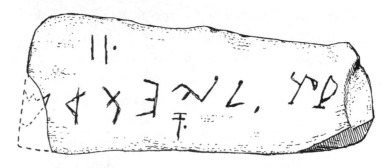

In Peru, the great naturalist/explorer Alexander Von Humboldt learned that books had indeed been made there in earlier ages. To date, four hundred signs or elements of an Incan writing system have been identified. Thought to be characters, they relate historical events, poetry, and myths of the ancestors. Also indicating a written script in this region are metal plaques and stelae with fifty-six different characters (von Däniken 1974, 35, 54). Explorers have come across other unidentified scripts in South America, such as the little-known alphabetic script in Matto Grosso, Brazil, which is associated with the high Tupi culture and, significantly, thought to be of Oceanic (i.e., Panic) origin. In Bolivia, too, the Panos (little people) cultivated the art of picture writing. One missionary saw a venerable man with a great book open before him reading about the wanderings of their forebears. The book, covered with intricate figures and signs, was shaped like nineteenth-century volumes in quarto, but made of plantain leaves. Are they the last remnant of a lost art?

Writing on banana leaves has mostly disappeared from the Andean lands. Various legends say an ancient ruler had prohibited writing because (1) it caused a plague, or (2) it weakened the memory, or (3) it contained prophecies of the regime's downfall. According to Oahspe, certain writing was indeed kept secret after the Flood.

> The |hins . . . made tablets and . . . engravings . . . of the children of Noe [variant of Noah] and of the flood and of the sacred tribes, Shem, Ham, and Jaffeth. . . . And these were the first writings since the flood, other than such as were *kept secret* [by the priests] amongst the |hins. (my italics)
>
> THE LORDS' FOURTH BOOK 1:1–6, 2:18–21*

In the next cycle, the only holy books were those

> given in secret to the tribes of |hins, of which the different nations of the earth knew nothing.
>
> OAHSPE, BOOK OF GOD'S WORD 30:22

Egypt's golden age, at least according to astronomical data at Giza, goes back to 14,000 years ago. Gaston Maspero, the great Egyptologist, thought the Book of the Dead and the pyramid texts were written long before the First Dynasty of 7,000 years ago, at which time the books were already old. Manetho, in classical times, said that Egypt already possessed more than 11,000 years of protohistory. Matching and exceeding this, the Egyptian tablet called "Old Chronicles" records thirty dynasties during a period of 36,000 years. Indeed, Martianus Capella reckoned that Egypt's astronomers had kept secret their knowledge for 40,000 years. This is the same horizon suggested by Richard Rudgley

---

*This was fifteen thousand years ago, that is, in Thor's time, corroborating Philo who said that Taautos (variants: Thoth and Thoor [*sic*] in Egypt) conceived letters and opened the way for written documents.

(1999), who cites "extremely archaic" artifacts (up to 34,000 years old) in Siberia that record both the solar year and the phases of the moon. Such dates resonate with the antediluvian patriarch Enoch. Chapters 72 to 82 in the Book of Enoch record amazingly modern knowledge of astronomy, including intercalary days and the orbits of the Sun and Moon.

You were probably taught in school that Hipparchus of Nicaea was the first astronomer to map the heavens (ca. 130 BCE). Heavens, no! There is nothing to reconcile the extremely old dates for the first astronomers with today's conventional wisdom, teaching us that the Fertile Crescent is the cradle of civilization or that the Egyptian calendar is no more than 6,000 years old. The Egyptians said that Taut or Thoth (a.k.a. Thor, 15 kya) gave them both the alphabet and astronomy. As a matter of fact, workers outside the box have found evidence of astronomical sophistication in Egypt almost three times as old at a place in the Nubian Desert called Nabta Playa. One stone sculpture found there looked very much like a map of the Milky Way as seen from the northern galactic pole (Kenyon 2008, 69). One of its megalithic sight lines stands in relation to the galactic center, indicating a vernal heliacal rising fully 19,700 years ago.

The Guataman counterpart of those Egyptian sages, the Mayas, as indicated in Codex Vaticanus, also kept their calendar system for some 20,000 years, a great deal earlier than assumed by most archaeologists. Probing the time in which the Mayan astronomers made their first calculations, protohistorian Paul Von Ward counted back thirteen Mayan cycles (of 5,163 years each) to a time 67,119 years ago (2011, 169). Clearly, the Mayan observatories have been much too conservatively dated, for they were rebuilt by successive generations, simply incorporating earlier complexes into their buildings. When we dig into the Mesoamerican pyramids and temple platforms—lo and behold—inner pyramids are found.

The first men described in the Popol Vuh were far-seeing (literally): "They examined the four corners, the four points of the arch of the sky, and the round face of the earth." The apertures of the round tower at the

Yucatan Chichen Itza Observatory seem to coincide with readings of the planets and constellations.* Their successors, the Aztecs, left pictograms showing stargazers looking at the heavens through a tubular instrument.

Telescopes

"He excelled in mathematics . . . in making telescopes and micro-scopes" some 12 kya (The Lord's Fifth Book 7.2) Chinese, Hebrew, Indian, Persian, and Babylonian astronomers measured the size, velocity, and orbit of the planets. But how? Did they have tele-scopes? The ancient Persians and Indians were far advanced in learn-ing, as proved by the stars and planets they named and mapped. The ruined temples of these lands suggest that in the time of their build-ing, astronomers knew nearly as much about the heavens as we do today. Within their Temple of the Stars (Oke-I-git-hi) were

> mirrors and *lenses* and dark chambers . . . so constructed
> that the stars could be read as well in the day as at night.
> (my italics)

> OAHSPE, THE LORDS' FIFTH BOOK 2:16

A ground lens, now at the British Museum, was found in the ruins of Nineveh (Iraq) the likes of which today can only be manufactured with a special abrasive based on cerium oxide. Those Chaldeans knew with exactness the mean motions of the Moon and could fore-tell the appearance of comets. Their cuneiform tablets and cylinders recorded the phases of Venus as well as the location of an outer planet that could not be observed without telescopes of some sort. Predynastic finds in Egypt as well, according to Russian research, include crystal lenses of the utmost precision.

As for the New World, researchers say that the ancient Peruvians used a mirror called *quilpi* made of concave and convex glasses, "an

---

*Those astronomers kept a precise calendar year for Earth and Venus, the exact length of the solar year correct to three places of decimals and the Moon's orbit to four decimal places.

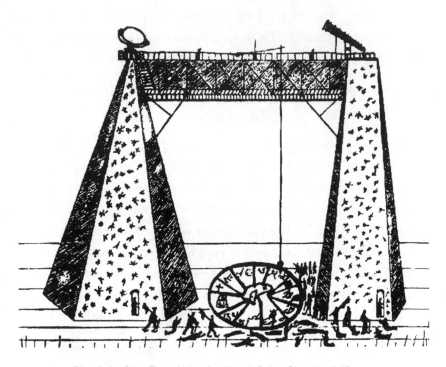

Fig. 8.6. Star Temple in the time of the Shepherd Kings.
From Oahspe, The Lords' Fifth Book

optical instrument for seeing far away" (Charroux 1971, 214). Following these clues, French researcher and explorer Robert Charroux (1974, 22, 66) found, among Peru's controversial Ica Stones, drawings of astronomers observing the heavens with the aid of a telescope. He dates the Ica Stones to thirty thousand years ago. In Bolivia as well, Tiahuanaco's astronomers apparently studied the stars using the equivalent of a telescope. One hears of similar ancient lenses in China, Jordan, Libya, Brazil, Central Australia, and Megalithic Europe.

Is there a link between these ancient astronomers, as widely separated as Mexico, Mesopotamia, China, and Egypt? Philip Coppens shows a map linking "the four most important archeological sites in the world: Giza, Stonehenge, Teotihuacan, and Tiahuanaco [which] when mapped on the world, reveal that they are laid out in the formation of the constellation of Orion" (2013, 197). And while staid archaeology dates those Tiahuanacan ruins to a mere three thousand years ago,* Alexander Marshack of Harvard's Peabody Museum, having examined those artifacts, saw a meaningful pattern of dots, zigzags, and circles that could not have been mere decorations or even hunting tallies. They looked much more like notations, with information on the phases of the Moon and the constellations. Some were dated as old as 30 kya, matching Charroux's date for the Ica Stones as well as the work of archeology Professor George Michanowsky who, in the 1950s, deciphered one of these engravings. It represented a region of the sky where a supernova had exploded 30 kya, giving birth to the Gum nebula, which, incidentally, could not have been observed with the naked eye.

Marshack was struck by similar notations on a twenty-thousand-year-old bone in Central Africa, the Ishango Bone, which had apparently been used to track the phases of the Moon. Its notches, arranged in groups, look like a mathematical table. The artifact was not unlike engraved bones left by Europe's Cro-Magnon people, such as the Kelna Bone (Czechoslovakia) and the Gontzi ivory (Ukraine). Ancient astronomers at Cro-Magnon Lascaux, France, painted a lunar calendar similar to the Ishango Bone, 15 kya or even 20 or 30 kya, according to Charroux (1974, 83).

Likewise, India's *Mahabharata* indicates that the skies were closely watched around 15,000 BP, perhaps corroborated by a map of the heavens found at Bohistan in the Himalayan foothills showing the position of the heavenly bodies 13,000 years ago. Those ancient scientists predicted eclipses and could calculate the precession of the equinoxes. The man known as Asuramaya is a kind of Hindu Methuselah, for he was their antediluvian sage, astronomer extraordinaire, and author of *Surya-Siddhanta*. Discovered in Benares, this work is perhaps the world's most ancient astronomical treatise.

---

*Also conservatively dated, although trumpeted as the "oldest" solar observatory in the Americas, are Peru's ruins at Chankillo, with thirteen towers that align in such a way that they are oriented to the Sun at sunrise and sunset (Ornes 2007, 13).

The Sumerians, for their part, had knowledge of Mars's satellites. These astronomers accurately measured the rotation of the Earth and the Moon with little deviation from modern figures. They counted periods of revolution in seconds (not days). Among their clay tablets was found one containing a fifteen-digit number: 195,955,200,000,000. Now known as the Nineveh constant, it indicates that the Sumerians were able to calculate the revolutions of all the planets in our solar system. They also drew pictographs of the planets revolving around the Sun—eight thousand years before the vaunted Copernican Revolution! But were the Sumerians the *first* astronomers? Why do we stop tracking paleoastronomy at the Fertile Crescent?

I have to ask whether the Sumerian savants can be traced back to Asia Minor. It does not make sense to date the world's earliest astronomers to Sumerian times, five thousand or six thousand years ago, when we know of Turkish observatories like Gobekli Tepe (which was also a ritual center, a sanctuary) dated to 12,000 BP. This Star Chamber in southeast Turkey's highlands (north of Sumer on the Euphrates), has finely carved reliefs on massive megalithic columns. It is aligned north-south (as is the later Great Pyramid of Egypt), betraying a precise knowledge of geodesy seven thousand years *earlier* than the presumed beginning of exact science. Nor is it coincidental that Gobekli Tepe, accidentally discovered in 1994, was a temple complex with round and oval buildings.

The Lord provided . . . oracle houses, in which he could speak face to face with mortals, through his angels . . . persuading them to industries . . . [and] teaching them about the stars.

OAHSPE, THE LORDS' FIFTH BOOK 6:3–5

(TIME FRAME ROUGHLY 11 KYA)

The round *tholoi* structures ("beehive chambers") discovered in Turkey and Iraq answer to the description of those oracle houses. These windowless domes were well-covered (intentionally dark), for they were constructed specifically to hold communion with their mentors in the

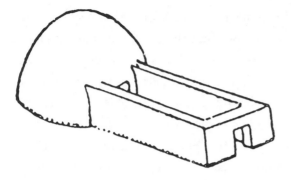

Fig. 8.7. Tholos of Iraq

spirit world. All along the sweep of the Caspian, beehive structures dotting the landscape stand as the signature of the wise men who convened with invisible helpers in their dark chambers.

In spirit chambers, the sitters were instructed by their unseen host to build certain scientific instruments, such as the *gau*, from which the word *gauge* is derived. The *gau* was a measuring device, a triangle with plumb and level combined, by which was discovered the roundness of Earth: "By the gau was the earth proven to be a globe." Also by using this instrument the temples were built and the stars were observed. Other measuring instruments with Panic names suggest they are at least 24,000 years old:

> Sam'tu was an instrument for measuring; tu'fa was a compass, used to measure circles and circumference
>
> OAHSPE, BOOK OF SAPHAH:
> TABLET OF SE'MOIN, INTERPRETATION 28, 32

Fig. 8.8. A gau. This simple device was the basis of ancient technology.

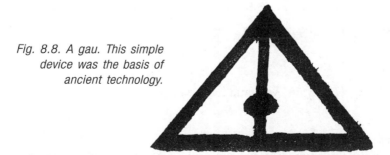

The land of Turkey, bridging East and West, was one of the last outposts of the old masters. Here we find the first known astronomers, as well as wheat domestication and town life, not to mention the key founders of Europe. A terraced town with canals and irrigation works, the Hittite settlement of Çatalhüyük is among the oldest known cities in the world, boasting the world's earliest mural and landscape paintings, the oldest finely woven cloth, fired pottery, and cuneiform writing. The early Hittites enjoyed an enviably harmonious society featuring democratic councils and female equality. These institutions signal the sacred legacy of the sons of Noah, a distinctly "separate" people. Indeed, the Hittites practiced the levirate* "to preserve the seed" of their own holy race. (Çatalhüyük, incidentally, is near Mt. Ararat, where Noah's Ark is said to have beached.)

As long ago as 9000 BP (with deeper levels still unexcavated), Çatalhüyük was filled with comfortable brick-and-timber houses. Sorry, no stonework in this virtually "Stone Age" town. In fact, the famous site contains the oldest known examples of metallurgy and smelting in the Near East. Did Higher Intelligence, the angelic host with whom these people communed inside their *tholoi,* teach them their sophisticated industries and technologies? Ancient Turkey, after all, had the first wheel,† the first coins, fine jewelry, comfortable homes, and polygonal masonry. Among their ruins were found excellent metalwork (copper cylinders), superb craftsmanship in obsidian, and wooden articles produced with "lathelike" perfection.

Their colleges taught

> mathematics and chemistry and minerals and assaying and the rules
> for inventing chemical combinations.
>
> OAHSPE, BOOK OF WARS 21:10

Adjoining Turkey, in Armenia (Medzamor), metallurgy again goes back,

---

*To marry your brother's widow.
†Contradicting the textbooks, which still say Sumerians invented the wheel 5 kya.

oxymoronically, to the "Stone Age," the site boasting thirty-thousand-year-old iron workshops producing eighteen varieties of bronze (Charroux 1974, 45, 117–18), long before the Mesopotamian "founders of civilization" began mining copper and making bronze.

Extremely early metallurgy has also been identified in southern Africa, where there are reports of 47 kyr copper mines in Rhodesia, 43 kyr iron mines in Swaziland, and extensive mining more than 20 kya at Broken Hill (Corliss 1978, 362). And in Egypt, long before the official "Iron Age," iron is mentioned in the Pyramid Texts. Chert mining in the Nile Valley also goes back tens of thousands of years (Rudgley 1999, 174). In Africa, Ham, son of Noah, taught mining, metallurgy, crafts, and also the intricacies of the world beyond:

To Ham | allotted the foundation of the migratory tribes of earth . . . allotted to teach the barbarians.

OAHSPE, THE LORDS' FIFTH BOOK 3:3

It is undeniable that some great civilization exercised its influence on various peoples . . . and was so strong as to even affect the barbarians.

PETER KOLOSIMO, *TIMELESS EARTH*

Leaders of men were these flood survivors, the children of Noah—the sacred tribes of Ham, Shem, Jaffeth, and their offshoots. They spread into India, Persia, Arabia, Egypt, Africa, and China. They taught the wild tribes. And they were men, not gods or spirits but a priestly race, the first on Earth to acquire knowledge from the angels and disseminate it to the unlearned in their midst.

I think those traveling Hamites are the same as the Gumba who, say the Kenyan Kikuyu, taught them the art of smelting and ironworking. And since the sons of Noah in America (Guatama) were old cousins to these teachers, we are not surprised to find that Kikuyu beadwork is identical to that found among the Plains Indians. And, yes, iron implements have been found at moundbuilder sites in

Virginia, Missouri, and Michigan (Steiger 1974, 44–50). In Ohio as well, near Paint Creek, smith shops and furnaces for smelting were excavated, and they are similar to workshops in North Carolina and New York. In addition, 12 kyr copper and bronze have been unearthed in Illinois (Kolosimo 1973, 6–7).

In the dead cities of South America as well, especially at sites with *tholoi*/beehivelike structures resembling those of the Near East, we again come across "anomalous" (too early) metalwork, like platinum objects, a metal known to melt only at 3300°F. How were such high temperatures achieved by these early people? Curiously, Heinrich Schliemann found in Turkey (at Troy, which he famously unearthed) pottery shards of the same chemical composition as those found at Bolivia's Tiahuanaco, along with metal objects of an identical amalgam of platinum, aluminum, and copper. Iron mines, according to *Mercurio Peruano,* were anciently worked here, perhaps 16 kya. The Tiahuanacans were master metallurgists, their techniques—for silver-plating, filigree, and damascene work—among the lost arts.

Consider also the high quality of ancient textiles eight or nine thousand years ago, which may allow us to link areas as disparate as Tiahuanaco and Turkey, opening our eyes to that lost horizon—those dispersed cousins. The ruins of Turkey's Çatalhüyük contained very finely woven cloth and pieces of carpet of supreme workmanship, whose quality was rivaled by fine textiles in North America,* Panama, Mexico, and Peru. Unsurpassed in technique and beauty, brilliant-colored Peruvian shrouds and tapestries are minutely stitched: five hundred threads per inch! Older yet than Çatalhüyük are 30 kyr textiles found at a location north of Turkey, in a cave in Georgia. These were spun of flax, traditionally used to make linen. The knotted fibers were dyed violet, red, black, and turquoise. (Women in Iran, Portugal, and Ireland still dry and spin flax to be woven into linen.)

---

*Mining and textiles are in the assemblage of ancient Wisconsin sites (at Oconto and Osceola), dating back at least 11 kya, while Florida's bog mummies (8,300 kya) were also found in association with finely woven cloth (Dewhurst 2014, 98, 295).

Egypt's ancient flax tradition was famous throughout the world for its fine linen; the mummies are all wrapped in it. Every single thread was composed of 360 minor threads twisted together. But were the Egyptians the inventors of linen, or did they draw on the arts of a still older culture? It seems significant that the overseers of linen and the chiefs of the royal textile works in Egypt were typically "dwarfs" (little people, that is, Noachic). In the same vein, the "fairy folk" (read: little Ihins) of the British Isles were famous for their delicate linens and "webs of the finest cloths." Also small in stature were Jordan's Natufians who, more than 10,000 years BP, dressed in elegant fabrics made from spun yarn. Incredibly fine textiles are repeatedly found in little people enclaves, for the Ihins were the first to make garments, tens of thousands of years ago.

> Storing in their cities and mounds ample provisions of food and clothing for the winter . . . [t]hey tilled the ground and brought forth flax and hemp, from which to make cloth for covering the body.
> OAHSPE, FIRST BOOK OF THE FIRST LORDS 3:12;
> THE LORDS' FIRST BOOK 2:13

The tradition became part of Cro-Magnon culture. No, they did not invent it; they inherited it. Their artifacts include spun, dyed, and knotted flax. To explain these precocious Stone Age industries, anthropologists (as we saw in chapter 3) suggest that cold weather spurred restless innovations; scholars even contended that Cro-Magnons changed history by inventing the sewing needle.* But Cro-Magnon was only carrying on an older culture, one branch of which goes back to Russia's Sungir site, perhaps 40,000 years old, where (frozen) fabric manufactured into trousers, pullovers, and coats has been found—perhaps part of the Uighur textile industry.

---

*The moundbuilders in Ohio also had sewing needles and woven cloth.

Brushing aside these intriguing finds (and their too-old dates), archaeology remains loyal to the SM, which says such industries could only have developed around 6,000 years ago. Indeed the primitive status of Mesolithic man is a requirement of Darwinian evolution. What would happen to that precious paradigm if all these protohistorical discoveries saw the light of day? After all, a great deal has been written about the Superior Ancestor in the past fifty years. Much has been uncovered: "Stone Age" people also performed medical operations, including amputations, cranial surgery, and organ transplants. The ancient Hindus removed cataracts and performed C-sections. The ancient Egyptians and Peruvians had brain surgeons, just as the moundbuilders near Cincinnati performed cranial surgery. The ancient Aleuts of Alaska removed abcesses of the eye.

LOST LANDS

The fleets that left vestiges of culture and language over the world did not come from the centers we usually associate with Greeks and Phoenicians, but from other earlier lands now lost. . . . In a previous advanced age of navigation . . . seafarers sailed all over a younger world, mapped its oceans and charted its skies.

CHARLES BERLITZ,
MYSTERIES AND FORGOTTEN WORLDS

Although the first world map is generally attributed to Anaximandros of Miletos in the sixth century BCE, the Sea Peoples, older than the Egyptians and anterior to the so-called dawn of history, were familiar with all the continents, their navigators using spherical trigonometry to determine exact locations. This was a worldwide seafaring civilization that preceded the maritime Sumerians of the Fertile Crescent and were perhaps the last—not the first—of these golden-age thalassocrats. Hawaiian legend recalls the Mu folk as master shipbuilders and skilled mariners. "These ancient people of Mu were great navigators and sailors who took their ships all over the world" (Churchward 2011, 25).

Without a blink Dame History has written off such early trans-oceanic contacts as merely a fluke, probably the result of accidental landings (see the "random" card, as discussed in the introduction and chapters 1 and 2), and this crumbling isolationist school of thought is still the ruling paradigm. When "much-too-early" buildings were found in Chile, "the site was obviously a fluke . . . the accidental remnant of shipwrecked [raft-wrecked?] castaways. . . . [But] men were on the move around the world much earlier than mainstream scholars would have us believe" (Frank Joseph 2013, 85).

Americanist Norman Totten protested that

> the most authentic ancient oral knowledge is dismissed out of hand when it counters current academic opinion, particularly the out-moded though still dominant paradigm of radical isolationism. . . . I have grown weary of the arrogance of those . . . who feel that their published reputations should constitute a fixed view of the past. (Mahan 1983, 2)

But the times are changing: British anthropologist Richard Rudgley has persuasively argued that these "Stone Age" explorers discovered all the world's major lands. In India, the maritime Vedics navigated around Africa and across the Atlantic, mining copper and tin in the Americas. A terra-cotta seal from Mohenjo-Daro, Pakistan, along the Indus, shows a large, high-prowed ship with a spacious on-deck cabin, hinting at "an ancient science of cartography and navigation that explored the world and charted it accurately" (Hancock 2002, 126, 669). Indeed, Valmiki's *Ramayana* describes the early people of India as mighty navigators whose ships traveled from the Western to the Eastern oceans, and from the Southern to the Northern seas, "in ages so remote that the sun had not yet risen above the horizon." Compelling parallels between this Indus Valley civilization and the Southeast Amerinds, as Joseph Mahan sees it, speak for "an ancient culture of truly worldwide influence" (Mahan 1983, 155).

Vast copper works along Lake Superior run down to sixty-foot depths, the distribution of their products suggesting a great maritime and commercial people who traded copper, tin, bronze, and silver during the "Stone" Age. More than a billion pounds of high-grade copper was removed by those prehistoric miners and distributed widely throughout America,* if not the world.

> The people of learning built great sailing boats.
>
> OAHSPE, LORDS' THIRD BOOK
> REFERRING TO A TIME 18 KYA IN THE NEW WORLD†

A good map of South America was drawn 15,000 years ago—the Piri Reis world map, based apparently on even earlier sources. Models of four-masted ships have been found in Brazil, resembling ancient Cretan ships that could carry some eight hundred passengers; in both languages they are called by the same name—*cara-mekera*. Likewise, 17 kyr cave art in southwestern France depicts a large ship. Around 24,000 BP,

> there were thousands of cities and hundreds of thousands of inhabitants . . . and they sailed abroad on the seas.
>
> OAHSPE, THE LORDS' FIRST BOOK 2:19–20

In fact, Alexander Marshack has found 35 kyr notations that indicate knowledge of world geography. Yet even farther back, 40 kya, there were great cities in all five divisions of the Earth.

> And man built ships and sailed over the ocean in all directions, around the whole world.
>
> OAHSPE, SYNOPSIS OF SIXTEEN CYCLES 1:21

---

*Michigan copper and Illinois galena have been found in Florida.
†The quote matches Cherokee lore, which says the Little People traveled all over the world.

The latest round of Australian archaeology also points to oceanic voyages some 45 kya.

Early humans may well have been seafarers as much as forty thousand years ago.

GRAHAM HANCOCK, *UNDERWORLD*

We do not believe for a minute that these mighty navigators were naked hunter-gatherers; for these same people had mastered the techniques of agriculture, which is the cornerstone of civilization. Forty thousand years ago,

the earth was tilled and it brought forth abundantly everything that was good for man to eat.

OAHSPE, SYNOPSIS OF SIXTEEN CYCLES, 1:21

This, of course, refers only to the civilized tribes of the Paleolithic, for people of learning have always coexisted on the planet with tribes of the most primitive type. Nevertheless, the official version has *everyone* before 10,000 BP as rude hunter-gatherers and cave dwellers. Agriculture is presumed to have finally arisen around that time in the Fertile Crescent, then spread to China, moving into Europe some eight thousand years ago.

But if farming began in the Fertile Crescent, the Sumerians themselves said it was brought to them from the sacred mountain of Du-Ku, in the region of Turkey. As we saw on page 348, science, metallurgy, ceramics, and clothing of Mesopotamia were all predated in Turkey. Wheat and oats were already well-established crops in Turkey thirteen thousand years ago. And these people, in turn, appear to have had prior experience in agriculture, an art that had been perfected elsewhere. This brings us right back to Churchward's date of 17,000 BP for agriculture among the North Asian Uighurs, who made their way west through the land of Turkey (Asia Minor).

Some regions in the Fertile Crescent that traditional prehistorians cite as the cradle of agrarian life, and hence of civilization itself, look "singularly uninviting" to an ancient horticulturalist. More likely, the practice of farming was brought to these places from elsewhere (alluvial plains) and doggedly pursued in the new landscape despite harsh conditions. Such uninviting areas include the rugged terrain of the Taurus and Zagros Mountains and the Upper Zab's (Iraq) "rocky limestone hillocks, not really suitable for farming. . . . If farming really were the stimulus for the creation of civilization, why were the [supposedly] earliest examples of it located in such agriculturally difficult places?" (Kreisberg 2012, 57).

The Fertile Crescent's touted "Neolithic Revolution" (around 10,000 BP) blithely ignores the practice of agriculture in Thailand fully thirteen thousand years ago; in fact, yam and taro cultivation in Indonesia go as far back as seventeen thousand years. The proto-Malay folk at Kota Tam*pan* (ancestors of the Lenggong little people) practiced farming and manufactured porcelain containers eleven thousand years ago. Only settled agriculturalists were likely to develop the ceramic arts (no pottery in the Mesopotamian Fertile Crescent until 8500 BP). It is also likely that Middle Stone Age microliths were fitted onto a sickle for harvesting work. We find these "pygmy flints" at the same spots settled by our protohistorical farmers: Russia, Palestine, Southeast Asia, Philippines, China, Japan,* and Egypt.

Egypt also shows signs of very early (eighteen-thousand-year-old) cultivated wheat and barley, most notably at Wadi Kubbaniya, with its tuber remains and grinding stones. Legends of the Nile Valley recall the Egyptian god Osiris introducing domesticated crops in the "First Time." Not surprisingly, the oldest horticultural people of the Nile

*Japan's delicate ceramics of the Yamato (Jomon little people) are about 18 kyr. Pottery is almost exclusively associated with settled, agricultural people. The Jomon, according to Graham Hancock (2002), cultivated rice, beans, and gourds. As old as Jomon, and possibly older, are Eastern European ceramics found at Dolni Vestonice, Czechoslavakia.

and Libya, according to Hecataeus, were the *pugmalos* (pygmies).

So let's connect the dots. The earliest agriculturalists were little people—Noachics. In Palestine, sixteen-thousand-year-old emmer wheat grains have been found at Nahal Oren. Here, too, it seems to be a short-statured people, the aforementioned Natufians, who introduced the arts of horticulture. Their forebears, the sacred little people known as Ihins, were strict vegetarians (no hunting!) and the first farmers in the world, even before the Flood.*

According to The Secret Commonwealth†, the little people had their own agriculture before the Flood when cities and great plantations filled the Earth—before being effaced.

> They tilled the soil . . . [but all was] dissipated by the dread hand of war . . . their cities are destroyed.
>
> OAHSPE, SYNOPSIS OF SIXTEEN CYCLES 3:9

The SM tells us that agriculture began in the Americas around 7 kya, even though beans and peppers were grown in the Andes at least 11 kya. There are even hints of horticulture before the Flood in Peru, for the pre-Inca creation myth has Pachacamac planting the first corn, yucca, vegetables, and fruit before creating the present race. In other words, those agriculturalists belong to the Second Age, which "ended in cataclysmic deluge." Archaeologist George O'Neill regarded the Peruvian Chachas as a once-vast civilization that built cities and contoured agricultural terraces, which undoubtedly maintained a very large population. Terracing itself may be an index of these pre-Neolithic farmers: there are terraced hillsides at Machu Picchu, Tiahuanaco, Ollantay, and Pisac. Advanced techniques in terracing (along with weaving and metallurgy) have been found at other ancient Andean sites, such as the Gallinago civilization, and at Viru Valley in Peru.

---

*On Pan itself, Melanesians 11 kya grew yams and sago. Even antediluvian farmers left evidence of 28 kyr cultivated grains in the Solomon Islands (Frank Joseph 2006, 132).
†Old Book of the Irish.

In Mexico, Mayan engineers set up an efficient system of irrigation and drainage, enabling them to grow crops to feed millions. Monte Alban near Oaxaca, flanked on all sides by terraced steps, lay atop an artificially leveled "mountain" (read: mound) There are two thousand terraces here and many others at such sites as Cacaxtla, where archaeologists marvel at extraordinary prehistorical murals. Agricultural terracing in Mexico's Puebla Valley dates from very early times, while radar and aerial imaging confirms the same in Belize, Quintana Roo, and the Maya Mountains.

Corn

There has even been talk of corn in Mexico dating back to 80,000 BP, evidence presented by Jeffrey Goodman (1981, 179) in the form of pollen grains extracted by drill core under Mexico City. Similarly, grinding tools, dated 70,000 BP, have been unearthed in California. Maize culture alone gives us "food for thought." The pattern of dispersal through the Americas was north and south from Guatemala (Guatama-la). The oldest cultivated corn in the Western Hemisphere is located in the highlands of Guatemala. Beans and squash also seem to have originated in Central America, spreading both north and south. Here in the Americas, it was the race founders (not the "gods") who brought corn, a boon spoken of in many native traditions. They declare that First Man brought maize. The Pueblo peoples of the American Southwest, who migrated there from Central America, are unlike other desert tribes, for they are preeminently agricultural, known for their clever irrigation systems. The Hopi say their ancestors domesticated corn in the First World. In Zuni tradition, it is Corn Maiden who accompanied the First People to the surface of the Earth. This Emergence usually stands as a metaphor of the Flood and its survivors.

Protohistorical America has been vastly underrated. The remarkable mound works of ancient North America, as Sir John Lubbock in

Prehistoric Times saw it, indicate a population both large and stationary . . . which must have derived its support, in a great measure, from agriculture." Now according to orthodox history, agriculture in the Americas is no more than seven thousand or eight thousand years old. But if 8000 BP is the earliest date for New World farmers, that is as far down as they've dug. Lower fossil beds remain largely unexplored, for they are "way too old" (or so it is thought) for remains of humanity's developed industries.

In oral history it always seems to be the little people who bring agriculture, especially after the Emergence (read: Flood times). In fact, it is right in the area of tiny graves in Tennessee that archaeology has located some of the earliest American cultigens—squash and sunflower. This six-acre graveyard contained the (extremely short, three-and-a-half-foot high) remains of almost 100,000 souls. Significantly, all the oldest mounds in North America are located in alluvial regions, affording ample scope for agricultural pursuits. This rich alluvial soil was once the seat of a numerous and permanent population who supported themselves by the cultivation of maize.

Let archaeologists say the first-ever cities on Earth came about 7000 BP; we know a lost horizon lies buried deep in the unexplored Earth. After the Flood, thousands of cities sprang up everywhere. Even before the Flood,

The Lord called them [the Ihins] together, saying: Come dwell together in cities. . . . And the Ihins . . . built mounds of wood and earth . . . hundreds and thousands of cities and mounds built they.

OAHSPE, FIRST BOOK OF THE FIRST LORDS 3:8

By 18,000 BP there were four million moundbuilding Ihins spread throughout Tennessee, Kentucky, Ohio, and Kansas, and along the shores of Lake Michigan and Lake Superior. We have many traces of this lost civilization, particularly in the extensive trade networks of the

Mississippi Valley culture. When Europeans first settled America, they came upon thousands of mounds of varying heights throughout the Mississippi and Ohio waterways, some of which had been whole cities, protected by earthen walls. All this indicated "a population as numerous as . . . the Nile or the Euphrates . . . Cities of several hundred thousand souls have existed in this country" (Brackenridge 1818).

There was the great king Hoajab . . . [whose] capital city was Farejon-kahomah, with thirty-three tributary cities, having tens of thousands of inhabitants, on the plains of He'gow [southeastern Ohio]. The next great king was Hiroughskahogamsoghtabakbak, and his capital city was Hoesughsoosiamcholabonganeobanzhohahhah, situated in the plains of Messogowanchoola* and extending eastward to the mountains of Gonzhoowassicmachababdohuyapiasondrythoajaj, including the valleys of the river Onepagassathalalanganchoochoo, even to the sea, Poerthawowitcheothunacalclachaxzhloschista-combia† [Lake Erie].

OAHSPE, FIRST BOOK OF GOD 25:7–9

Little People on the Mounds

Through diligent research, physicist and encyclopedist of anomalies William Corliss rescued an 1837 report of burial grounds in Cochocton, Ohio, situated on a mound and containing the remains of wooden coffins. These people "were very numerous, and must have been tenants of a considerable city" (Corliss 1978, 681). The bones belonged to a race of tiny people from three feet to four-and-a-half feet in height. Similar nineteenth-century reports found "buried nations, unsurpassed in magnitude . . . Ohio is nothing but one vast cemetery of the beings of past ages . . . prehistoric white men" (Atwater 1973).

---

*Indiana, North Ohio, and Pennsylvania.
†Now you can see how long those agglutinative names could get!

Pennsylvanian grave-mounds ("tumuli")—extensive, stone-lined, and filled with copper ornaments—are on a par with the Ohio burial grounds. One especially interesting Pennsylvania site is the Meadowcroft rock shelter in the western part of the state. Archaeologist Jeffrey Goodman announced in 1981 that material recovered there could be as old as twenty thousand years. The item in question was the fragment of a basket. Goodman's date is in line with historian Harold Gladwin's "pygmy groups" in the Americas, going back as long ago as 25,000 BP (in Texas and the Southwest). These were the earliest people in the Southwest and they were also basketmakers, short and slender. (Jeffrey Goodman 1981)

The refugees from Pan carried their moundbuilding tradition not only to America but to every land they settled. After the Flood, moundbuilding people appeared all over the world: in Jarmo, Iraq, for example, they built mounds twenty-three feet high. Coming from the lost lands, all the world's moundbuilders were of a similar breed and, wherever they went, their works were of a comparable type.

The mystery of the moundbuilders' origin is only a mystery if we ignore these facts. Unwilling to accept their common origin, historians indulge in wild guesswork. They came from the stars, they came from Atlantis, they came from Africa, or they came from the sea—which comes closest to the truth—for they were boat people refugees from Pan.

As common survivors of the sunken continent, it should come as no surprise that these cousins resemble one another, both physically and culturally. Their "similarity of works almost all over the world indicates that [they] sprang from one common origin" (Atwater 1973). Indeed, it was their very presence that ensured the progress of humanity wherever these sons of Noah made landfall. Be it a mound in North America, a human-made hill in South America, a tumulus in Scandinavia, a barrow in England, a sidhe in Ireland, a tepe in

Turkey or Greece, a tel in Western Asia, or a kurgan in the North, these structures all had the same origin.

> These are My chosen that live in mounds. . . . They are the people of learning. They survey the way for the canals, they find the square and the arch; they lead the Ihuans [the Indians] to the mines.
>
> OAHSPE, THE LORD'S THIRD BOOK 1:15

The Ihins, in short, are a lost race. For the most part, clear memory has not outlived deep antiquity. I am not, however, a member of the we-shall-never-know club, and neither, presumably, is archaeology. But given the conservative streak in all the sciences, the pundits of the past are loath to revise the playbook, sworn to uphold the (now obsolete) cradle of civilization in the Fertile Crescent—seven thousand years ago.

America, as is her wont, presents the keenest challenge. The lost race theory, accounting for the North American moundbuilders, stimulated one of the greatest scientific controversies of the nineteenth century. Favoring the lost race theory in the 1820s was Caleb Atwater, the postmaster at Circleville, Ohio. Atwater examined many skeletons in the mounds, concluding they were a race apart from the indigenous type. The Indians, he would point out, are a tall people; the early moundbuilders were short.

When non-Indian articles, such as glass or iron items, were recovered from the mounds, opponents of the lost race theory, most notably the Smithsonian Institute, pegged them as fakes or errant items of European manufacture. But a misplaced trinket here and there cannot throw us off the trail of the moundbuilders and their finely wrought artifacts—handsome tools, fine pottery, and works of art unknown among the hunter tribes of North America. Distinctly un-Indian were their inscribed tablets, hieroglyphic writing, telescopic tubes, astronomical calendars, metallurgy, glass beads, batik, and weaving. The infrastructure was also beyond the skeptics' ability to explain away—extensive causeways, canals, fortifications, temple pyramids, as well as the science of surveying. All these were signs of a cosmopolitan civilization.

Still, the assumption held fast that Indians were the only previous inhabitants of America;* hence, their ancestors must have built the mounds. Even today we call them "Indian mounds." And most of them *are* Indian mounds, at least the newer ones. Nevertheless, the bones of a different race lie underneath them, distinct from the surviving native populations and with arts unknown to them. But this kind of reverse evolution, if you will, did not sit well with the deans of science who were led in the nineteenth century by ethnologist Henry Rowe Schoolcraft. They scorned the lost race theory as pure bunk. Anything that contradicted the gradual linear ascent of man, a smooth and constant climb upward (see chapter 2), would be neither politically correct nor academically de rigueur.

Schoolcraft took the pulpit, arguing that there is nothing in the structure of our Western mounds that a semi-hunter and semi-agricultural population, like that of the existing Indians, could not have produced. Yet, besides the aforementioned metalwork, writing, weaving, pyramids, canals, and so forth, there was another flaw to this argument of Schoolcraft and his colleagues. The Indians themselves did not claim ancestry from the builders of the early mounds. Most of the tribes had no knowledge of the mounds' origins.

The lost race theory, nonetheless, continues to be assailed up to the present moment. For more than 150 years, experts have been saying that the moundbuilders were no "mythical" tribe of people who became extinct prior to the advent of the Indians. Well, it is understandable that nineteenth-century man, who had scarcely just begun the work of archaeology, might have clung to old-school ideas and doctrines. But for twenty-first-century man, who has looted the world's buried treasures, and with so much discovery under his belt, the continued whitewash is inexcusable. Races have come and gone.

---

*Clearly, there were political reasons in the age of manifest destiny for portraying native culture as shallow and inconsequential.

Civilizations rise and fall. Each cycle, overcome with darkness and the degeneracy of its ruling elite—and of its ruling dogmas—is swept away, like the dead leaves of winter.

Then comes spring, the new dawn, and the winds of change. In the last chapter of this book we will take a peek at that new dawn, and the paradigm shift waiting in the wings.

CHAPTER 9

DEMOCRACY

On the Eve of Fraternity

To keep man from interfering with man, this hath been great labor.

OAHSPE, BOOK OF INSPIRATION 11:11

Even though civilizations, with the help of cosmic cycles of darkness and light, rise and fall, there is yet a sense that each new round brings us a little higher. Isn't this what we mean by progress?

First on earth, monarchies, then republics, then fraternities, the latter order of which is now in embryo, and shall follow after both the others. Behold how hard it is for an ignorant man to conceive of a state without a master, or for the people of a republic to understand a state without votes and majorities, and a chief ruler. Yet such shall be the fraternities.

OAHSPE, BOOK OF DISCIPLINE 11:11–12

In September 1787, the United States Congress sent the newly minted Constitution to the individual states for ratification. By December it was ratified by Delaware, New Jersey, and Pennsylvania; then Georgia and Connecticut in January. And by June 1788, Maryland, South Carolina, New Hampshire, and Virginia had also approved the Constitution, making up the requisite two-thirds vote. By the end of the following winter, on March 4, 1789, Congress convened and the

U.S. Constitution went into immediate effect. Freedom's Scroll had been written and approved. The great event is hailed in Oahspe's Book of Inspiration, chapter 15:

> The gods sanctified that day of the . . . sealing of His compact [the American Constitution] as the day of the Holy Seal. . . . Remember this day, and keep it holy to the end of the world, for hereat was the *beginning* of the liberty of man! (my italics)

Yes, it was only the beginning. "We are only on the threshold of the democratic state," declared Manley Hall in *The Secret Destiny of America* (2000). "Not only must we preserve that which we have gained through ages of striving, we must also perfect the plan of the ages, setting up here the machinery for a world brotherhood of nations and races." The key word here, of course, is *brotherhood*. It is one thing to establish a republican constitution, but quite another to fulfill the secret destiny of the ages, which is nothing more or less than the brotherhood of man!

Fig. 9.1. Nachwach, the Hopi symbol of brotherhood

Certain ironies of that new American government were apparent early on. This freedom-loving republic, rebel-child of a repressive parent, this young America was now itself fast becoming the instrument of a ruling class. This is what Thomas Jefferson (a founding father who had prophesied humanity's perfectibility and the ultimate triumph of self-government) had feared more than anything else. Reviewing the office of the presidency from his diplomatic post in France, Jefferson commented snidely: "Their president is a bad edition of a Polish king." But, of course, he is also the gent who famously said, "That government is best which governs least." Later, in a letter to John Taylor, he would write, "Our government has swallowed more of the public liberty than even that of England." His rival Alexander Hamilton, fancy man and federalist leader, was a New Yorker who had written much of the

Constitution himself with the help of Gouverneur Morris and James Madison, while Jefferson and John Adams, the real intellects, were (conveniently?) away in Europe.

Still, it had been necessary to frame a potent charter for the new nation. All the "perils of democracy" (Winston Churchill's phrase) would somehow be met in the challenge of the future.

In the year of ratification, every detail of policy—from taxation to the slavery issue—had been fiercely debated. A multitude of disparate opinions were in the air; still, consensus worked in favor of the elite. (And isn't that the theme of this book?) The slave, or Negro, who was legally "three-fifths" of a person, was never mentioned directly in the U.S. Constitution. Men were created equal, but that did not include women or blacks.

The peculiar institution of slavery would have to take care of itself. Slaves, housewives, farmers, laborers, artisans, small shopkeepers, students, teachers, Native Americans, religionists of all stripes, minorities, frontiersmen, and all the lower classes were now subsumed under a central power controlled by politicians, aristocrats, captains of industry, shippers, manufacturers, bankers, land speculators, lawyers, and clergymen of rank—all with an interest in upholding class society and property rights.*

Woodrow Wilson, the twenty-seventh president of the United States, said: "The government of the United States . . . is a foster child of special interests . . . [it] has gotten into the hands of bosses. . . . An invisible empire has been set up above the forms of democracy . . . American industry is not free. . . . We have come to be one of the worst ruled, one of the most completely controlled and dominated governments in the civilized world."

Looking back on the colonial situation and summarizing the anti-federalist stand, Wilson's contemporary, Winston Churchill, in *The Age of Revolution,* opined:

---

*Nevertheless, the states were mindful of their godly trust, and would quickly write the first ten Amendments, the Bill of Rights, ratified in 1791. But, I often wonder, should freedom be legislated? Or is it a God-given essential? Why must I depend on a beneficent authority to grant freedom to me?

To the leaders of agrarian democracy, the backwoodsmen, the small farmers, the project seemed a betrayal of the Revolution. They had thrown off the English executive . . . [but] created another instrument no less powerful and coercive. They had been told they were fighting for the Rights of Man and the equality of the individual. But they saw in the Constitution an engine for the defence of property against equality. They felt in their daily life the heavy hand of powerful interests behind the contracts and debts which oppressed them.

The famous patriot Patrick Henry disliked the Constitution altogether, saying, "It squints toward monarchy." Henry, along with soon-to-be president James Monroe and others, perceived the defects of a "democracy" managed out of New York (the nation's capital was still on Manhattan Island) by a slender power elite.

Nonetheless, when George Washington was inaugurated as the first president of the United States at Federal Hall in New York City, in his perfect dignity he refused pay for the office, accepting only reimbursement for expenses incurred during his administration. However, Washington did sign the first central bank's charter, setting the scene for monopolies and financial corruption. Jefferson was bitterly opposed to the concept, which Hamilton favored, winning the support of Southern politicians by moving the capital from New York to the more southerly Washington, D.C.

The problem was that the children of kings and aristocrats did survive in the great American republic, and they became the leisure class. Autocracy survived despite the shift to republicanism. Oligarchy survived. Plutocracy survived, making the world safe for lucrative corporate investment. Republicanism did not rid the world of these things. A remark by commentator Jay Tolson in 2001 on the fledgling democracy argues that "the competitiveness and individualism released by the Revolution began to produce something quite different from a genteel republic: a rough-and-tumble democracy with a vigorous capitalist economy."

Capitalism, we know by now, is the thing that ends in a ratio of one billionaire to a million people in poverty, or one millionaire to tens of thousands who are struggling. Who really wields power in this wonderful democracy? The capitalist. German historian Oswald Spengler held that democracy inevitably lapses into plutocracy (rule by the wealthy). Our thinking, after all, is manipulated by public opinion, which is manipulated by the media, which is manipulated by politics, which is manipulated by business interests.

Today, economic power in "democracies" has been concentrated to an unprecedented degree: 90 percent of the economy is in the hands of the top 500 companies, and 80 percent of that group is in the hands of the top 100 companies. Takeovers continue to concentrate that wealth until only two dozen financial institutions control ownership of those companies' stocks—a cozy financial elite.

Similarly, rich countries today consume 80 percent of the world's natural resources. If this is democracy, only one out of five of the world's people is reaping its benefits. And doesn't history warn against this? The Inca's advanced civilization suffered from the corruption of a privileged class. The empire's benevolent despots closely controlled six million people, ruling through a paramilitary chain of command. Greece, too, was governed by a tight aristocracy. And when Egypt fell, a mere 4 percent of the populace owned all the land, just as Babylon went by the wayside when 3 percent of the population controlled all the wealth. When ancient Persia collapsed, 2 percent dominated everything, and when Rome fell, a scant two thousand people owned the entire civilized world. And the Dark Ages followed.

Is there a lesson to be learned from the past?

As long ago as thirty-five hundred years, the prophet Capilya of India admonished: "To labor for one's self at the expense of the state is to rob the state; to hoard up possessions is to rob the poor. . . . After all, is not the earth-life but the beginning, wherein we are as in a womb, molding our souls into the condition which will

Fig. 9.2. Capilya of India, a contemporary of Moses drawn in the dark (automatically) by J. B. Newbrough.

come upon us after death? |n which case we should with alacrity . . . appropriate the passing time to doing righteous works to one another."

OAHSPE, BOOK OF BON 7:13–14

Although the sharing of wealth was taught by the sages of the past, democracy today, rather than leveling the field, has widened the gap between the rich and poor. Our own beloved sixteenth president, Abraham Lincoln, prophesied in his last year of life:

I see in the future a crisis approaching. . . . As a result of war, corporations have been enthroned, and an era of corruption in high places will follow. The money power will endeavor to prolong its reign by working on the prejudices of the people until all the wealth is aggregated into a few hands and the republic is destroyed. (Martinez 2007)

Less than a hundred years later, the thirty-second president, Harry Truman, echoed that sentiment: "When there are too few people at the top who control the wealth of the country, then we must look out."

Truman, as you may recall, ended World War II. During that war a conversation was held in the erudite parlor of Mr. and Mrs. Alfred North Whitehead. One of their guests, referring to the relative youth of the American republic, opined that "We Americans seem to have no folk mythology." Whereupon Mr. Whitehead replied, "You Americans are creating your myths now."

Lucien Price, who miraculously recorded these extensive dialogues, informs us that this comment then "led to a lively discussion of some of our myths." "One of them is democracy," said Mrs. Whitehead wickedly. Her husband then remarked, "The political concepts on which your American society is founded are a kind of myth. . . . The founders of your republic [began] the political myth. . . . The right of the common man to a good life [was] interpreted as the right of a few exceptional individuals . . . to exploit the resources . . . [making] themselves inordinately rich. When I say 'exceptional,' I do not mean that they are superior. In every other relationship of life besides the making of money they quite often are inferior. But . . . they were the ones who ran away with your political myth . . . the false and vulgar idea that anybody in America could get rich" (Whitehead 1956, 217–18).

CAPITALISM

Alas, the two extremes—riches and poverty—have made the prospect of a Millennium a thing of mockery. For one rich man there are a thousand poor, and their interests ceaselessly conflict with one another. When Capital strikes Labor, Labor cries out in pain, but Capital strikes him with a heartless blow. Nation is against nation; king against king; merchant against merchant; consumer against producer; yes, man against man, in all things upon the earth. Because the state is rotten, the politician feeds on it; because society is rotten, the lawyer and court have riches and sumptuous feasts.

OAHSPE, THE VOICE OF MAN 1:34–37

As we near the end of this book it is clear that our theme has been the monopoly of ideas enjoyed by the SMs and the intellectual elites, not only of academia but also of the new age. How democratic is that? Yet democracy and monopoly are kissing cousins. Albert Einstein mused,

> Private capital tends to become concentrated in a few hands . . . [creating] an oligarchy of private capital, the enormous power of which cannot be effectively checked *even by a democratically organized society.* The members of legislative bodies are selected by political parties, largely financed . . . by private capitalists. (my italics) (1954, 157–58)

For his own part, Einstein, something of a recluse and prophet himself and famous for his casual and offhand appearance, once said, "I am strongly drawn to a frugal life." Regarding frugality: in the fraternities, which we will shortly touch on, everyone has just enough, and no one too much.

"Capitalism," said a Cuban citizen, after some recent changes on the island nation, "makes people hard. Things are better than before. But I'm afraid" (Iyer 2013, 15). In our day, as democratic capitalism spreads to less progressive countries, rather than a triumph of self-rule, it has in some places only brought a change of masters, from dictators to financial overlords—same dogs, different collars. Foreign investors and multinational corporations step in, right along with bureaucrats from the old regime, all at the expense of the working population. What is democracy in these lands? The pleasure of eating at McDonald's? Of owning a cell phone? Of wearing designer jeans?

Capitalism brought them freedom, poverty, and the mafia.
LEIF ERIKSEN, *THE WOMAN FROM BRATISLAVA*
(THE QUOTE IS IN REFERENCE TO THE NEW SLOVAKIA)

Russia's market reforms in the early 1990s (after the collapse of the USSR) brought a nouveau riche, the Mafiosi, and a crash privatization program; more than half the workforce moved into the private sector. All this, as one Russian-American sees it,

> cost ordinary Russians dearly. The lack of purchasing power in the impoverished population and the reduction of state subsidies brought entire branches of the economy to a halt.
>
> . . . Fewer and fewer had the money to buy [goods]. Millions fell below the poverty line. . . . Teachers, doctors, officials, police were not paid for months at a time. . . . The intelligentsia lost faith in democracy. Crime rose. (Goldfarb 2007, 32)

Privatization and market economy may bring prosperity to a few but decline and unemployment to the many. And with the privatization boon in America comes outsourcing. What's so democratic about outsourcing? It only spikes unemployment at home, erodes the great American middle class, and creates a permanent underclass.

> *They shipped entire job sectors abroad and then railed at the demise of the middle class.*
>
> STEVE MARTINI

All this has gone in lockstep with tremendous financial power in the hands of the military, which sees fortunes spent on "security" and the armed forces. In this connection, Einstein once dubbed militarism "the plague spot of civilization. The maxim . . . has been security through superior military power, whatever the cost . . . [but] the idea of achieving security through national armament is . . . a disastrous illusion" (Einstein 1954, 159).

The quote comes from Einstein's contribution to Eleanor Roosevelt's TV program on the hydrogen bomb, on February 13, 1950. Mrs. Roosevelt herself made bold to say, "Private profit is made out of the dead bodies of men. The more we see of the munitions business . . . the

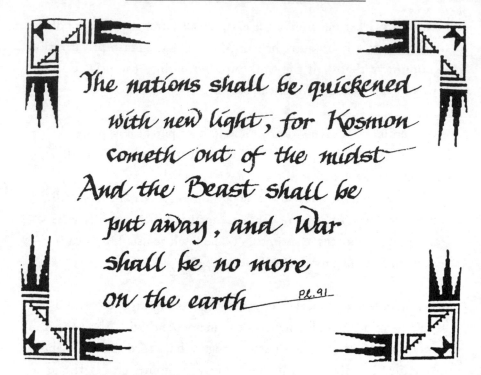

Fig. 9.3. Plate 91: The end of war, as prophesied in Oahspe

more we realize that human cupidity is as universal as human heroism. . . . If we are to do away with the war idea, one of the first steps will be to do away with all possibility of private profit" (Cook 2000, 239).

> The boast of a government is not of its virtue and goodness, and its fatherly care of the helpless, but of its strength in arms, and its power to kill.
>
> OAHSPE, GOD'S BOOK OF ESKRA 36:63

Today, the military budget is truly staggering:

- One fully equipped pair of NVGs (night vision goggles): $65,000
- One Tomahawk cruise missile: $840,000
- One SR-71 "Blackbird" reconnaissance flyer (the ultimate spy plane, in service 1964–1990): $34 million

- One Chinook, big war helicopter: $35 million
- One C-37A Gulfstream V, used by several agencies for homeland security: $60 million
- One F-35 fighter jet, a lethal hunter: more than $160 million
- One C-17 Starlifter: $202 million (a cargo aircraft with a 169-foot wingspan made for troops and equipment)
- One nuclear-powered aircraft carrier: $762 million
- One U.S. Navy fast-attack submarine: $2 billion
- One B-2 stealth bomber: $2 billion

Militarism is a common feature of [societal] breakdown and disintegration.

ARNOLD TOYNBEE, *A STUDY OF HISTORY*

We are spending close to half the world's budget on defense and "star wars," bankrupting our own country in the process. About a third of U.S. taxes underwrite the military, covering costs of the wars and veterans' benefits. During the worst economic crisis since the Great Depression, politicians continued to spend tax dollars on the war in Afghanistan at a rate of $2 billion each week, while they made devastating cuts at home.

As Einstein saw it, "The state should be our servant and not we its slaves. The state transgresses this commandment when it compels us . . . to engage in war, the more so since the object and effect of this slavish service is to kill people belonging to other countries or interfere with their freedom of development" (Einstein 1954, 96).

Throughout the chapters of this book I have assailed some of our most cherished theories, calling them state mythology. Why is the state so interested in maintaining the status quo? Specifically, how is it that Darwinian survival-of-the-fittest and manifest destiny (American expansionism) emerged in the same moment (the mid-nineteenth century)? Coincidence?

THE ECONOMICS OF WAR

Behold the famous States/Harrying Mexico/With rifles and with knife.

RALPH WALDO EMERSON

Now that most of the Indian campaigns were over, America waged its first war of foreign conquest (the Mexican War [1846–1848]). The war was launched on the pretext that Providence most manifest had ordained the expansion of a superior race into backward places such as Mexico, where the lofty ideal of democracy may now take root. With the delicate balance of fifteen slave states and fifteen free states, the slave South (having annexed Texas) was now seeking further horizons for slavery's extension. Was this not the very motive for the Mexican War—to provide the slaveholding interest with expanded territory and influence?

Nothing could have suited the aims of imperialism more than Darwinian "survival of the fittest" and the philosophy of manifest destiny. Might is right, though it is cloaked now in the mantle of progress, enlightenment, and democracy. But it was still plunder and unnecessary wars and exploitation that humanity made for itself, and continues to make, whenever our so-called democracies support dictators. Has the United States of Amnesia forgotten that Saddam Hussein was our friend before he was our enemy? Backing pro-Western powers around the world—for the sake of trade and dominion—is not the same as backing freedom and democracy. America (usually the CIA) in the twentieth century was covertly instrumental in undermining or ousting the democratic leadership in Iran, Dominican Republic, Chile, Brazil, Haiti, Guatemala, and Honduras.

The question I am getting at is this: Is it true that Western society has achieved the "final form of human government" (as declared by Francis Fukuyama)? Hasn't it been the conceit of each civilization, in turn, to fancy its own attainments the crown of history? Our own civilization is no exception to the rule, trumpeting the present forms, specifically democracy, as the very apex and pinnacle of progress, little hoping or dreaming of a better way to come.

The American prophet and seer J. B. Newbrough once questioned his spirit teachers on this subject, asking, "Ours is a republic; wherein can it be improved, or is it really the highest form of earthly government?" Whereupon his controls could not refrain from scoffing at the naïve question: "The highest! O vain, vain man! Many, many changes shall follow your republic; many governments arise and go out of memory before the highest shall be reached! Your government is far behind the requirements of the people. Even now are curses rising up amidst you against tyranny" (Newbrough 1874, 52).

In a 2013 poll, Gallup asked people from many countries, "Is your government highly corrupt?" The highest score (94 percent replying yes) came from the Czech Republic. Next was Ghana at 89 percent, and third was the United States at 73 percent ("Briefing," *Time,* November 4, 2013, 10).

So who are we trying to fool? The road ahead is long, with more bridges to cross before we can boast of the perfect society, before the darkness of despotism has become a thing of the past, before calling ourselves brothers and sisters, before election fraud, outdated laws, scandals, and wars are gone forever, before the new paradise is won, and before society, as prophesied, proceeds smoothly without votes and rulers!

FRATERNITY

They shall have no leaders . . . or rulers. The progress of man is toward this.

OAHSPE, BOOK OF JUDGMENT 1:42 AND 35:2

Nor will we desire a leader. Newbrough, who was the human instrument, the amanuensis, for transcribing Oahspe,* tried to "make them comprehend that institutions like colleges have no real leader for each department,

---

*My biography of Newbrough and his times is titled *The Hidden Prophet* (2009).

but only a teacher. And that some such model must be ours. . . . Are we not to develop ourselves? Or are we to be led?" He himself begged off a leadership role, which the new faithists were expecting of him. "I want the members to depend on themselves. . . . We must steer clear of a leader. We must learn *fraternity*. . . . We must learn to counsel together"* (Bates letters 1884). Isn't this the ultimate meaning of *grassroots*?

The suggestion here is that government is fast becoming obsolete. As I write these words in October 2013, the U.S. government is in shutdown mode. During this month *Time* magazine ran a commentary titled "Congress Is Bad for the Economy" (Foroohar 2013, 19), arguing that partisan politics and infighting have brought the country to a standstill. But let us remember that self-rule means just that. Do bureaucracies really keep us in line? Do they represent the common man? Aren't we capable of running our affairs by ourselves? John Stossel thinks so: "Politicians don't run the country. America isn't a car that will crash without a president at the helm. America is run by millions. . . . The complex thriving giant that is the United States mostly runs itself" (Stossel 2006, 105).

In July 2013, I read in the news that Congress was considering establishing national parks on the Moon. Around the same time, one columnist wrote about a

> paralyzed Congress . . . on track to crush its record of fewest bills passed per year. . . . [W]hat Congress has passed this session are barely laws . . . they are mostly to do things you shouldn't even need Congress for: appointing a chief financial officer for D.C., picking a diameter for baseball's Hall of Fame commemorative coins, and naming a bridge in St. Louis for Stan Musial. (Stein 2013)

What does self-rule consist of? It all depends on becoming organic, meaning interconnected. This is to say that things done reciprocally are

---

*Just as the Afghan *jirga* and council of elders in tribal communities do throughout the developing world.

more important than personal strivings (or special-interest agendas).

> *When men no longer help each other, civilizations decline*
> *and collapse.*
>
> ROBERT CHARROUX, *MASTERS OF THE WORLD*

This is part of the paradigm shift everyone is talking about. It means going from overrated individualism (or clubbiness) to a collaborative mind-set. This new age of consciousness is not as individual or metaphysical or "inner" as many have portrayed it. Einstein, who abhorred being "exhibited like a prize ox," put it like this: "The cult of individuals is always, in my view, unjustified. To be sure, nature distributes her gifts unevenly among her children. But . . . it strikes me as unfair, and even in bad taste, to select a few of them for boundless admiration." Then he really gets to the point: "The contrast between the popular estimate of my [own] powers and achievements and the reality is simply grotesque" (Einstein 1954, 4).

> Judgment is rendered against government wherein it . . . neglecteth
> providing means for the development of the talents created with *all*.
> (my italics)
>
> OAHSPE, BOOK OF JUDGMENT 34:11

> *Let no man be idolized.*
>
> ALBERT EINSTEIN, *IDEAS AND OPINIONS*

If the ancients deified their heroes, we moderns are not too different, given our obsession with celebrities and superhyped icons of science and technology. The great American hype only succeeds in obliterating a general sense of innate worth. No wonder the hidden woe of Americans is the want of "self-esteem," the feeling of inadequacy, of inferiority or ordinariness. Celebritism is not really part of a true democracy, but a vestige of elitism.

Yet Newbrough predicted, "I am convinced that organization for good works will succeed in this cycle* . . . good works, charitable or educational. I believe all other organizations will come to naught" (1883). That included both national government and organized religion.

> When we view our own beloved country, where all forms of Christianity have struggled as if in a moral race, and observe that all the affairs of our government are a stupendous aggregate fraud, and that there is scarce an officer, from a senator to a judge, that can be believed under his solemn oath, we may well pause and ask whether our religion has such a proper restraining and elevating influence as to deserve our further support. (Newbrough 1874, 60)

And that statement was made fully 140 years ago!

In 1892, the year after Newbrough's death, the reform-minded in America were fulminating against "the law-breaking corporations and iniquitous business combines [that] controlled America . . . a plutocracy having been substituted for democracy" (Boyer 1995, 94).

> Lawmakers trade in projects and schemes for their own profit and glory. . . . Let this be a guide to you, O man, in prophesying the change and the overthrow of governments: According to the square of the distance a government is from, righteousness, so is accelerated the pace of its coming change or destruction. . . . And remember, the percentage of inspiration that comes to you from the lowest grade is doubly degraded in the cities and great capitals. Know, O man, that all cities built by men, sooner or later fall into destruction. . . . In time, all holiness passes away from there; and when your God abandons that city for a day, taking away his holy angels, the people fall into anarchy, or run with brands of fire and burn down the city.
>
> OAHSPE, BOOK OF JUDGMENT 34:6–7, 20, AND 15:10–13

In some ways, the prophesied fraternity of the future resembles that

---

This cycle refers to the Kosmon Era, begun in 1848 and expected to take off by 2048.

of the past, established by our enlightened forebears. The Zarathustrian law, given more than eight thousand years ago, enjoined that the city shall not exceed two thousand souls. And in that city,

> the oldest, wisest, best man shall be the high rab'bah . . . [and he] shall be the priest and ruler of the city; and the sins of the people of the city shall be upon his head.
>
> OAHSPE, BOOK OF GOD'S WORD 24:5 AND
> FIRST BOOK OF GOD 6:1–5

This is not too different from the Indian Way. Today, of all the extant tribes in America, the Hopi, the "Little People of Peace" (as the missionaries once called them), reflect most faithfully this ancient rule. It is God-made, not human-made, laws that they honor. The village chief, as of old, is their spiritual leader. His job is to constantly exemplify the proper observance of things and encourage his people to conduct their lives mindfully, and to look after the welfare of his people. Actual force is seldom required, for public opinion among the Hopi is the most effective form of control. To the casual visitor, government is so lacking in form as to seem nonexistent. "But back of this disarmingly artless administration lies organization and discipline that might well be emulated outside of Hopi country" (James 1956).

Other traditional societies in today's world follow suit: Daoud Hari in *The Translator* tells us about the villages in Darfur "where the omdas and the sheikhs . . . have earned the respect of the people they live among. It is a very different kind of democracy, with the people voting for their local leaders not with ballots, but rather with their attitudes of respect for those who stand out in their service to their communities." Isn't this like the service of the rab'bah of old—"voting" without ballots?

You can't solve a problem in higher mathematics by voting on it.

ERIC LERNER, *THE BIG BANG NEVER HAPPENED*

As I have contended throughout these chapters, the rigged consensus of the experts, the elite, has proven to be an uncertain guide to truth. Although the pseudosciences in my purview have certainly amassed a great chorus of supporting voices, let us not be fooled by the majority vote. Remember the little story Abu-Ishak Chisti told at the front of this book? All fifty witnesses pronounced the man dead! The point is, if everyone is using the same erroneous premise or agenda, of course their results will be "consistent," and there will be "consensus." But it is a bogus consensus.

The man called Buddha, a.k.a. Sakaya, was once asked,

> "Why not decree according to the majority vote?" Sakaya answered: "That is the lower light, being the light of men only. . . . It is incumbent on every man in the community who enters the discussion, to speak from the higher light . . . without regard to policy or consequences. And the same law shall be binding on the rab'bah; and though nine men out of ten side the other way, yet the rab'bah's decree shall stand above all the rest."
>
> OAHSPE, GOD'S BOOK OF ESKRA 28:12–14

How then shall we become organic and fraternal if the highest wisdom in the state, or in a community, is not with the majority, but with a small minority?

> Therefore let . . . each group select its wisest man . . . and his title shall be Chief.
>
> OAHSPE, JEHOVIH'S KINGDOM ON EARTH 6:32

Before embracing the majority rule so glorified by democratic societies, consider these remarks by two great Americans: "Mankind will in time discover that unbridled majorities are as tyrannical and cruel as unlimited despots" (John Adams). And this: "When were the good and the brave ever in a majority?" (Henry David Thoreau). More recently Alfred North Whitehead opined: "The publishers of newspapers, instead of appealing to a select group who will take excellence, must dilute and distribute their article so that it will appeal to all classes, and that results in a leveling down to some lower common denominator . . . everything has to be pepped up and made exciting" (Whitehead 1956, 59).

A republic cannot follow the highest Light; it follows the majority. And a majority is, and was, and ever shall be, the lesser light. Therefore, a republic is not the all highest government.

OAHSPE, BOOK OF JEHOVIH'S KINGDOM ON EARTH 26:2–3

In *The Hidden Prophet* I gathered together Newbrough's thoughts on the subject of votes, laws, and majorities in America. These, he came to believe, had resulted, not in welfare for all but in a system that despised and even criminalized the helpless and disenfranchised. "The laws in most cities," Newbrough decried, "treat the little waif [homeless street kid] as an outlaw. In some cities no one has a right to take the little thing in and care for it."*

Neither did J. B. come to believe that democracy was the ideal form of government, as he has one character in his Fort Sumter novel declare: ". . . A republican government can never raise the common people to a high state of civilization. The more elevated [ones] are constantly pulled down to the level of the great multitude." And in his own voice, this "hidden prophet" stated, "Another thing I find will take some time to eradicate. That is, votes and majorities, that disposition for the majority to make the minority do thus and so. In fact, a *fraternal* government is so new that it will take some time to get people to understand it" (Bates letters 1884).

---

*Speaking of such laws, gun sales and laws on firearms are based on the doctrine of antic- ipated, prospective damage. But with so much violence in the streets and in the schools, are we sure the cure (being well armed) is better than the disease? Does it make sense to pass laws allowing us to arm almost anyone with lethal weapons, yet pass other legisla- tion outlawing hunting bullfrogs with a firearm (Illinois); throwing a snowball (in Bel- ton, Missouri); pumping your own gas (in New Jersey and Oregon); playing bingo while drunk (Kern County, California); defacing a milk carton (Massachusetts); and having an artificial lawn (Palm Harbor, Florida)?

Ministering to others will heal our woundedness.
ABRAHAM VERGHESE, *CUTTING FOR STONE*

Wherein one man is weak, let two or more unite; a simple thing, by which even the stars of heaven can be turned from their course!
OAHSPE, BOOK OF OSIRIS 3:7

The unfoldment of an age of harmony and universal fellowship was envisioned by Martin Luther King Jr., in his acceptance of the Nobel Peace Prize fifty years ago:

> I refuse to accept the idea that man is mere flotsam and jetsam in the river of life that surrounds him. I refuse to accept the view that mankind is so tragically bound to the starless midnight of racism and war that the bright daylight of peace and brotherhood can never become a reality.

Presciently, King wrote his own eulogy: "Tell them I tried to feed the hungry. Tell them I tried to clothe the naked. Tell them I tried to help somebody." His hero, Mahatma Gandhi, the greatest man of India, the rab'bah of the land though unsupported by any outward authority, had been entirely devoted to the upliftment of his people.

> *I have a dream . . . that the brotherhood of man will become a reality in this day . . . and with this faith I will go out, and carve a tunnel of hope through a mountain of despair.*
> MARTIN LUTHER KING JR. (DETROIT, JUNE 23, 1963)

Since the assassination of both Gandhi and Martin Luther King, Jr., few have come along to replenish the pure faith of justice and brotherhood. Now the ball is in our court. The court is the world and the starting point is trust. There is no other way.

It is not in numbers but in unity, that our great strength lies.

THOMAS PAINE

Having achieved a free society, it only remains to achieve a good society. The motto of the French Revolution was "Liberty! Equality! Fraternity!" Yes, liberty first, then equality. Last to come, the hardest, will be fraternity—mutual trust and unity.

Only by harmony and the union of many can any great good come unto the generations of man.

OAHSPE, BOOK OF JUDGMENT 36:30

Separation and isolation remain the lot of people who lack the vision, the vision of fellowship. "Man," mused Einstein "can find meaning in life, short and perilous as it is, only through devoting himself to society." He thought the ideal community would develop in its members

a sense of responsibility for his fellow-men, in place of the glorification of power and success in our present society. . . . The student is trained to worship acquisitive success. . . . One should guard against preaching to the young man success in the customary sense . . . for a successful man receives a great deal from his fellow-men, usually incomparably more than corresponds to his service to them. . . . The value of a man should be seen in what he gives and not what he is able to receive. . . . The profit motive . . . is responsible for an instability . . . of capital which leads to increasingly severe depressions. Unlimited competition leads to a huge waste of labor and to crippling of the social consciousness. . . . Our whole system suffers from this evil. (Einstein 1954, 8, 62, 156–58)

In the day thou hast rendered judgment against thyself for not practicing thy highest light, thou art as one departed from a coast of breakers toward mid-ocean . . . like one turned from perishable things toward the Ever Eternal. And when thou hast joined with

Fig. 9.4. Coast of breakers

others in a fraternity to do these things: then thou has begun the second resurrection.*

OAHSPE, BOOK OF DISCIPLINE 13:11

All these prophecies, which I have been quoting, envision the future community coming together spontaneously with

> a few here and there, capable of the All Light. And these shall . . . form a basis for My kingdom on earth. . . . And they shall pledge themselves unto one another . . . holding their possessions in common. . . . To live for the sake of perfecting themselves and others in spirit, and for good works. . . . And they shall become an organic body in communities of tens and twenties and hundreds and thousands. . . . Come thou . . . and inherit the wilderness of this land. And they shall bloom as a new paradise before thy hand.

OAHSPE, BOOK OF JUDGMENT 1:32–42

---

*The *second resurrection* refers to the realm of coordinated effort. The "oceanic" metaphor is particularly apt in this contest. Though we may all be individual drops in the ocean, we are in it together and there is no longer a way of distinguishing one drop from the other.

The following prophecy appears at the end of the Book of Jehovih's Kingdom on Earth (chapter 26:10–24). It is also the end of Oahspe. It is written in "ante-script," that is, though set in the future, it reads in the past tense:

And thus it had come to pass, that . . . on earth, man was without a government. . . . And this was the next higher condition that came upon the earth after republics. The angels . . . had said to [man]: "Do not bother your heads much about passing new laws . . . [even though] many men shall rise up, saying: If the government would make a law of peace; or, if the government would prohibit the traffic and the manufacture of this curse or that curse. But we say to you, all these things shall fail. *Do not trust in the ungodly to do a godlike thing.* The societies shall fail; the Peace Society shall become a farce . . . for they have fallen under the lower light. . . . Under the name of liberty, they shall claim the right to practice ungodliness. But you shall come out from among them, and be as a separate people in the world.

And so it came about that the people were admonished by God and his angels . . . speaking in the souls of mortals. And those of the Spirit believed; but those of the flesh disbelieved. Wider and wider apart, these two peoples separated. And the believers . . . practiced righteousness, rising higher and higher in wisdom and purity. But the disbelievers went down in darkness; were scattered and lost from the face of the earth. Thus, Jehovih's kingdom swallowed up all things in victory; His dominion was over all, and all people dwelt in peace and liberty.* (my italics)

---

*Though this utopian prophecy comes from and applies to our own era, it was also known to a few ancient sages. Plutarch, in *Isis and Osiris,* for example, prophesied that at the end of time, the whole Earth would become homogenous and humanity would live a blessed life.

"The Time Has Come"

In 1957, the elderly hereditary chief of the Yuchi Indians in North America wrote: "The signs are plenty. . . . The time has come, the Sun and Moon have bloomed and ripened, bringing a soft light to the hearts of everyone and the hidden secrets together for everlasting brotherhood. When this is done, my work will be finished. The tough roots of the fire tree will be torn apart where the great secret of brotherhood was hidden by false prophets." (Mahan 1983, 4)

In the coming utopia, the flame of universal brotherhood melts the shackles of self-serving nationalism. Where republicanism and democracy have fostered nationalism, leading to xenophobia or racism or excessive ethnic pride, it has left true brotherhood behind.

You shall also consider this, O man: All governments are tending toward oneness with one another [globalization]. . . . Consider, then, what wisdom is between governments: to make themselves reciprocal toward one another.

OAHSPE, BOOK OF JUDGMENT 34:23–24

CONCLUSION

In writing this book, I have noticed that democracy's obsession with political correctness (PC) reaches into every phase of our existence. In a blunt example, you do not assassinate a vicious dictator because that is not PC. No, instead you invade his country and thousands of innocent civilians wind up dead as "collateral damage." As for matters of the intellect, to be politically correct in this day and time, your theory must not include a Higher Power. Instead, something called "humanism" appears and takes the spotlight, the name little more than a euphemism for atheism.

Our PC vocabulary has given us euphemisms and circumlocutions that are nothing but a fake, cloaked in the mantle of decency and respect. What's more, PC is transient, unreliable: what's PC today was not PC a generation ago and won't be PC in the next generation, reflecting "our national gullibility for fads and scientism" (Gross 1978, 266). One of the best examples is global warming. While it's now PC, the climate verdict was "nuclear winter" a generation or two ago!

Matthew Chapman, the great-great grandson of Charles Darwin, in his book *Trials of the Monkey* (2001, 45–46) let fly his opinion of PC in the course of recounting his interesting bus trip to the American South: "A Redskin just got on board. Oh, no. I'm sorry. *Native American!* Oh, no, sorry again, I just heard it ought to be *First Nation Person.* How I despise this Index Expurgatorius of 'inappropriate' words. . . . As a writer it feels like someone's putting a hand in my toolbox. [Amen.] The whole concept seems Orwellian to me, a way of changing appearance without changing substance, a hypocritical ploy . . . by which self-satisfied closet racists and bigots can safely hide behind a set of linguistic rules. . . . When did you last hear one black man say to another, 'Whassup, African-American?'"

I have seen government agencies change their name just to get away from the bad image so richly earned by their former but unchanged selves. PC and euphemisms are there to dodge, not resolve, our prejudices and failures. When the sly substitute replaces the unvarnished truth, we are in for a sharp wake-up call. It is not just in political or academic circles that the truth has been overwhelmed by the spineless switch. . . . It is everywhere.

> In a materialist age like ours nothing is real except what is false.
>
> MALCOLM MUGGERIDGE, *A THIRD TESTAMENT*

One secret I have learned: mortals may not be able to find the truth all by themselves, but they are inherently capable of *recognizing* it. And that is sufficient. In a perfect world we are receptive to the truth (like a built-in Geiger counter), rather than having to explain everything. This book has been a critique of current "explanations." I believe that in the coming age of understanding, theories and explanations will lose much of their relevance. Science today has become the God of explanation. "In our scientific-industrial West, everything must be explained by *reasons,* even if they later prove to be not so reasonable" (Gross 1978, 35). This same approach has us questioning everything, challenging everything. We even challenge the givens, which should be axiomatic, as if that were the smart thing to do— givens, like the fiery beginnings of Earth (chapters 3 and 4); or the perfect innocence of the newborn babe; or our very soul, our link to the Infinite (chapters 6 and 7).

In every chapter of this book I have railed against explanations of things that never happened. How have we gotten into this predicament? Simply enough: because we have refused to deal with the unseen.

Gods and Lords mold the inhabitants of the earth as clay is molded in a potter's hand. O man, what is thy folly! How hast thou found such cunning ways to put off thy Creator [whose] star is within thy soul? Feed it, O man.

OAHSPE, BOOK OF APOLLO 5:14

James Churchward (who said all this a hundred years ago) reminded us in *The Lost Continent of Mu* (2011) that

> man, the most complex of all forms of life, and the most perfect . . . was created for a special purpose . . . [with] a force or soul for the purpose of ruling the earth. This great gift has been bestowed on no other form of life, which proves conclusively that man is a separate and distinct creation, possessing a divine force.

But in these secular times, this "divine force" is not acceptable, not PC, and so nature (and "humanism") have been exalted in place of it.

> Yet mankind reflects a higher origin and possesses capabilities that transcend anything found in nature. . . . Why this insistence upon insupportable theories? Sophisticated denials that claim to have the support of science attempt to explain spiritual experience within the materialistic framework. The result is a mythologized science. (Hunt and McMahon 1988, 100, 122)

The hard-nosed, rank materialism of today's intelligentsia, desanctifying human life (their craft is sometimes called "scientific fundamentalism") was put in motion by the theories of Darwin, Marx, and Freud—the great mythographers of our era—and atheists all. But Darwin, in time, will be replaced by true creationism; Freud by true spiritualism; and Marx by fraternities. I see the coming paradigm shift as something like the prophesied battle between dark and light (dark = materialism; light = the unseen power). Until we twig to the subtle but potent underpinning of the seen world, we will continue going in circles for the answers we claim to seek and continue to explain things that never happened!

It is misleading, even bullying, to polarize rationalism against belief in the Unseen, a misstep taken ever since the so-called Enlightenment. Today it is clumsily called science versus religion. While the assumption, the ruling paradigm, is that what we can see and measure trumps all, the truth is that the great Force lies in the unseen. This is a given that should have been established by the atomic age.

Objective, impartial science? Paradoxically the secular scheme is not the empirical science it poses as, but is itself taken on faith: the God-free zealot both assumes and demands a universe without a Designer. In the end, it resolves into pantheism—nature worship and cosmos worship—revering the purely physical creation. Yet it is the unseen ephemeral Force that shapes, nourishes, and sustains all!

Most of us have underestimated the role of spirit in this earthly adventure—the unsinkable human spirit that will outlive all our pubescent theories.

What controls our destiny? Is it our simian instincts? DNA? The constellations? Our unconscious? Our past lives? Who or what is in control? Not knowing the answer, we embrace alarmist ideas; such the-

Immortal self am I

SPE   Fig. C.1. Spe: the spirit
part, the indweller

ories, it should be noted, typify societies on the verge of collapse. Panic especially comes alive in the soil of guilty conscience. We are waiting for the other shoe to drop, half-expecting some dreadful payback for lives played out in self-interest.

But there is no greater threat than our own illusions, one aspect of which is our susceptibility to easy answers—catastrophism, ETs who brought civilization, and so forth. These entertaining solutions are part and parcel of today's anti-intellectual streak, our famous two-minute attention span. We let others do our thinking and go along with their easy or cute or exciting answers.

If we leave it to our most popular theories to "save the planet," I am afraid we are doomed. But we are not doomed, for we have already begun to move on. No, not "improving" the paradigm or patching up old theories, but remembering Einstein's words: "No problem can be solved from the same consciousness that created it." In other words, we need to think outside the box. We search in vain for a Band-aid solution to our problems. The solution is transformation. We cannot survive without the truth and the trust it brings.

> We are now in an epoch where change is more drastic than that which slew the nineteenth century. . . . We are on the threshold of an age of liberation, a better life for the masses . . . a new form of society.
>
> ALFRED NORTH WHITEHEAD,
> DIALOGUES OF ALFRED NORTH WHITEHEAD

I would not have written this book had I not felt the "ripeness" of our age, the gameness of our times, with our growing awareness that we have all been indoctrinated and that these theories and dogma are somehow out of step with the vital pulse of the people. In any case, we are on the verge of truth. It has been coming for hundreds of years, during which "public affairs were infused with a political spirit that was both democratic and scientific in a way that people believed heralded a bright new age" (Milton 1996, 199).

The name of that age is "Kosmon," though the Hindus call it "Satya Yuga" (Age of Truth). But it only comes when the Age of Half-Truths is fully played out. Erasmus in his day cried out: "They will smother me beneath six hundred dogmas; they will call me heretic and they are nevertheless Folly's servants. They are surrounded with a body-guard of definitions, conclusions, corollaries, propositions explicit and implicit. . . . They are looking in utter darkness for that which has no existence whatever" (quoted in Koestler 1959, 111).

THE THIRTY-THREE-YEAR CYCLE

A fruitful approach to weather rhythms is revealed in the study of cyclical time. The prophetic numbers—all multiples of 11—can help us track recurrent hurricanes, blizzards, earthquakes, volcanic eruptions, and so forth. First of all is the well-known 33-year (11 x 3) rhythm: modern science has rediscovered the "spell" in the 33-year meteor cycle. In Oahspe's Book of Knowledge (4.18) it is said that man

> found that every thirtythird year was alike on the earth as to heat and cold, and from these he discovered the nebulous regions within the vortex of the earth, and the cause of the variations in the times of falling meteors.

In addition, it is known that influenza viruses tend to mutate approximately every 33 years. Also the PDO (Pacific Decadal Oscillation), which is a shift in weather phases of the North Pacific, occurs every 30 years or so, affecting "the ways that nature has to move heat around the Earth" (Spencer 2010, 15–16). Then, too, in the pioneering work of Eduard Bruckner (1862–1927), climatic phenomena, especially the tides, also show synchronized phases in a cycle of 33–35 years. Only recently have Bruckner's climate cycles, which survey the past thousand years, begun to receive support and confirmation. (I thank James McGill for pointing this out to me on his website, studyofoahspe.com.)

There is also the website http://tamino.wordpress.com, which stalks the elusive connection between the solar cycle and temperature on Earth, and found a 33-year cycle of alternating cooling and warming phases, according to the amount of cloud cover.

Taking hurricanes as an example of the 33-year cycle called a "spell":

HURRICANES ON THE SPELL

YEAR	PLACE OF HURRICANE	CYCLE	YEAR	PLACE OF HURRICANE
1900	Texas, kills 6,000	+33	1933	Tampico, Mexico, hit hard
1928	Florida, Puerto Rico, 5,000 killed	+33	1961	Belize (Hattie)
1932	Cuba, 2,500 killed	+33	1965–1967	Cuba and Haiti (Inez); Bahamas and Florida (Betsy)
1935	Florida, 400 killed	+33	1968	Florida (Gladys)
1955	North Carolina to New England (Connie and Diane)	+34	1989	East Coast (Hugo)
1960	Antilles, East Coast US (Donna)	+32	1992	Southern US, Bahamas (Andrew)
1973	Eastern US, Cuba (Agnes/Cecilia)	+32	2005	Southern US (Katrina/ Wilma, Rita, Stan)

Storms (hurricanes, flooding rainfalls, etc.) cannot be blamed on global warming. Instead, I lean toward the visionary work of Tom Veigle (www.earthvortex.com), which presents the vortexian model of an aging Earth and urges the use of this new science to bust up hurricanes and meteors and to bring warmth to cold regions and rainfall to drought-stricken regions. All things, James Churchward once mused, point to a day when we will have perfect control over all the Earth's elements and many of her forces—a state to which we are now advancing. The day is coming when we will manage the vortexian energies to regulate temperatures and prevent destructive storms—an enlightened application of vortexyan science.

THE PROPHETIC NUMBERS

UNITS OF TIME AND THE 3,000-YEAR CYCLE

NUMBER OF YEARS	ANCIENT NAME	COMMENT
11	Ode	Sunspot cycle
33	Spell	Same as meteor cycle; one "generation"
66	Beast	2 spells
99	Wave	3 spells; centennial
121	Semoin	1/3 tuff or ode squared
200	Half-time or Dan	Six generations; double wave
363	Tuff ("Circle")	Solar year; 11 spells
400	Time	Baktun (Mayan); 144,000 days
666	Period	"The number of the beast"
3,000	Cycle	Danha; 100 generations

Every 3,000 years the Light of dawn and the Voice come to provide deliverance from the enemies of the Great Spirit. We are now in a new dawn called Kosmon. Although Kosmon began in the midnineteenth century, the few centuries preceding paved the way for it. Arthur Koestler, for example, calls the sixteenth century "that century of awakening":

> The pulse of all humanity was quickening as if our planet, after traversing on its journey through some somnolent and bemused zone

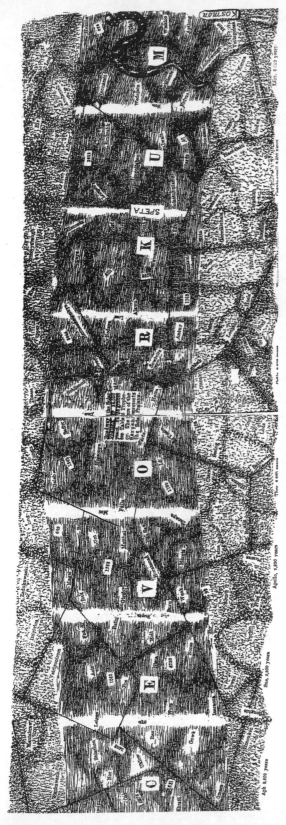

Fig. B.1. C'vorkum. The 3,000-year danha is shown to be a regular time interval between periods of light.

The sun, with his family, plieth in a large circuit, which is divided into 1,500 arcs, the distance of which for each arc is about three thousand years, or one cycle.

OAHSPE, SYNOPSIS OF 16 CYCLES 1:2

The student will draw a curved line, representing the travel of the great serpent for three thousand years [i.e., for one danha cycle]. . . . Make one [chart] for every great division of the earth.

OAHSPE, BOOK OF COSMOGONY AND PROPHECY 7:17

of the universe, was now emerging into a region bathed in vivify-
ing rays, or filled with cosmic Benzedrine. . . . It seemed to act as
a stimulant . . . manifesting itself as a thirst of the spirit, an itch of
the brain, a hunger of the senses. (Koestler 1959, 144, 147)

Climate shifts are also related to the long arc cycles of 3,000 years.
In the past 12,000 years, according to studies of lake sediments, there
were four periods, peaking every 3,000 years, that saw a great number
of intense floods.

This entire system of prophetic numbers—from the ode to the
cycle—came out of Egypt, as described in Oahspe's Book of Bon:

> For events of prophecy there was a calendar called the ode,
> signifying skytime. . . . One ode was equivalent to eleven years;
> three Odes, one Spell; eleven spells, one Tuff. Thothma, the
> learned man and builder of the great pyramid, had said: "For
> the heat or the cold, or the drought or the wet, no matter which,
> the sum of one eleven years is equivalent to the sum of another
> eleven years. . . ." Whoever will apply these rules to the earth shall
> prophesy truly regarding drought, famine, and pestilence. And if
> he applies himself to find the light and the darkness of the earth,
> these rules are sufficient.
>
> OAHSPE, BOOK OF BON 14:4

The ancient tables of prophecy discovered the rhythms of the Earth
in these multiples of 11 (33, 66, 99, etc.); and the prophets and mathematicians called the 33-year cycle the First Rule of Prophecy. This was
subdivided by 3, into 11 years. And this was called the Second Rule of
Prophecy. The Third Rule was 99 years. Thus do the currents of vortexya fall under three primaries: 11, 33, and 99.

BEWARE DROUGHT

On a Cooling and Drying Planet

Recently, some British genius engineered a plan to send balloons aloft to spray sun-deflecting particulates in the air to reverse global warming. Environmental groups said, "Whoa, *drought* could result from such a release."

Meanwhile, nuclear power production uses up huge amounts of water. During the 2007 drought, nuclear reactors across the American Southeast were threatened with temporary shutdown due to the awesome amounts of cooling water needed to operate them (billions of gallons are used to cool and condense the steam that turns the turbines). A typical reactor sucks up 33 million gallons of water a day.

We are too close to a water crisis to afford the luxury of heeding bogus doctrines and pseudoscientific oracles. Disregarding the drying trend would be as bad a mistake as blaming it on global warming. This ignores the fact that warmer and drier is inconsistent with the well-known sunspot cycle, in which warmth correlates with moisture (not dryness), and cold (not warmth) correlates with dryness, as at the dessicated poles. "Decreased rainfall in most places has accompanied the cooling. . . . The total area covered by sandy deserts has significantly increased since the Climatic Optimum [7,000 years ago]" (Imbrie and Palmer 1979, 179).

Today, scorched farmlands and disappearing wetlands affect freshwater fisheries, driving food prices higher. In many parts of the world, water levels are dropping in lakes, reservoirs, and wells (White 1980, 363). Half of the world's population faces ongoing shortages of freshwater. Predicted are unprecedented droughts and desertification of the breadbaskets. Parts of Africa—the Sahel, Ruanda, and Ethiopia—are at risk. China is also seeing dire water shortages. Neither is the outlook in India good for the twenty-first century. Spain has been called the future Sahara of Europe.

Australia's eastern coast, along with Central America, the Mideast, and Central Asia are also drying. As for the United States, hardest hit are California, Texas, Nevada (especially Las Vegas), and the Southeast. Water wars, disputes over water rights, tugs-of-war between neighboring states, and court battles are anticipated amid shortfalls and dry hurricanes worldwide. Emergency drought plans have already been drawn up.

On this aging and drying planet, the prophetic numbers are hard at work in cycles of drought. "Scientists now believe the eleven-year sunspot cycle has a strong connection with drought on the earth" (White 1980, 358). Indeed, dry times in Africa have followed the ode: 1973 saw severe drought; hundreds of thousands of Africans perished in the resulting famine. Then, in 1984, again drought struck Africa (1973 + 11 = 1984). There is also one ode separating droughts in India: 1792 saw the last of the "Skull Famine," so named after severe drought led to cannibalism in Bombay and other cities. Then 1792 + 11 = 1803, when drought, war, and locust plague caused thousands of deaths in India. After these calamities ended in 1804, add 33 (11 x 3) years = 1837–1838: India's northwest region is hit by severe drought and 800,000 perish. Then add a double ode of 22 years = 1860: Drought once again desiccates the New Delhi region. A double ode is also seen to separate two periods of devastating drought in China: 1928 + 22 = 1950, killing 10 million. The rhythms of heat radiation from the sun actually follow a 22-year cycle. George E. Hale of Mt. Wilson Observatory uncovered a magnetic cycle of 22 years, which was confirmed by Charles G. Abbot of the Smithsonian Institute. The Sun's magnetic field changes dramatically over its 11- and 22-year cycle.

THE FOUR HUNDRED YEARS
OF THE ANCIENTS

Muslims believe that each prophet (Adam, Noah, Abraham, Moses, Jesus, and Muhammad) is attended by twelve imams, covering the prophet's reign. We perceive the "time" count (the 400-year unit) embedded in the thinking of these Arab sages when we consider their doctrine of the "twelve imams." With one imam appearing per generation (one-third of a century), twelve generations of imams would give us the 400 years, the same as the Mayan *baktun* or the Egyptian unit called "time."

History's scroll often submits to the inexorable rhythm of the *baktun:* early German migrations across Europe lasted a *baktun,* i.e., from 200 to 600 CE; just as the Muhammadan raids on India ran for 400 years, from 600 to 1000 CE. In the Americas, from the fifth to the ninth century CE (400 years), sixteen kings ruled the Mayan Copan of Honduras. The classic Maya flourished at their peak between 435 and 830 CE, just as Mesoamerican civilization was previously unified under the Olmecs for 400 years (900–500 BCE).

Japan's golden age lasted from 794 to 1192—398 years. China's Han Dynasty lasted 400 years. The Ottoman Empire is also considered to have lasted about four centuries. In 1517, the Turks took possession of Jerusalem; exactly 400 years later, in 1917, they were ousted by the British, who were then given the mandate over Palestine.

The year 399 BCE saw the judicial murder of Socrates; 400 years later came the birth of Christ who would become the next great martyr for humankind. And advancing one more "time," the pope claimed the keys to Christ, in 400 CE. Later, 1215 + 401 = 1616: in 1215 the Inquisition was officially established by Pope Innocent III; 401 years later came the draconian Papal Edict of 1616, which, among other things, outlawed the Copernican doctrine (heliocentric) and would become Galileo's lifelong nemesis. Moreover, two years before Copernicus's theory was published, Ignatius Loyola founded the Jesuits. That was in 1541. Add 400 = 1941, the year in which the absolute power of pope and church was usurped by the fascist states of Mussolini and Hitler.

Henry Adams's education (*The Education of Henry Adams*) taught him one thing: "There must be a unit from which one could measure motion down to his own time. . . . The laws of history only repeat the lines of force or thought . . . [which have] the motion of a cannonball. . . . One could watch its curve." Ambling through history, Adams perceived one trajectory beginning in 1500 . . . and in "1900, the continuity snapped," starting a new trajectory: four hundred years.

The Dark Ages lasted 400 years (675–1075) and so did the Middle Ages (1075–1475), according to Arnold Toynbee, whose *Times of Troubles* conform amazingly to the 400-year *baktun* in most of the civilizations he surveyed: Egyptiac, Sumeric, Sinic, Syriac, Asian, Hindu, Hellenic, and Russian Orthodox. Just so, periods of "universal peace" have also tended to last a *baktun*: Egyptiac: 2060–1660 BCE as well as 1580–1175 BCE; Orthodox Christian: 1372–1768 CE; Russian Orthodox: 1478–1881 CE. From the sacred histories of some 3,500 years ago, we hear of this 400-year cycle of peace:

Let the student compare the faithists of Capilya in India with the Cojuans of the same country; and the Faithists of Moses in Egupt with the Eguptians of the same country. The Faithists of both countries advanced but their persecutors both went down to destruction. The peace of the Faithists held *four hundred years;* and then both peoples began to choose kings, which was followed by 900 and 99 years of darkness. (my italics)

OAHSPE, BOOK OF COSMOGONY AND PROPHECY 8:11

(SEE ALSO GENESIS 36:31)

Why did the faithists of Moses fall after 400 years? This, we learn, is the duration of light in every new cycle after the Earth has entered the beam of danha:

The scale then riseth for four hundred years, more or less; and after that, wars and epidemics come upon the people. . . . Such then is the general character and behavior of man during a cycle . . . he riseth and falleth in all these particulars as regularly as the tides of the ocean.

OAHSPE, BOOK OF COSMOGONY AND PROPHECY 7:8

COMPARISON OF AMERIND AND CHINESE VOCABULARY

The Chinese system of writing, as well as that of ancient America, arose independently of the so-called Mesopotamian (Sumerian) "cradle of civilization," yet all three languages are of the agglutinative type. Most Native American languages, like Algonquian, are agglutinative, akin to Chinese (single words can be as long and descriptive as phrases or even sentences).

FAR-FLUNG LINGUISTIC COUSINS

ALGONQUIAN	CHINESE	TRANSLATION
mai'ah	ma*	expression
p'boa†	m'boa	winter; destroyer
hagni	ah'gni	fire; to burn (same in Sanskrit)
go'ongwe	oe'gwong	love-offering
ni'oh'ghoo	ni'ghoo	prayer
shu	su	enlightened; prophetic
yope'ang	yoke'eng	sacred star
haden	haden	sky, heaven (*hades* in Greek)
haw'git	git'haw	Sun, the all-heat (same in Arabic)
hogawatha	hogawata	mastodon

---

*The term is *mai* in the mother-language (Panic).

†"Winter" is *p'boa* in Panic. The term *Panic* refers to both the language and the culture of Pan (Mu, Lemuria)—the lands lost in the Pacific to the Great Flood of Waters (my next book!).

One Sumerian tablet at Nineveh records the wisdom of King Ashurbanipal, who declared he was "initiated into the secrets of writing by the god of scribes" and thus could understand the "stone carvings of the days before the flood." Is this not a clue to an antediluvian relationship of these far-flung "cousins": Sumerians, Amerinds, and Chinese?

ANCIENT WRITING IN THE MEDITERRANEAN

It may have been the Egyptians, not the Phoenicians, who established purely consonantal writing about 6,000 years ago. Cuneiform and the 4,000-year-old Phoenician* script are actually Johnny-come-latelys in the scheme of things. Hints of earlier knowledge of writing in the Mediterranean include:

1. France: Finds at Glozel include 100 stone objects, "bricks," incised with letters and dating to the Magdalenian era (more than 14,000 BP). They contain some Phoenician and Greek letters. These were slandered because they were "archeologically unacceptable . . . [it] would imply a sweeping re-evaluation of established theories" (Berlitz 1972, 171). Nevertheless, this alphabet includes the letters C, H, I, J, K, L, O, T, V, W, and X! (Charroux 1971, 35).

2. Spain: Writing and books of the Tartessos culture date to 9000 BP (according to Strabo's *Geography*) and they may correspond to rock inscriptions still seen today in southwest Spain. These Iberians kept records before the Sumerians and, as some

---

*Twenty-five feet below the ruins of Ras Shamra, Syria—the depth indicating great antiquity—are found pins, bracelets, and other jewelry imported from the Caucasus, the Balkans, and the Rhine Valley. The Canaanites and the later Phoenicians were evidently heirs to some great civilization of the remote past.

speculate, the script may have been picked up by the seafaring Phoenicians. "Tartessos possessed thousands of ancient manuscripts dating back to before 10,000 BCE" (Chouinard 2012, 32).

3. The Canary Islands: Large lava slabs with undeciphered hieroglyphics (Charroux 1971, 105) relate to Tifinagh (Berber), a written language of the Tuareg (Moroccan Sahara), which is similar to the later Phoenician. The rock inscriptions in the Canaries seem to be alphabetic (letters), not glyphic (picture writing).

Fig. F.1.Inscription from the Upper Magdalenian (ca. 13 kya) found at Dordogne, France

Fig. F.2. Pebbles with characters similar to letters of the alphabet, like these, were found in the Cro-Magnon caves of Mas d'Azil in southern France.

BIBLIOGRAPHY

Abrams, Michael. 2011. "Brain Map Shows You Think Like a Worm." *Discover,* January: 36.

Acheson, Amy. 2008. "Is the Big Bang Dead?" In *Forbidden Science,* edited by Douglas Kenyon. Rochester, Vt.: Bear & Co.

Adler, Jerry. 2006. "Of Cosmic Proportions." *Newsweek,* September 4: 46.

Alexander, Eben. 2012. "The Science of Heaven." *Newsweek,* November 26: 24–26.

Allen, Sue. 2007. *Spirit Release.* Winchester, UK: O Books.

Allen, Thomas B. 1994. *Possessed.* New York: Bantam Books.

Alley, Richard B. 2004. "Abrupt Climate Change." *Scientific American,* November: 64.

Anderson, Mark. 2005. "The Universe, Appearing at a Reactor Near You." *New Scientist,* June 11: 8.

Arguelles, Jose. 1987. *The Mayan Factor.* Rochester, Vt.: Bear & Co.

Asimov, Isaac. 1971. *The Stars in Their Courses.* New York: Doubleday.

———. 1978. *Quasar, Quasar, Burning Bright.* New York: Doubleday.

Atlantis Rising Staff. 2008. "The Enigma of the Polar Dinosaurs." *Atlantis Rising* 67, January–February: 11.

———. 2010. "Are Sunspots Causing Global Cooling?" *Atlantis Rising* 93, May–June.

Atwater, Caleb. 1973. *Antiquities Discovered in the State of Ohio,* vol. 1. Cambridge, Mass.: Harvard University Press.

Augustine. 1967. *City of God.* New York: Brown, 77.

"Bangkok Activists Call for an End to Climate Change." 2009. *WIN* magazine, Fall: 6.

Bates letters. 1884. Faithist Archives.

Battersby, Stephen. 2005. "Oh, for a Theory of Everything." *New Scientist,* September 24: 6–8.

Battersby, H. F. Prevorst. 1979. *Man Outside Himself.* Secaucus, N.J.: Citadel Press.

Begley, Sharon. 2009. "Climate Change Calculus." *Newsweek,* August 3: 30.

———. 2012. "Solar Flair," *Smithsonian,* December: 83–87.

Bell, Robin E. 2008. "The Unquiet Ice." *Scientific American,* February: 60.

Berlitz, Charles. 1972. *Mysteries of Forgotten Worlds.* NewYork: Dell Books.

Berman, Bob. 2009. "Strange Universe." *Astronomy,* April: 16.

Bernstein, Morey. 1965. *The Search for Bridey Murphy.* New York: Doubleday.

Blatty, William. 1983. *Legion.* New York: Pocket Books.

Bok, Bart J., and Lawrence E. Jerome. 1975. *Objections to Astrology.* Buffalo, N.Y.: Prometheus Books.

Boule, Marcellin. 1923. *Fossil Men.* London: Oliver & Boyd.

Bowman, Carol. 2001. *Return from Heaven.* New York: HarperCollins.

Boyer, Paul. 1995. *When Time Shall Be No More.* Cambridge, Mass.: Harvard University Press.

Brace, C. Loring. 1979. *The Stages of Human Evolution.* Englewood Cliffs, N.J.: Prentice-Hall.

Brackenridge, H. H. 1818. "On the Population and Tumuli of the Aborigines of North America." *Transactions of the American Philological Society.*

Brandon, Ruth. 1984. *The Spiritualists.* Buffalo, N.Y.: Prometheus Books.

"Briefing." 2013. *Time,* September 23.

Bros, Peter. 2008. "Inquisition: The Trial of Immanuel Velikovsky." In *Forbidden Science,* edited by Douglas Kenyon. Rochester, Vt.: Bear & Co.

Bulkeley, Kelly. 1995. *Spiritual Dreaming.* New York: Paulist Press.

Buskirk, Michael Van. 1978. *Astrology: Revival in the Cosmic Garden.* Christian Apologetics Research and Information Services.

Cannell, Alan. 2010. "The Dmanisi Hominids," Part I. *Pleistocene Coalition News* no. 2, no. 6. November.

Castle, Kit, and Stefan Bechtel. 1999. *Katherine, It's Time.* New York: Harper & Row.

Cavendish, Richard. 1967. *The Black Arts.* New York: G. P. Putnam's Sons.

Cerminara, Gina. 1967. *Many Mansions.* New York: Signet.

Chapman, Matthew. 2001. *Trials of the Monkey.* New York: Picador.

Charroux, Robert. 1971. *One Hundred Thousand Years of Man's Unknown History.* New York: Berkley Books.

———. 1974. *The Mysteries of the Andes.* New York: Avon Books.

Chase, Truddi. 1987. *When Rabbit Howls.* New York: Jove Books.

Chernicoff, Stanley. 1995. *Geology: An Introduction to Physical Geology.* New York: Worth Publishers.

Childress, David H. 1998. *Lost Cities of Ancient Lemuria and the Pacific.* Stelle, Ill.: Adventures Unlimited Press.

Chouinard, Patrick. 2012. *Forgotten Worlds.* Rochester, Vt.: Bear & Co.

Chown, Marcus. 2005. "Mystery Rays Could Be Sign of Cosmic Strings." *New Scientist,* June 4: 16.

Churchill, Winston. 2005. *The Age of Revolution.* New York: Barnes and Noble.

Churchward, James. 1931. *The Children of Mu.* New York: Paperback Library, Inc.

———. 1968. *The Second Book of the Cosmic Forces of Mu.* New York: Paperback Library, Inc.

———. 2011. *The Lost Continent of Mu.* Jack E. Churchward's edition. Huntsville, Ark.: Ozark Mountain Publishing.

Cicero, Marcus Tullius. 1997. *The Nature of the Gods, and On Divination.* Amherst, N.Y.: Prometheus Books.

Cohen, Barry E., Esther Giller, and Lynn W., et al., eds. 1991. *Multiple Personality Disorder from the Inside Out.* Baltimore: Sidran Press.

Cohen, Daniel. 1973. *How the World Will End.* New York: McGraw-Hill.

Cook, Blanche Wiesen. 2000. *Eleanor Roosevelt: The Defining Years 1933–1938,* vol. 2. New York: Peguin Classics.

Coppens, Philip. 2013. *The Lost Civilization Enigma.* Pompton Plains, N.J.: New Page Books.

Corliss, William. 1978. *Ancient Man.* Glen Arm, Md.: The Sourcebook Project.

———. 1980. *Unknown Earth.* Glen Arm, Md.: The Sourcebook Project.

Crabtree, Adam. 1985. *Multiple Man.* New York: Praeger Books.

Crapanzano, Vincent, and Vivian Garrison, eds. 1977. *Case Studies in Spirit Possession.* New York: John Wiley & Sons.

Crichton, Michael. 2004. *State of Fear.* New York: HarperCollins.

Cummins, Geraldine. 1965. *Swan on a Black Sea.* London: Routledge & Kegan Paul.

Curry, Andrew. 2008. "Ancient Excrement." *Archaeology,* July/August: 42–45.

Darwin, Charles. 1964. *On the Origin of Species.* Cambridge, Mass.: Harvard University Press.

———. 2004. *The Descent of Man.* New York: Penguin Books.

Davison, Wilma. 2006. *Spirit Rescue.* Woodbury, Minn.: Llewellyn Publications.

Davis, Kenneth C. 2002. *Don't Know Much about the Universe.* New York: Perennial Books.

Dawkins, Richard. 1996. *The Blind Watchmaker.* New York: W. W. Norton.

Day, Michael. 1988. *Guide to Fossil Man.* Chicago: University of Chicago Press.

Dean, Malcolm. 1980. *The Astrology Game.* Don Mills, Ontario: Nelson Foster & Sons.

de Camp, L. Sprague. 1970. *Lost Continents.* New York: Ballantine Books.

———. 1996. *Time and Chance, an Autobiography.* Hampton Falls, N.H.: Donald M. Grant Publisher.

Denton, Michael. 1986. *Evolution: A Theory in Crisis.* Bethesda, Md.: Adler & Adler.

Dewhurst, Richard J. 2014. *The Ancient Giants Who Ruled America.* Rochester, Vt.: Bear & Co.

Diamond, Jared. 1992. *The Third Chimpanzee.* New York: HarperPerennial.

Donnelly, Ignatius. 1882. *Atlantis: The Antediluvian World.* New York: Harper.

Doyle, Arthur Conan. 1930. *The Edge of the Unknown.* New York: G. P. Putnam's Sons.

Drake, Raymond. 1968. *Gods and Spacemen of the Ancient East.* New York: Signet Books.

———. 1974. *Gods and Spacemen in the Ancient Past.* New York: Signet Books.

Drayson, Alfred Wilks. 1890. *Untrodden Ground in Astronomy and Geology.* London: K. Paul, Trench, Trubner & Co., Ltd.

Ebon, Martin. 1971. *They Knew the Unknown.* New York: The World Publishing Co.

———, ed. 1978. *Parapsychology.* New York: Signet Books.

Eddington, Arthur. 1937. *Science and the Unseen World.* New York: Macmillan Publishing.

Edey, Maitland, and Don Johanson. 1990. *Blueprint: Solving the Mystery of Evolution.* New York: Penguin Books.

Editors, *Scientific American.* 2008. "Climate Fatigue." *Scientific American,* June: 39.

Editors, *Psychic* magazine. 1972. *Psychic.* New York: Harper & Row.

Einstein, Albert. 1954. *Ideas and Opinions.* New York: Wings Books.

Eiseley, Loren. 1962. *The Immense Journey.* New York: Time Reading Program Edition.

———. 1972. *The Unexpected Universe.* New York: Mariner Books.

Eldredge, Niles. *Reinventing Darwin: The Great Debate at the High Table of Evolutionary Theory.* New York: Wiley, 1995.

Fagan, Brian M. 1999. *World Prehistory: A Brief Introduction.* New York: Longman.

———. 2001. *The Little Ice Age: How Climate Made History.* New York: Basic Books.

———. 2010. *Cro-Magnon: How the Ice Age Gave Birth to the First Modern Humans.* New York: Bloomsbury Press.

Feliks, John. 2013. "Debunking Evolutionary Propaganda," Part 3. *Pleistocene Coalition News* 5, nos. 4 and 5: 10–12. www.pleistocenecoalitionnews .com (accessed Oct. 2013).

Field, William O. 1955. "Glaciers." *Scientific American,* September: 84–92.

Fiore, Edith. 1987. *The Unquiet Dead.* New York: Dolphin.

Fix, William. 1984. *The Bone Peddlers.* New York: Macmillan Publishing.

Fodor, Nandor. 1974. *An Encyclopedia of Psychic Science.* Secaucus, N.J.: Citadel Press.

Foroohar, Rana. "Congress Is Bad for the Economy." *Time,* Oct. 14, 2013: 19.

Frazier, Kendrick, ed. 1986. *Science Confronts the Paranormal.* Buffalo, N.Y.: Prometheus Books.

Freud, Sigmund. 1896. "The Aetiology of Hysteria." Translated by James Strachey and read before the Society for Psychiatry and Neurology, Vienna, April 21.

Gamow, George. 1948. *Biography of the Earth.* New York: Mentor Books.

Gauquelin, Michel. 1967. *The Cosmic Clocks.* Chicago: Henry Regnery Co.

Ginzberg, Louis. 1961. *Legends of the Jews.* New York: Simon & Schuster.

Glass, Justine. 1969. *They Foresaw the Future.* New York: Putnam.

Glausiusz, Josie. 2007. "Meet the Ancestors." *Discover,* May: 70.

Goldfarb, Alex. 2007. *Death of a Dissident.* New York: Free Press.

Goldsmith, Donald. 1997. *The Hunt for Life on Mars.* New York: Penguin Books.

Goodman, Felicitas. 1981. *The Exorcism of Anneliese Michel.* New York: Doubleday.

Goodman, Jeffrey. 1978. *We Are the Earthquake Generation.* New York: Seaview Books.

———. 1981. *American Genesis.* New York: Berkley Books.

———. 1983. *The Genesis Mystery.* New York: Times Books.

Gore, Rick. 1997. "The Dawn of Humans." *National Geographic,* July: 96–102.

"Gravity's Pull." 2003. *Time,* Feb. 18: 4.

Greene, Brian. 2012. "The Mystery of the Multiverse." *Newsweek,* May 28: 20–25.

Gribbin, John R., and Stephen Plagemann. 1974. *The Jupiter Effect.* New York: Walker & Co.

Groothuis, Douglas R. 1986. *Unmasking the New Age.* Downers Grove, Ill.: InterVarsity Press.

Gross, Martin L. 1978. *The Psychological Society.* New York: Random House.

Gugliotta, Guy. 2009. "Forest Primeval." *Smithsonian,* July: 14.

Guirdham, Arthur. 1982. *The Psychic Dimensions of Mental Health.* Wellingborough, Northamptonshire, UK: Turnstone Press.

Haddon, Alfred. 1911. *The Wanderings of Peoples.* Cambridge, UK: Cambridge University Press.

Hall, Manley. 2000. *The Secret Destiny of America.* Philosophical Research Society.

Hancock, Graham. 2002. *Underworld.* New York: Crown Publishers.

Hapgood, Charles. 1999. *The Path of the Poles.* Kempton, Ill.: Adventures Unlimited Press.

Hayden, Thomas. 2009. "What Darwin Didn't Know." *Smithsonian,* February: 41–48.

Hawking, Stephen. 2005. *Discover.* 65.

Head, Joseph, and S. L. Cranston. 1961. *Reincarnation: An East-West Anthology.* Wheaton, Ill: The Theosophical Publishing House.

Heilprin, John. 2013. "Physicists Say They Have Found a Higgs Boson." http://news.yahoo.com/physicists-found-higgs-boson-101311366.html (accessed March 14, 2013).

Heinberg, Richard. 1989. *Memories and Visions of Paradise*. Los Angeles: Jeremy P. Tarcher.

Helvarg, David. 1999. "On Thin Ice." *Sierra Magazine,* November/December: 38.

Hendricks, Wanda. 1980. "Vortex Energy," Part 3. *Kosmon Voice.*

Hertsgaard, Mark. 2012. "The Pasta Crisis." *Newsweek,* December 17: 34.

Hitching, Francis. 1982. *The Neck of the Giraffe*. New Haven, Conn.: Ticknor and Fields.

Hogan, Craig J., et al. 1999. "Surveying Space-Time with Supernovae." *Scientific American,* January: 45–50.

Hooton, Earnest. 1963. *Up from the Ape*. New York: Macmillan.

Horner, Christopher. 2008. *Red Hot Lies*. Washington, D.C.: Regnery Publishing.

Hornyanszky, Balazs, and Istvan Tasi. 2009. *Nature's IQ*. Badger, Calif.: Torchlight Publications.

Hotz, R. L. 2013. "Puzzling Hominid Had Human Traits." *Wall Street Journal,* April 12: A2.

Howard, Virginia. 1971. *The Messenger*. Self-published.

Howells, William. 1993. *Getting Here*. Washington, D.C.: The Compass Press.

Hoyle, Sir Fred, and N. C. Wickramasinghe. 1984. *Evolution from Space*. New York: Simon & Schuster.

Hunt, Dave, and T. A. McMahon. 1988. *The New Spirituality*. Eugene, Ore.: Harvest House.

Huse, Scott M. 1983. *The Collapse of Evolution*. Grand Rapids, Mich.: Baker Book House.

Imbrie, John, and Katherine Palmer. 1979. *Ice Ages: Solving the Mystery*. Short Hills, N.J.: Enslow Publishers.

Iyer, Pico. 2013. "Cuban Evolution." *Time,* July 18: 15, 69.

Jaffe, Aniela. 1971. *From the Life and Works of C. G. Jung*. New York: Harper & Row.

———. 1978. "The Psychic World of C. G. Jung." In *Parapsychology,* edited by Martin Ebon. New York: Signet Books, 30, 88–100.

James, Harry. 1956. *The Hopi Indians.* Caldwell, Ind.: Caxton.

James, William. 1903. *The Varieties of Religious Experience: A Study in Human Nature.* New York: N. W. Longmans, Green and Company.

Jastrow, Robert. 1977. *Until the Sun Dies.* New York: Norton.

Johnson, Philip. 1991. *Darwin on Trial.* Downers Grove, Ill.: InterVarsity Press.

Jones, John S., and William T. Smith. 2007. "Warming Not Man-Made." *Washington Times,* November 25.

Joseph, Frank. 1997. "Underwater City Found Near Japan." *Ancient American* 3, no. 17. March/April: 200.

———, ed. 2006. *Discovering the Mysteries of Ancient America.* Franklin Lakes, N.J.: New Page Books.

———. 2013. *Before Atlantis.* Rochester, Vt.: Bear & Co.

Joseph, Lawrence E. 2007. *Apocalypse 2012.* New York: Morgan Road Books.

Jung, Carl G. 1964. *Man and His Symbols.* New York: Dell Books.

———. 1964. *Memories, Dreams, Reflections.* London: Collins and Routledge & Kegan Paul.

———. 1973. *Collected Letters,* vol. 1. Princeton, N.J.: Princeton University Press.

———. 1974. *Dreams.* Princeton, N.J.: Princeton University Press.

———. 1978. "The Psychological Foundations of Belief in Spirits." In *Parapsychology,* edited by Martin Ebon. New York: Signet Books, 101–17.

Kay, John. 2007. "Science." *Financial Times.* October 9.

Keats, Jonathon. 2013. "Earth's Explosive Origins Revealed." *Discover,* January 2: 41.

Keith, Arthur. 1929. *The Antiquity of Man.* London: Williams & Norgate Ltd.

Kelly, Ivan W., and Don H. Saklofske. 1986. "Alternative Explanations in Science." In *Science Confronts the Paranormal,* edited by Kendrick Frazier. Buffalo, N.Y.: Prometheus Books.

Kenyon, Douglas, ed. 2008. *Forbidden Science.* Rochester, Vt.: Bear & Co.

Keyes, Daniel. 1982. *The Minds of Billy Milligan.* New York: Bantam Books.

King, David. "UK Scientist Gagged." *Asheville Global Report* no. 269.

Klein, Joe. 2013. "This Is How a Nation Unwinds." *Time,* June 17: 22.

Kluger, Jeffrey. 2001. "A Climate of Despair." *Time,* April 9: 30–36.

Koestler, Arthur. 1959. *The Sleepwalkers.* New York: Macmillan Co.

Kolosimo, Peter. 1973. *Not of This World*. New York: Bantam Books.

———. 1975. *Timeless Earth*. New York: Bantam Books.

Krauss, Lawrence. 2008. In reply to Letters, *Scientific American*. July: 12.

———. 2012. "How the Higgs Boson Posits a New Story of Our Creation." *Newsweek*, July 16: 5.

Kreisberg, Glenn, ed. 2012. *Mysteries of the Ancient Past*. Rochester, Vt.: Bear & Co.

Krippner, Stanley. 1975. *Song of the Siren*. New York: Harper & Row.

Kruglinski, Susan. 2006. "Plants Grab Control of the Global Greenhouse." *Discover*, May: 15.

Kurtz, Paul and Andrew Fraknoi. 1986. "Scientific Tests of Astrology Do Not Support Its Claims." In *Science Confronts the Paranormal*, edited by Kendrick Frazier. Buffalo, N.Y.: Prometheus Books.

Laird, Gordon. 2002. "Losing the Cool." *Mother Jones*. March/April: 76–78.

Landsburg, Alan, and Sally Landsburg. 1974. *In Search of Ancient Mysteries*. New York: Bantam Books.

Langley, Noel. 1967. *Edgar Cayce on Reincarnation*. New York: Hawthorn Books.

Lasch, Christopher. 1978. *The Culture of Narcissism*. New York: W. W. Norton.

Leakey, Richard. 1973. "Skull 1470." *National Geographic*. June: 819–29.

Leek, Sybil. 1974. *Reincarnation: The Second Chance*. New York: Stein & Day.

Lemley, Brad. 2005. "No Easy Way Out of the Greenhouse." *Discover*, October: 30–31.

———. 2009. "Why Is There Life?" *Discover*, November: 64–69.

Lemonick, Michael D. 1999. "Up from the Apes." *Time*, August: 58.

———. 2001. "Life in the Greenhouse." *Time*, April 9: 24–29.

———. 2013. "The Missing Universe." *Time*, April 22: 15.

Lerner, Eric J. 1991. *The Big Bang Never Happened*. New York: Random House.

———. 2004. "Bucking the Big Bang." *New Scientist*, May 22: 20.

Lethbridge, T. C. 1965. *ESP: Beyond Time and Distance*. London: Sidgwick & Jackson.

Lindsay, Jack. 1971. *Origins of Astrology*. London: Frederick Muller Ltd.

L. K. 2009. "Sky Survey Finds Dusty Universe." *Astronomy*, June: 22.

Logan, Cynthia. 2008. "Dr. Quantum's Big Ideas." *Forbidden Science,* 273.

Lovtrup, Søren. 1987. *Darwinism: The Refutation of a Myth.* New York: Croom Helm Ltd.

Lubenow, Martin. 2005. *Bones of Contention.* Grand Rapids, Mich.: Baker Books.

Lucretius. *On the Nature of the Universe.* London: Penguin, 1994.

Macdougall, J. D. 1996. *A Short History of Planet Earth.* New York: John Wiley & Sons.

———. 2004. *Frozen Earth.* Berkeley: University of California Press.

MacLaine, Shirley. 1986. *Dancing in the Light.* New York: Bantam Books.

Mahan, Joseph P. 1983. *The Secret.* Columbus, Ga.: Star Printing Co.

Markman, Ronald, and Dominick Bosco. 1989. *Alone with the Devil.* New York: Doubleday.

Marsa, Linda. 2013. "Earth Goes to Extremes." *Discover,* Jan. 2: 22–23.

Marshack, Alexander. 1991. *The Roots of Civilization.* Mt. Kisco, N.Y.: Moyer Bell Ltd.

Marshall, Carolyn. 2005. "Schwarzenegger Issues Plan to Reduce Greenhouse Gases." *New York Times,* June 2.

Marsolek, Patrick. 2012. "Searching for the Higgs." *Atlantis Rising,* May/June: 62.

Martin, Malachi. 1992. *Hostage to the Devil.* San Francisco: Harper SanFrancisco.

Martinez, Susan B. 2007. *The Psychic Life of Abraham Lincoln.* Franklin Lakes, N.J.: New Page Books.

———. 2009. *The Hidden Prophet: The Life of Dr. John Ballou Newbrough.* Creative Space Publishing Platform. www.CreativeSpace.com.

———. 2011. *Time of the Quickening: Prophecies for the Coming Utopian Age.* Rochester, Vt.: Bear & Co.

———. 2013a. *The Lost History of the Little People.* Rochester, Vt.: Bear & Co.

———. 2013b. *The Mysterious Origins of Hybrid Man.* Rochester, Vt.: Bear & Co.

Mayer, Robert. 1988. *Through Divided Minds.* New York: Doubleday.

McGowan, Kat. 2013. "Firestorm Over New Psychiatry Bible." *Discover,* January 2: 44.

Milton, Richard. 1996. *Alternative Science.* Rochester, Vt.: Park Street Press.

Mirsky, Steve. 2013. "Anti-Gravity." *Scientific American,* December: 86.

Mitton, Simon. 2005. *Conflict in the Cosmos.* Washington, D.C.: Joseph Henry Press.

M. M. 2009. "Extreme Planet." *Astronomy,* February: 26.

Mone, Gregory. 2011. "Two Degrees of Separation." *Discover,* June: 46.

Moote, Lloyd. 1972. *The Seventeenth Century: Europe in Ferment.* Lexington, Mass.: D. C. Heath.

Morehouse, David. 1998. *Psychic Warrior.* New York: St. Martin's Press.

Morley, George. 1962. "Path of Light." Kosmos Church Archives.

Montagu, Ashley. 2000. *Man, His First Two Million Years: A Brief Introduction to Anthropology.* New York: Delacorte Press.

Muggeridge, Malcolm. 1983. *A Third Testament.* New York: Ballantine Books.

Nadis, Steve. 2012. "When Universes Collide." *Discover,* 72.

Neimark, Jill. 2011. "Stone Age Romeos and Juliets." *Discover,* Jan. 2: 67.

Newbrough, John B. 1874. *Spiritalis.* Long Creek, S.C.: Tri-State Press.

———. 1883. *Medium and Daybreak.* Spiritual Journals. March 2.

New Scientist Staff. 2005. "It Pays to Be Green." June 4: 5.

Ornes, Stephen. 2007. "Sun Kings." *Discover,* June: 13.

Oxnard, Charles E. 1984. *The Order of Man.* New Haven, Conn.: Yale University Press.

Panek, Richard. 2005. "Two against the Big Bang." *Discover,* November: 49–52.

Palmer, Douglas. 2010. *Origins: Human Evolution Revealed.* Cambridge, UK: Mitchell/Beazley.

Palus, Shannon. 2011. "Lightning Unleashes Antimatter Storms." *Discover,* April: 18.

Park, Alice. 2013. "Head Case." *Time,* June 3: 16.

Pearce, Fred. 2005. "Forests Paying the Price for Biofuels." *New Scientist,* November 19: 19.

Peck, M. Scott. 2005. *Glimpses of the Devil.* New York: Free Press.

Pendick, Daniel. 2009. "Is the Big Bang in Trouble?" *Astronomy,* April: 48–51.

Perkins, James S. 1977. *Experiencing Reincarnation.* Wheaton, Ill.: The Theosophical Publishing House.

Pollack, H. N. 2003. *Uncertain Science, Uncertain World.* New York: Cambridge University Press.

Price, Harry. 1940. *The Most Haunted House in England.* London: Longmans, Green & Co.

Quinn, Noreen. 1975. *She Can Read Your Past Lives.* New York: Pillar Books.

Raup, David M. 1986. *The Nemesis Affair.* New York: W. W. Norton.

Rennie, John. 2002. "15 Answers to Creationist Nonsense." *Scientific American,* July: 78–85.

Rhine, J. B. 1956. "Did You Live Before?" *The American Weekly,* April 8.

Richet, Charles. 2003. *Thirty Years of Psychic Research.* Translated by Stanley Debrath. Whitefish, Mont.: Kessinger Publishing Co.

Ridley, Matt. 2012. "A Relief to Darwin: The Eyes Have It." *Wall Street Journal,* November 3–4: C4.

Robinson, Arthur, and Zachary Robinson. 1997. "Science Has Spoken: Global Warming Is a Myth." *Wall Street Journal,* Dec. 4.

Robinson, J. Hedley, and James Muirden. 1979. *Astronomy Data Book.* New York: John Wiley & Sons.

Rogers, R. R. 2007. "Tracking an Ancient Killer." *Scientific American,* February: 50.

Ruddiman, William F. 2005. "How Did Humans First Alter Global Climate?" *Scientific American,* March: 47–53.

Rudgley, Richard. 1999. *The Lost Civilizations of the Stone Age.* New York: The Free Press.

Runcorn, S. K. 1955. "The Earth's Magnetism." *Scientific American,* September: 152–63.

Russell, Walter. 1947. *The Secret of Life.* Waynesboro, Va.: Universe of Science and Philosophy.

Sanduleak, N. 1986. "The Moon Is Acquitted of Murder in Cleveland." In *Science Confronts the Paranormal,* edited by Kendrick Frazier. Buffalo, N.Y.: Prometheus Books.

Schiaparelli, G. V. 1889. *De la rotation de la terre sous l'influence des actions geologique.* St. Petersburg: N.p.

Schoch, Robert. 2012. *Forgotten Civilization.* Rochester, Vt.: Bear & Co.

Schoenewolf, Gerald. 1991. *Jennifer and Her Selves.* New York: Donald I. Fine, Inc.

Schreiber, Flora R. 1973. *Sybil.* Chicago: Henry Regnery Co.

Scott, Beth, and Michael Norman. 1985. *Haunted Heartland.* New York: Barnes & Noble.

Shea, Neil. 2005. "The First Americans?" *National Geographic,* May: 2.

Sherman, Harold. 1972. *You Live after Death.* Greenwich, Conn.: Fawcett Publications.

Simpson, George Gaylord. 1953 (reprinted in 1969). *The Major Features of Evolution.* New York: Columbia University Press.

Smith, Carole. 1998. *The Magic Castle.* New York: St. Martin's Press.

Smith, Susy. 1970a. *Ghosts Around the House.* New York: The World Publishing Co.

———. 1970b. *Widespread Psychic Wonders.* New York: Ace Publishing.

"Special Report." 1999. *Scientific American,* January: 53–55.

Spencer, Roy W. 2010. *The Great Global Warming Blunder.* New York: Encounter Books.

Steiger, Brad. 1974. *Mysteries of Time and Space.* New York: Dell Books.

Stein, Benjamin, and Kevin Miller. 2008. "Expelled: No Intelligence Allowed." Directed by Nathan Frankowski. Distributed by Amazon.

Stein, James. 2012. *The Paranormal Equation.* Pompton Plains, N.J.: New Page Books.

Stein, Joel. 2013. "Bill Me Later." *Time,* August 12.

Stearn, Jess. 1973. *The Search for a Soul.* Greenwich, Conn.: Fawcett Crest.

Stevenson, Ian. 1987. *Children Who Remember Past Lives.* Charlottesville, Va.: University Press of Virginia.

Stillman, William. 2006. *Autism and the God Connection.* Naperville, Ill.: Sourcebooks Inc.

Stix, Gary. 2008. "Traces of a Distant Past." *Scientific American,* July: 56–60.

Stone, Alex. 2007. "The Birth of Dark Energy." *Discover,* April: 12.

Storr, Anthony. 1996. *Feet of Clay.* New York: Simon & Schuster.

Stossel, John. 2006. *Myths, Lies and Downright Stupidity.* New York: Hyperion.

Switek, Brian. 2010. *Written in Stone: Evolution, the Fossil Record and Our Place in Nature.* New York: Bellevue Literary Press.

Sylvester, Paul. 2001. *Discover,* February.

Tarduno, John A. 2008. "Hotspots Unplugged." *Scientific American,* January: 73–74.

Tattersall, Ian, and Jeffrey Schwartz. 2000. *Extinct Humans.* New York: Westview Press.

Taylor, G. R. 1983. *The Great Evolution Mystery.* New York: Harper & Row.

Tenodi, Vesna. 2013. "Problems in Australian Art and Archeology." *Pleisto-cene Coalition News* 5, no. 1: 15–17. www.pleistocenecoalitionnews.com (accessed November 11, 2014).

Teresi, Dick. 2011. "Lynn Margulis: Q & A." *Discover,* April: 67–71.

Tolson, Jay. 2001. "Founding Rivalries." *US News & World Report,* February 26.

Tomas, Andrew. 1973. *We Are Not the First.* New York: Bantam Books.

Trench, Brinsley LePoer. 1974. *Temple of the Stars.* New York: Ballantine Books.

Toynbee, Arnold J. 1961. *A Study of History.* New York and London: Oxford University Press.

Tucker, Jim B. 2005. *Life before Life.* New York: St. Martin's Press.

Twigg, Ena. 1973. *The Woman Who Stunned the World.* New York: Manor Books, Inc.

Tymn, Michael E. 2013. "Was William James a Wimp?" *The Journal for Spiritual and Consciousness Studies, Inc.,* January: 1–5.

Valentine, Tom. 1975. *The Great Pyramid.* New York: Pinnacle Books.

Vecchi, Gabriel A., and Brian J. Soden. 2007. "Global Warming and the Weakening of the Tropical Circulation." *Journal of Climate* 20: 4316.

Velikovsky, Immanuel. 1965. *Worlds in Collision.* New York: Delta Books.

Verrill, A. H. 1943. *Old Civilizations of the New World.* New York: New Home Library.

von Däniken, Erich. 1974. *Gold of the Gods.* New York: Bantam Books.

von Franz, Marie-Louise. 1964. "The Process of Individuation." In Carl G. Jung, *Man and His Symbols.* New York: Dell Books.

Von Ward, Paul. 2011. *We've Never Been Alone.* Charlottesville, Va.: Hampton Roads.

Wadler, Arnold. 1948. *One Language.* New York: The American Press for Art and Science.

Walsh, Bryan. 2013. "Defending the Waterfront." *Time,* June 24: 10.

Wambach, Helen. 1979. *Life before Life.* New York: Bantam Books.

Webster, James. 2009. *The Case Against Reincarnation.* Surrey, UK: Grosvenor House Publishing Limited.

Webster, Richard. 1995. *Why Freud Was Wrong.* New York: Basic Books.

Weiner, Jonathan. 1989. "Glacier Bubbles Are Telling Us What Was in Ice Age Air." *Smithsonian,* 86.

Weisman, Alan. 1977. *We Immortals.* New York: Pocket Books.

West, John Anthony. *The Case for Astrology*. New York: Edward-McCann, 1970.

Wheatley, Margaret J. 1994. *Leadership and the New Science*. San Francisco: Berrett-Koehler Publishers.

White, John. 1980. *Pole Shift*. New York: Doubleday.

Whitehead, Alfred North. 1956. *Dialogues of Alfred North Whitehead*. New York: Mentor Books.

Whitton, Joel, and Joe Fisher. 1986. *Life between Life*. New York: A Dolphin Book.

Wickland, Carl. 1924. *Thirty Years among the Dead*. London: Spiritualist Press.

Wilcock, David. *The Source Field Investigations*. New York: Dutton, 2011.

Williams, Nancy. 2008. "The Super Collision Threat." *Atlantis Rising*, September/October: 23, 59.

Wilson, Clifford. 1975. *The Chariots Still Crash*. New York: Signet Books.

Winchester, Simon. 2003. *Krakatoa*. New York: HarperCollins Publishers.

Wright, Frederick G. 1912. *Origin and Antiquity of Man*. Oberlin, Ohio: Bibliotheca Sacra Co.

INDEX

on victims of sudden death, 300
on vortexian currents' effect on man, 213
on the vortexian lens, 197, 198
on vortexya, 58, 59f, 204, 205
on worlds without end, 65
Zarathustra speech in, 220–21
on Zarathustrian law, 381
Occam's razor, 26
Oeschger, Hans, 177
Oppenheimer, J. Robert, 1
oracle houses, 346–47
Original Sin, 295
oscillation of the Earth
apparent ice ages and, 15, 122–24, 127
magnetic evidence of, 127–28, 157–58
Oahspe on, 154
polar dinosaurs and, 124–26
pole shift due to, 15, 122–24, 126–27, 129–30, 139, 157
process of, 127
Velikovsky's concept of pole shift, 129f, 130
warm-weather species in the far North, 124, 156
Outer Darkness, 255f, 256, 302
out-of-body experience (OBE), 28, 246–48, 269

Pacific Decadal Oscillation (PDO), 164f, 175, 193, 233, 395
Paine, Thomas, 385
paleomagnetics, 127–28
Pan
Central American sites named after, 326–27
end of, 329
as source of writing, 336–37, 339
Panchatantra fable, 293–94
paradigm shift to come, 391–92, 393–94

Paulson, Genevieve, 297
Peck, M. Scott, 268
peer review, 20–21
Perkins, James, 297, 317
Persian astronomy, 220
phobias, 304
Pike, James, 309
Piper, Leonora, 4, 5, 6
plutocracy, 369, 380
polar bears, 193
pole shift, 15, 122–24, 126–27, 129–30, 139, 157. *See also* oscillation of the Earth
political correctness (PC), 389–90
poltergeists, 243–44, 245–46
possession or spirit obsession
aura or dam against, 256
brain vs. soul and, 257, 258
by demons, 257–58, 267, 268
drugs and, 252
by drujas, 311, 322–23
gangs in Egypt, 322–23
knowledge of unlearned languages and, 264–66
"like unto like" principle and, 307–8
misdiagnosis of, 251–52
MPD and, 267, 268
psychosis due to, 254
reincarnation as, 293, 297, 300–314, 320
sloth of will and, 256–57
trauma and, 256
by twin spirits, 249–50, 310–11
the unconscious and, 264–65
prophecy, ancient
Algonquin, 214
astronomy-based, 203, 206–7
Babylonian, 205–6
of drought and disease, 202
Greek, 207–9, 220

BOOKS OF RELATED INTEREST

The Lost History of the Little People
Their Spiritually Advanced Civilizations around the World
by Susan B. Martinez, Ph.D.

The Mysterious Origins of Hybrid Man
Crossbreeding and the Unexpected Family Tree of Humanity
by Susan B. Martinez, Ph.D.

Time of the Quickening
Prophecies for the Coming Utopian Age
by Susan B. Martinez, Ph.D.

Forbidden History
Prehistoric Technologies, Extraterrestrial Intervention,
and the Suppressed Origins of Civilization
Edited by J. Douglas Kenyon

The Ancient Giants Who Ruled America
The Missing Skeletons and the Great Smithsonian Cover-Up
by Richard J. Dewhurst

Göbekli Tepe: Genesis of the Gods
The Temple of the Watchers and the Discovery of Eden
by Andrew Collins

DNA of the Gods
The Anunnaki Creation of Eve and the Alien Battle for Humanity
by Chris H. Hardy, Ph.D.

Slave Species of the Gods
The Secret History of the Anunnaki and Their Mission on Earth
by Michael Tellinger

INNER TRADITIONS • BEAR & COMPANY
P.O. Box 388
Rochester, VT 05767
1-800-246-8648
www.InnerTraditions.com

Or contact your local bookseller